URBAN THEORY
BEYOND THE WEST

Since the late eighteenth century, academic engagement with political, economic, social, cultural and spatial changes in our cities has been dominated by theoretical frameworks crafted with reference to just a small number of cities. This book offers an important antidote to the continuing focus of urban studies on cities in 'the global north'.

Urban Theory Beyond the West contains 20 chapters from leading scholars, raising important theoretical issues about cities throughout the world. Past and current conceptual developments are reviewed and organized into four parts: 'De-centring the city' offers critical perspectives on re-imagining urban theoretical debates through consideration of the diversity and heterogeneity of city life; 'Order/disorder' focuses on the political, physical and everyday ways in which cities are regulated and used in ways that confound this ordering; 'Mobilities' explores the movements of people, ideas and policy in cities and between them; and 'Imaginaries' investigates how urbanity is differently perceived and experienced. There are three kinds of chapters published in this volume: theories generated about urbanity 'beyond the West'; critiques, reworking or refining of 'Western' urban theory based upon conceptual reflection about cities from around the world; and hybrid approaches that develop both of these perspectives.

Urban Theory Beyond the West offers a critical and accessible review of past and current theoretical developments, providing an original and groundbreaking contribution to urban theory. It is essential reading for students and practitioners interested in urban studies, development studies and geography.

Tim Edensor teaches cultural geography at Manchester Metropolitan University, UK. His research interests include tourism, materialities and mobilities.

Mark Jayne is a Senior Lecturer in human geography at the University of Manchester, UK. His research interests include consumption, the urban order, city cultures and cultural economy.

URBAN THEORY BEYOND THE WEST

A world of cities

Edited by Tim Edensor and Mark Jayne

Routledge
Taylor & Francis Group

LONDON AND NEW YORK

First published 2012
by Routledge
2 Park Square, Milton Park, Abingdon, Oxon, OX14 4RN

Simultaneously published in the USA and Canada
by Routledge
711 Third Avenue, New York, NY 10017

Routledge is an imprint of the Taylor & Francis Group, an informa business

British Library Cataloguing in Publication Data
A catalogue record for this book is available from the British Library

Library of Congress Cataloging in Publication Data
Edensor, Tim, 1957–
 Urban theory beyond the West / Tim Edensor and Mark Jayne.
  p. cm.
 Includes bibliographical references and index.
 1. Urbanization–Philosophy. 2. City planning–Philosophy. 3. Urban
geography. 4. Political geography. 5. Political ecology. I. Jayne, Mark,
1970– II. Title.
 HT113.E44 2011
 307.7601–dc22
                                                          2011013425

ISBN: 978-0-415-58975-8 (hbk)
ISBN: 978-0-415-58976-5 (pbk)
ISBN: 978-0-203-80286-1 (ebk)

Typeset in 10/12.5pt Bembo by
Graphicraft Limited, Hong Kong

Printed and bound in Great Britain by
CPI Antony Rowe, Chippenham, Wiltshire

CONTENTS

LIST OF PLATES

LIST OF FIGURES

LIST OF TABLES

NOTES ON CONTRIBUTORS

Hooshmand Alizadeh is Assistant Professor in Urban Design and Head of the Urban Planning and Design Group at the University of Kurdistan, Iran. His research interests include public spaces, urban form, urban regeneration and the built form of Kurdish cities, and he has published in journals including *Cities, Urban Morphology, Kurdish Studies, International Journal of Kurdish Studies, International Journal of Environmental Research, Journal of Asian Architecture and Building Engineering.*

Yooil Bae joined Singapore Management University as an Assistant Professor of Political Science, School of Social Sciences, in 2008. He was previously a Post-doctoral Fellow at the National University of Singapore and he received his doctorate from the Department of Political Science at the University of Southern California. He specializes in comparative politics, urban and regional political economy, environmental politics and the politics of Northeast Asia. His work examines the diffusion of ideas and transfer of policy ideas across national borders, state–civil society relations and urban political economy in Northeast Asian countries, and has appeared in *The Pacific Review, International Journal of Urban and Regional Research* and *Korean Journal of Policy Studies*. He is currently working on how middle power countries, not advanced economies, diffuse their experiences and developmental models to the third world through various programmes such as Overseas Development Assistance.

Filip De Boeck is Professor and Coordinator at the Institute for Anthropological Research in Africa (IARA), a research unit of the University of Leuven, Belgium. Filip is actively involved in teaching, promoting, coordinating and supervising research in and on Africa. Since 1987 he has conducted extensive field research in both rural and urban communities in D.R. Congo. His current theoretical interests include local subjectivities of crisis, postcolonial memory, youth and the politics of culture, and the transformation of private and public space in the urban context in Africa. Together with Alcinda Honwana he edited *Makers and Breakers: Children and Youth in Postcolonial Africa* (Oxford: James Currey, 2005). Other recent book publications include *Kinshasa: Tales of the Invisible City*, a joint book project with photographer Marie-Françoise Plissart (Ghent/Tervuren: Ludion/Royal Museum of Central Africa, 2004). Most recently, De Boeck directed *Cemetery State*, a 70-minute documentary film about a graveyard in Congo's capital.

Melissa Butcher is a Lecturer in the Department of Geography, Open University, UK. The focus of her research is transnational mobility, cultural change and conflict in diverse urban spaces, emphasizing questions of identity and belonging. Before joining the OU, Melissa taught in universities in Ireland and Australia, and she has also worked as a journalist, a development education specialist and an intercultural consultant and trainer in the private, government and community sectors. She has lived and worked in Asia, primarily India and Australia, and is currently writing a book on transnational mobility and cultural change. Her previous publications include *Dissent and Cultural Resistance in Asia's Cities* (with Selvaraj Velayutham, Routledge, 2009), *Ingenious: Emerging Youth Cultures in Urban Australia* (with Mandy Thomas, Pluto Press, 2003) and *Transnational Television, Cultural Identity and Change: When STAR Came to India* (Sage, 2003).

Swati Chattopadhyay is an Associate Professor in the Department of History of Art and Architecture at the University of California, Santa Barbara, USA. She directs the Subaltern-Popular Workshop, and is the Editor Designate of the *Journal of the Society of Architectural Historians*. She is the author of *Representing Calcutta: Modernity, Nationalism, and the Colonial Uncanny* (Routledge, 2005, paperback 2006), and her forthcoming book on contemporary urbanization and popular culture is titled *Unlearning the City*. She is also the guest editor of an issue of *Urban History* (forthcoming) and co-guest editor of a special issue of *PostColonial Studies* (Nov 2005).

Tim Edensor is a Reader in Human Geography at the Department of Environmental and Geographic Sciences, Manchester Metropolitan University, UK. He has written widely on a wide range of research interests that include urban and rural cultures, spaces of tourism, national identities, walking and driving, rhythmanalysis, haunting, urban materiality, football and contested cultural memory. Tim is author of *Industrial Ruins: Space, Aesthetics and Materiality* (Berg, 2005), *National Identity, Popular Culture and Everyday Life* (Berg, 2002) and *Tourists at the Taj* (Routledge, 1998), as well as editor of *Geographies of Rhythm* (Ashgate, 2009) and co-editor of *Vernacular Creativity* (Routledge 2009). He is currently researching landscapes of illumination.

Yasser Elsheshtawy is Associate Professor of Architecture at the United Arab Emirates University. His research interests are environment-behavioural studies, urban planning in Middle Eastern cities, informal urbanism and architectural theory. He has published in numerous journals such as *Journal of Architectural*

Education, International Journal of Urban and Regional Research, Built Environment and *Architectural Theory Review.* In addition he authored and edited a series of books published by Routledge which include *Dubai: Behind an Urban Spectacle* (2010), *The Evolving Arab City* (2008) (winner of the 2010 International Society for Planning History award) and *Planning Middle Eastern Cities* (2004, 2010).

Slavomíra Ferenčuhová is Assistant Professor in the Department of Sociology, Masaryk University, Czech Republic. Her research interests include urban planning and governance, especially in the Central Eastern European countries, (post)socialist urban ethnographies and identities, and history of academic urban studies. Her PhD dissertation on urban planning under state socialism and its transformations after 1989 in the Czech Republic, based on the case study of the city of Brno, will be published as a book in 2011 (Masaryk University Press).

Jorge Inzulza-Contardo is a Chilean architect and Lecturer in Urban Planning at the University of Chile. Jorge has a MA in Urban Development and completed a PhD in Planning and Landscape at the University of Manchester (UK). He has worked in the development of masterplans and management of urban projects for Santiago Metropolitan Region at the Ministry of Housing and Planning from 2003 until 2010. Jorge has lectured in architecture, urban planning and design topics at three universities of Chile (Catholic, Bío-Bío, Talca and Las Américas) and also at the University of Manchester. His research interests include gentrification, urban regeneration, urban design briefs, heritage and collective memory, with especial focus on Latin American contexts.

Deljana Iossifova is a Research Fellow at the School of Architecture and the Built Environment, University of Westminster, and teaches architecture and urban design at the University of Nottingham, UK. Trained as an architect at the Swiss Federal Institute of Technology (ETH Zurich), she has been in charge of a number of award-winning urban schemes and architectural design projects in various geographical contexts. She was a UNU-IAS PhD Fellow at the United Nations University, Institute of Advanced Studies (Sustainable Urban Futures Programme), and received her PhD in social engineering (public policy design) from Tokyo Institute of Technology. Deljana's main research interests include sociospatial transformation, multiple and place-related identities, and the sociocultural aspects of human adaptation to change. She is currently working on the HERA-funded project Scarcity and Creativity in the Built Environment (SCIBE).

Beatriz Jaguaribe is Associate Professor at the School of Communications of the Federal University of Rio de Janeiro. Her books include *Fins de Século: Cidade e Cultura no Rio de Janeiro* (Rocco, 1998), *Mapa do Maravilhoso do Rio de Janeiro* (Sextante, 2001), and *O Choque do Real: Estética, Mídia e Cultura* (Rocco, 2007). Her main research interests are representations of the city and contemporary urban experiences, inventions of the modern self in literature and the media, and the aesthetics of national imaginaries in Latin America. She is currently finishing a book on Rio de Janeiro for Routledge and is also working on a comparative study of public photography and propaganda during the Vargas and Peronist regimes in Brazil and Argentina.

Mark Jayne is a Senior Lecturer in Human Geography at the University of Manchester, UK. His research interests include consumption, the urban order, city cultures and cultural economy. He has published over 60 journal articles, book chapters and official reports including papers in *Progress in Human Geography*, *Transactions of the Institute of British Geographers*, *Environment and Planning A* and the *International Journal of Urban and Regional Research*. Mark is author of *Cities and Consumption* (Routledge, 2005), co-author of *Alcohol, Drinking, Drunkenness: (Dis)Orderly Spaces* (Ashgate, 2011) and *Rethinking Childhood, Families and Alcohol* (Continuum, 2011). Mark is also co-editor of *City of Quarters: Urban Villages in the Contemporary City* (Ashgate, 2004) and *Small Cities: Urban Experience Beyond the Metropolis* (Routledge, 2006).

Christina M. Jiménez is Associate Professor of History at the University of Colorado, Colorado Springs, USA. A native of San Jose, California she arrived at UCCS in 2000 after completing her PhD at the University of California, San Diego. Jiménez specializes in Mexican, Latin American, Latino and comparative urban history. Her work explores the dynamics of Mexican citizenship, urban politics and popular culture in the city during the nineteenth and twentieth centuries. She has published in *Urban History, Journal of Urban History* and *Black History Bulletin*, and is a contributor to *The Spaces of the Modern City*, edited by Gyan Prakash. She is currently working on a monograph titled *Making an Urban Public: How the City Revolutionized Citizenship in Mexico, 1880–1930*. She is also co-editor of *The Matrix Reader: Examining the Dynamics of Oppression and Privilege* (McGraw-Hill, 2008), an introductory reader for courses discussing race, gender, class and other social identities. Jiménez has received several prestigious fellowships at research centres such as the Shelby Davis Cullum Center at Princeton University and the Center for Cultural Studies at the University of California, Santa Cruz. With her co-editors, Jiménez co-organizes the annual national *Knapsack Institute: Curriculum Transformation Workshop* at UCCS, a workshop for educators teaching issues of diversity.

Cuttaleeya Jiraprasertkun is a Lecturer at the Landscape Architecture Division Faculty of Architecture, Kasetsart University, Thailand. Her research interests encompass both theoretical and methodological concerns of how to read and interpret space and place. She has published several papers in local and international journals which led to her receiving the Ellis Stones Memorial Award 2005 (First Prize) from the University of Melbourne. Her recent paper was published in the *Journal of Social Issues in Southeast Asia* (2007) entitled, 'Memory or nostalgia: the imagining of everyday Bangkok'. She is currently working on the case studies of community best practices for environment resource management in Thailand.

Nora Libertun de Duren's work focuses on the intersections of institutions, regional development and urban form. Her research has been published in a number of peer-reviewed journals, and she is the co-author of an edited volume on *Cities and Sovereignty*. Nora has taught urban planning theory and design at Columbia University, and at the University of Buenos Aires. Her work has been distinguished with various awards, including Fulbright, World Bank, MIT Presidential Scholarship, Harvard Fortabat scholarship and University of Buenos Aires Gold Medal. Nora holds an architectural degree from the University of Buenos Aires, a Master's in urban design from Harvard University, and a doctorate in urban planning from the Massachusetts Institute of Technology. In 2007 she was appointed Director of Planning of the NYC Department of Parks and Recreation.

Choon-Piew Pow is an Assistant Professor in the Department of Geography at the National University of Singapore. Trained as an urban/social geographer, he has an abiding interest in urban issues with a focus on Asia, in particular cities in China and Southeast Asia. His recent research examines how the development of housing enclaves in Shanghai has transformed the social and spatial organization of the city, leading to the formation of exclusive middle-class gated communities. Another current research project examines planning discourses and practices of urban sustainability and the ecological city. He has published in several international journals including *Urban Studies*, *Urban Geography*, *Social and Cultural Geography*, *Antipode* and *Asia Pacific Viewpoint* and is the author of the book *Gated Communities in China: Class, Privilege and the Moral Politics of the Good Life* (Routledge, 2009).

Dennis Rodgers is Senior Research Fellow at the Brooks World Poverty Institute, University of Manchester, UK, Visiting Senior Fellow in the Crisis States Research Centre, London School of Economics and Political Science, UK, and Associate Editor of the *European Journal of Development Research*. His research interests relate to issues of urban violence and urban governance in Nicaragua and Argentina, as well as questions of development epistemology and representation. He has published in numerous journals including the *Journal of Development Studies*, the *Bulletin of Latin American Research*, *Environment and Urbanisation*, *Critique of Anthropology*, *Security Dialogue* and the *Journal of Latin American Studies*. He is co-author (with José Luis Rocha) of *Bróderes Descobijados y Vagos Alucinados: Una Década con las Pandillas Nicaragüenses 1997–2007* (Envio, 2008), co-editor (with Gareth A. Jones) of *Youth Violence in Latin America* (Palgrave, 2009) and is currently completing a monograph based on 15 years of longitudinal ethnography with a Nicaraguan youth gang.

Scott Salmon is an Associate Professor of Geography and Urban Studies, and Chair of the Urban Studies Program, at the New School University in New York. His research interests centre on the contemporary transformation of urban imaginaries by processes of global competition – especially as they are reflected in the remaking of urban space and representations of the urban experience. He is currently working on a transnational comparison of the 'competitive' remaking of five global cities and also collaborating on a comparative analysis of the consequences of failed Olympic bids for Paris and New York.

Laura Shillington has a PhD in Geography from York University, Toronto, Canada and is a Research Associate in the Simone de Beauvoir Institute at Concordia University in Montreal. Her research focuses on everyday environmental struggles of marginalized households. At present, she is working on a project in Managua and Mexico City on everyday challenges of garbage and sewage. She is particularly interested in the strategies that women use to deal with environmental uncertainties in the home and how these shape and are shaped by household gender relations as well as broader power geometries.

Sofia Toufic Shwayri is an Associate Professor in International Planning and Development at Seoul National University (SNU) in South Korea. Prior to SNU, Dr. Shwayri was a Visiting Fellow in Lebanese Studies at Oxford University in St. Antony's College, where she organized an international conference, 'The Multiple Phases and Processes of Reconstructions in Lebanon', and before this she was an Assistant Professor at the Graduate School of Arts and Sciences at New York University. Sofia earned both her M.S. and Ph.D. in architecture at the University of California, Berkley (1997, 2002), where she was also a Research Fellow at the Centre of Middle Eastern Studies, an Instructor in Peace and Conflict Studies, and also worked in the Department of City and Regional Planning. Her interest in cities springs from growing up in wartime Beirut where she witnessed more than 15 years of simultaneous destruction and reconstruction. She is currently working on a book, *Civil War Beirut*. Her most recent publication is a chapter in *Flammable Cities*, edited by Jordan Sand, on Beirut titled 'The Beirut central district on fire: firefighting in a divided city with shifting frontlines, 1975–1976'.

AbdouMaliq Simone is an urbanist with particular interest in emerging forms of social and economic intersection across diverse trajectories of change for cities in the global south. Simone is presently Professor of Sociology at Goldsmiths College, University of London, and Visiting Professor of Urban Studies at the African Centre for Cities, University of Cape Town. His work attempts to generate new theoretical understandings based on a wide range of urban practices generated by cities in Africa, the Middle East and Southeast Asia, as well as efforts to integrate these understandings in concrete policy and governance frameworks. Key publications include: *In Whose Image: Political Islam and Urban Practices in Sudan* (University of Chicago Press, 1994), *For the City Yet to Come: Urban Change in Four African Cities* (Duke University Press, 2004) and *City Life from Jakarta to Dakar: Movements at the Crossroads* (Routledge, 2009).

ACKNOWLEDGEMENTS

The editors would like to thank Andrew Mould and Faye Leerink at Routledge for their patience and support and all of the contributors for ensuring that the compilation of this book was an enjoyable process.

Tim would like to thank Jon Binnie and Craig Young and the third-year geography students at MMU, and especially Uma Kothari, Kim and Jay.

Mark would like to thank his colleagues at the University of Manchester and 'down the road' at Manchester Metropolitan University, as well as Bethan Evans, David Bell, Gill Valentine, Slavi, the Wollongong Boys, the Leyshon's, Phil Hubbard, Sarah Holloway and Dr and Mrs Potts.

Special thanks to Daisy.

The authors, editors and publishers would like to thank copyright holders for their permission to reproduce the following material:

Figure 6.1 Seoul's Development (Photograph by Young-Wan Na, City of Seoul)

Figure 19.4 Location of the two future islands of 'Cité du Fleuve' in the Congo River, near Kingabwa (Photo courtesy of Cité du Fleuve)

1

INTRODUCTION

Urban theory beyond the West

Tim Edensor and Mark Jayne

This book has two main aims: to reflect on the progress made in studying cities throughout the world; and to act as point of departure in a re-imagining of cities that takes seriously urban theory beyond 'the West'. We respond to interdisciplinary concerns over the global disparities of academic knowledge and growing recognition by urban theorists of the need to appreciate the diversity of cities. To this end, chapters in this volume challenge core assumptions which frame urban theory and investigate the ways in which the study of cities has been dominated by parochial agendas, perspectives and assumptions. Accordingly, we contribute to broader theoretical agendas which highlight how making sense of urban life does not have to depend on pre-existing frameworks laid down by the 'Western' academy.

In this introduction we specifically focus on a range of exciting developments in urban theory that offer fruitful avenues towards such a project. Following Thrift (2000), who reminds us that 'one size does not fit all' and Robinson (2006), who insists on the imperative to look at the heterogeneity of urban practices, identities and processes, we demonstrate how a re-imaging of 'the city' is opening up the opportunity to contribute to broader intellectual debates that have largely bypassed urbanists. To be clear, urban theory has been slow in contributing to important advances in political, economic, social and cultural theories that have had a longer tradition of moving beyond theoretical agendas dominated by North American and European traditions. In these terms, urban theorists have tended to remain entrenched in conceptual and empirical approaches that have barely moved beyond the study of a small number of 'Western' cities which act as the template against which all other cities are judged.

Until very recently urban theorists have sidestepped any progressive engagement with labels such as 'West' and 'non-West' in contrast to disciplines across the social sciences which have challenged dualisms and typologies which fix particular spaces, such as 'first, second and third worlds', 'developed' and 'developing', 'North' and 'South', 'majority' and 'minority' and so on. With urban theory being dominated by a focus on a small number of cities mostly located in North America and Europe, imperatives to highlight the complex nature of urban life and thereby problematise these dualisms have not been forthcoming (see Chapter 5 and Chapter 18 for further critical discussion). As such, the possibility of making critical interventions into wider academic debates through championing an understanding of the increasing mixing of people, cultures and identities in most cities throughout the world has not been realized. For instance, while Dear and Leclerc (2003) highlight how the urban sprawl of Bajalta, from Southern California to Mexico, is a thoroughly postborder city, a transnational megalopolis that incorporates multiple shifting and intersecting spatialities in continual flux, a melding of 'non-Western' and 'Western' practices and processes, they fail to critically engage with the efficacy of such labels.

Urban theory has thus failed to systematically engage with such broader epistemological and ontological debates because of the persistence of situated theories that originate and reside in European and American academies – and resound through scholarly environments around the world as well. Such bias is also problematic given the salient historical context for many cities beyond 'the West' of colonial rule, where military, legal and governmental power held sway, ideological and cultural power was variously imposed, the urban fabric was often extensively redesigned, reorganized and divided into sections, and particular forms of development implemented and others discouraged. In Chapter 5, Chattopadhyay shows how the colonial city was a veritable laboratory in which policies of capitalist exploitation and state control were tested before being implemented, albeit usually in milder ways, in 'Western' cities. In addition to these extensive impacts, colonialism mobilized a series of discursive strategies through which space was claimed, contributing to a distorted geographical knowledge, the impacts of which continue to persist in contemporary imaginaries, not least in the 'Western' academy. For instance, Spurr (1993) highlights how colonizers variously aestheticized, classified, conceived as debased or ethereally insubstantial, idealized, naturalized and eroticized colonized urban spaces.

Disciplines such as anthropology, biology, history and geography were particularly complicit in proliferating these ways of understanding and classifying the 'other' in contrast to European identity. Studies of colonized spaces undertaken by 'Western' scholars tended to offer little or no mediation with local theorists. The lack of dialogue with colonized intellectuals assumed a universalist, scientific, rational expertise and a one-way flow of information built an imposing edifice of learning that added power and allure to European colonial society. Crucially, this colonial echo is still manifest in the contemporary marginalization of 'non-Western scholars' and their theoretical work in academic institutions, journals and textbooks. For example, Robinson (2003: 277) addresses the ways in which colonial approaches underpin persistent asymmetrical relationships across international scholarship, where 'Western' theorists ignore 'non-Western' thought but scholars throughout the world are expected to frame their work within 'the authorized Western canonical literature'. She claims that a 'very parochial scholarship has paraded the world in the clothes of universalism for some time' (ibid.: 275) and that this parochialism is expressed in urban theory through the imbalance between conceptions of the Western metropolis, implicitly construed as more developed, complex, dynamic and mature, in contradistinction to the 'non-Western' city.

Such ideological colonial rhetoric foregrounds a process through which 'advanced' European notions of urban progress can banish disorder, and produce a vibrant economic and civic life that will ultimately mirror the cities of the metropole. Echoing these colonial assertions of urban advancement, recent studies of cities beyond 'the West' have disproportionately focused on the notion that progress can be measured by universal stages of urban development, a discourse that reinstalls notions of irreducible difference by identifying those cities that are deemed to be at an 'earlier' stage. Indeed, Robinson (2006: 5) suggests that many previous studies of cities beyond 'the West' were 'doomed to be shipwrecked by developmentalism', citing Gugler (2004) for his focus on a narrow range of particular valuations of world city-ness. Not only has urban theory been overly focused on 'the West' but development studies have also focused on 'non-Western' cities by categorizing them as 'problems' in relation to 'Western' understandings of urban life. This 'developmentalism' limits the ways in which city life is imagined and represented, and as Robinson (2006: 4) remarks 'without a strong sense of the creativity of cities ... the potential for imagining city futures is truncated'. She (2002: 543) further argues that 'the explicit naming of the region or cities covered highlights the implicit universalist assumptions underpinning the often unremarkable localness of much writing on Western cities'. As Stenning and Hörschelmann (2008) emphasize, these assumptions also extend across European spaces regarded as less global or developed, such as post-socialist spaces, discussed in greater theoretical detail in a Czech and Slovak context by Ferenčuhová in this volume.

In a similar vein, Agrawal (2002) points to the tendency of scientistic codification, abstraction and separation of knowledge from its complex practical contexts as part of the way in which power structures knowledge production and consumption. Accordingly, Robinson (2003) advocates the need to de-colonize imaginations of city-ness in order to break free of the categorizing tendencies which dominate urban theory, suggesting that an emphasis on epochal or archetypal global cities has led to a dominant interest in the structural positions of cities (also see Bell and Jayne, 2009). A consequence of such 'measuring' has been that the distinctive ways in which actors and institutions are active agents in the making of cities have been ignored. Similarly, Panelli (2008) points to how geography is a 'modern' scholarly discipline that has historically contributed to Cartesian-based scientific knowledge and colonial politics, perpetuating the hegemony of Western academic knowledge through the sketching out of key fields and themes. The distortions of geographical focus continues to shape the scope of enquiry so that a 'modern-day parochialism shape[s] its research agenda' (Tolia Kelly, 2009: 2). As Dowling (2008) has indicated, the shaping and constraining of research are contextualized by the Western academy's own place within a neoliberal global economy: it is itself politically governed. Panelli (2008: 802) also points to 'the socially unequal peopling of geography departments, societies and publishing arenas' and the complicity of the university in constraining the topics studied, grants awarded, and journals published in reproducing marginality in the social sciences.

The persistence of such universalist ontologies and epistemologies is discussed by Raewyn Connell in her book, *Southern Theory*. Connell focuses on how theories of highly influential figures such as Coleman, Giddens and Bourdieu are replete with over-general assumptions which highlight their locatedness within the Western sociological theoretical tradition despite their claims to universality. For instance, Coleman claims that individuals purposively and continuously act to pursue their own interests across social realms, very much like idealized neo-liberal subjects, Giddens conceives of an irresistible and pervasive modernity without considering the diverse forms this modernity takes, and Bourdieu entirely ignores the influence of the colonial conflicts in Algeria on the Kabyle at the very time he was studying and utilizing them as the basis from which to create a 'universally applicable toolkit' (Connell, 2007: 44). Connell also identifies a key failing of what she terms 'Northern theory' in making claims to universality despite its specific geographical, historical and cultural origins. Such approaches either marginalize or subsume 'non-Western ideas', codifying and contextualizing them from a 'Western' standpoint, but more typically, they are ignored entirely, as if the 'non-Western' space being theorized is blank, awaiting 'Western' unmediated interpretation. Lacking reflexivity, these conceptions are unable to assess the specific conditions out of which they emerge.

Such critique constitutes 'a diverse range of responses to different colonialisms that have been differently experienced, encountered and dealt with in different times and places' and a common motive has been to question the continued salience of colonial knowledge (Nash, 2004: 115). Postcolonial geographical theorists have thus identified the historical effects of colonial forms of power and knowledge of space and place and discuss how this continues to be re-imagined. For example, Dipesh Chakrabarty (2000) questions the uncritically challenged tenets of 'Western' theory, calling for European scholarly knowledge to be 'provincialized' so that its situated emergence out of specific cultural, geographical and political contexts can be recognized in order to challenge universalism and provide space for theories that originate from around the world.

Critical scholars have thus responded to the distorted ways in which 'non-Western' cities have been portrayed in metropolitan accounts. In analysing Lagos, Thomas Hecker contends that 'Western' social science has focused on 'mechanistic accounts of spatial disorder, de-beautification, organized violence and crime, inter-ethnic strife, civil disorder, overcrowding, flooding, air and noise pollution, unemployment, widespread poverty, traffic chaos and risk-bearing sexual practices' (Ahonsi, cited in Hecker, 2010: 258). Such doom-ridden representation falsely suggests that Lagos is on the precipice of a total meltdown. Similarly, for Achille

Mbembe and Sarah Nuttall (2004: 348), African cities all too often end up epitomizing the 'intractable, the mute, the abject, or the other-worldly' in scholarship excessively dominated by functionalist, neoliberal and Marxist approaches; or in contrast where Africa serves as a venue for specialists on folklore, tribal patterns and witchcraft.

Such unreflexive generalizations have not only impacted on the study of cities 'beyond the West' but have curtailed the exploration of urban spaces and places beyond city centres, and marginalized the study of small cities. For example, Bell and Jayne (2006, 2009) point to the woeful neglect of small cities by urban theorists who, in seeking to conceptualize broad agendas and develop generalizable models – relating to epochal urbanism and an urban hierarchy – obscure as much as they illuminate. Indeed, given that the urban world is not made up of a handful of global metropolises but characterized by heterogeneity, research into small cities challenges an urban studies orthodoxy that labels them as of 'lesser' significance (Bell and Jayne, 2006; Jayne *et al.*, 2011). And the marginalization of cities beyond 'the West' is accompanied by a host of other distortions in urban research and theorization.

In this context, Nigel Thrift (2000) identifies four persistent myths about cities, two of which are relevant here. First, Thrift argues against the idea that cities are becoming globally homogeneous, for the actual diversity of the distribution of supposedly serial, homogeneous sites such as shopping malls and office blocks is more uneven than commonly asserted and in any case, the historical and cultural context into which apparently similar buildings and institutions are placed informs their adoption. Second, he draws attention to long-standing myths of urban exceptionalism. For example, Canclini (2008) highlights how within academic and popular discourse, certain cities have been celebrated and others pathologized. Berlin and Barcelona are identified as the epitome of 'urban planning innovation' whereas Mexico City is emblematic of uncontainable, polluted, unmanageable 'monstrosity'. Similarly, Paris is held to symbolize early modernity per se, New York is seen as the apex of urban high modernity, and Los Angeles is commonly cited as *the* postmodern city. As chapters throughout this book show nonetheless, there are numerous forms of urban modernity, yet the epistemic privilege bestowed on particular cities is a particularly unhelpful impediment to exploring the distinctive qualities of cities.

Challenging Western-centric urban theory

In thinking about how urban geographies might be reconfigured, Robinson (2003) echoes Chakrabarty (2000) in citing the need for scholars to acknowledge their own cultural and academic situatedness, and advises that 'Western' academic publishing should be more accessible to scholars around the world in order to counter what she calls the 'knowledge production complex'. She further contends that by adopting richer cosmopolitan spatial imaginations and ignoring categorization, urban hierarchies and developmentalism, a more inclusive understanding of the diversity of 'ordinary cities' and a broader range of themes and topics will emerge to extend our understanding of cities that is 'not limited to or fixated by the processes and places of the powerful' (Robinson, 2006: 90).

Robinson (2005: 709) thus champions the development of cosmopolitan theoretical perspectives through comparative urban research, substantively absent in urban theory since the 1970s, ensuring that 'there has been very little reflection on exactly how we go about speaking and theorizing across the very evident differences and similarities amongst cities'. Fortunately, a burgeoning interest in reinvigorating comparative urban research has recently emerged. For example, Nijman (2007a: 5) argues that there ought to be 'no single comparative method but rather a plurality of comparative approaches; it is equally clear that there are no universal or permanently fixed categories'. Similarly, Dear (2005: 251) suggests that 'since all ways of seeing are necessarily contingent and provisional, the best theoretical and applied urban geography will arise from a multiplicity of perspectives' which also includes a focus on cities 'in relation to their history and their networks' (Legg and McFarlane, 2008: 6–8).

Working towards such goals and in order to adopt comparative approaches that provides a method of 'fine-tuning a series of quantitative and qualitative comparative techniques', Kevin Ward (2008: 406) outlines a 'relational comparative' approach to understanding cities which stresses interconnected trajectories and identifies:

> how different cities are implicated in each other's past, present and future which moves us away from searching for similarities and differences between two mutually exclusive contexts and instead towards relational comparisons that uses different cities to pose questions of one another

While we elaborate on the relational dimensions of this cosmopolitan approach later in this chapter is important here to note Ward's (2010) argument that a relational comparative approach should theorize cities as dynamic aggregations of social relations and interactions that are often entangled with processes in other places at varying scales whilst recognizing grounded, local dynamics. In Chapter 8 Dennis Rodgers provides a fine-grained study of the highly distinctive national political factors that continue to impact on the evolution of Managua, Nicaragua, whilst suggesting how local processes and practices resonate with broader globalizing urban change, and in Chapter 7, Nora Libertun de Duren shows how the development of parks in Buenos Aries and New York was influenced by European urban design but also the imperative to draw urban and rural elites into the project of nation building.

Exemplifying such entanglements, and making theoretical, political and policy connections is Cindy Katz's (2004) examination of processes of development and global change through the perspective of children's lives in two seemingly disparate places: New York City and a village in northern Sudan in the 1980s and 1990s. Katz's Sudanese study shows how a large, state-sponsored agricultural programme trained children for an agrarian life centred on the family, a life that was quickly becoming obsolete. She draws comparisons with working-class families in New York City, noting the effects on children of a constantly changing capitalist environment with the decline of manufacturing and the increase in knowledge-based jobs, in which young people lacking skills and education faced bleak employment prospects.

Taking a different approach to considering how 'Western' urban theoretical orthodoxies might be contested and supplemented, Raewyn Connell (2007) draws attention to how four theoretical bodies of literature can contribute to a more dialogic, complex understanding of social theory. She refers to the vigorous debates amongst African intellectuals about the values of reinterpreting forms of folk knowledge as a basis for a distinctive African sociology. Connell also explores critical Quranic arguments of Iranian Islamic intellectuals in response to the baleful influences of (neo)colonialism, Westernized elites and conservative religious authorities. She also points to the development of innovative economic concepts, dependency theories, feminist approaches and critical cultural studies from Latin America. And finally, she describes how Indian scholars emphasize the lingering entanglements between colonizers and colonized in contemporary culture. In particular Connell cites the deconstruction of authoritative histories by the subaltern studies group, the forensic insights of Veena Das into the production of powerful, official forms of knowledge, and the critiques of Ashish Nandy that focus on the acquisition of cultural distinction by (over)reacting to colonialist hyper-masculinity and exaggerating religious identity, in his investigation of the harmful effects that colonialism continues to perpetrate on local identities and cultures.

Other geographical accounts also provide valuable ways of understanding the structuring of cities that cannot be addressed by Western dominated theories. For instance, Singh describes how the sacred symbol of *mandala*, which integrates the cosmos and human, has been superimposed upon the urban form of Varanasi. The main pilgrimage centres and routes in and through the city are complexly woven into diverse cosmological narratives and meanings by the spatial practices of the visiting pilgrims and enacted through their ritual passage, constituting what he terms a 'faithscape' that is inscribed on urban materiality (Singh, 1990: 141). Sacred and symbolic ritual paths within the city have proliferated, and the traversing of these routes integrates 'cosmogonic, mythic and experienced times', undergirding the affective and mnemonic significance of these rituals (Singh, 1987: 23). Varanasi also includes all other major sacred Hindu sites within India, symbolically encoded onto physical sites in the city. Moreover, transposed onto this distinctive urban geography are other pre-eminent holy cities, the 'four abodes of the Gods' located at the north, south, west and east of India, and the twelve *lingas* of light. Cosmologically then, 'Kashi is the paradigm of the sacred place', and symbolically embodies the whole of India; 'Kashi is a cosmopolis – a city that is a world' (Eck, 1991: 138). Whitehand and Gu (2006) similarly consider how Chinese cities were designed under the influence of Confucian cosmology or geomancy. Such concepts entirely bypass Western theoretical notions. In this book, contributors similarly reveal the very particular conceptions of space and place. In Chapter 17 Cuttaleeya Jiraprasertkun discusses the changing conceptions of urban place in Bangkok under conditions of urban sprawl, and in Chapter 9 Hooshmand Alizadeh unpacks distinctively Kurdish understanding of private and public space.

In a different vein, Achille Mbembe and Sarah Nuttall argue that rather than consistently reifying the distinctive otherness of cities beyond 'the West', we can de-provincialize them by identifying how multiple connections with other places help construct ideas of place. They insist that cities are dynamic and are continuously recomposed and reinvented in the present, and an acknowledgement of their indeterminacy, turbulence and provisionality in contexts where life changes rapidly means that understanding of the city cannot be fixed by documentation or empirical description. Accordingly, theoretical understanding always lags behind this ongoing practice and knowledge, in which 'the capacity to continually produce something new and singular, as yet unthought . . . cannot always be accommodated within established conceptual systems and language' (Mbembe and Nuttall, 2004: 349). This chimes with Robinson's (2005: 757) assertion that 'complexity and inherent changeability' means that efforts to establish the meanings of cities 'come quickly unstuck'. Mbembe and Nuttall further insist that a city is not simply a string of infrastructures and technologies but is also a site comprising images and architectural forms, fantasies, desires and pleasures, memories, sensations and manifold rhythms (Edensor, 2010) that emerge through daily encounters and experiences, all manner of cultures, strategies and contingencies that must be accounted for in theoretical discussions of urban life.

In foregrounding ideas that have emerged from beyond 'the West', Connell does not, however, call for the abandonment of 'Western' theory but rather conceives of it as a resource in aiding understanding rather than as a systematically applied diagnostic template. She argues that scholars should become adventurous, and open to 'dirty theory' as part of an omnivorous approach which chooses ideas that suit particular situations. This further highlights that it is crucial to acknowledge the polyvalent and polydiscursive characteristics of 'non-Western' theory that invariably emerges from fluid, multiple and dynamic settings rather than as part of a pure tradition of thought.

A world of cities: key themes

In responding to these challenges contributors to this book meld together grounded insights with theories from and beyond 'the West', in order to expand and multiply the ways in which cities can be understood. More broadly; chapters resonate with a re-imagining of urban theory, in which the city, often unquestioningly regarded as an identifiable and discrete entity, is now considered in line with Amin and Thrift's (2002: 1) suggestion that 'we no longer agree on what counts as a city . . . the city is everywhere and everything'. Similarly, Hubbard (2006: 1) insists that the city can be explored in multiple ways as 'a spatial location, a political entity, an administrative unit, a place of work and play, a collection of dreams and nightmares, a mesh of social relations, an agglomeration of economic activity and so on'.

Accordingly, we ally ourselves with an expanding body of innovative conceptual and empirical work that draws on diverse political, economic, social, cultural and spatial theories in order to better understand urban life. For instance, recent theoretical advances have included a focus on materialities (Latham and McCormack, 2004; Gandy, 2002a), relational urbanism (Massey, 2007; McCann and Ward, 2010; Jones, 2009), mobilities (Urry, 2007), rhythm (Edensor, 2010), assemblages (McCann and Ward, 2011), emotions and embodiment (Davidson et al., 2005) and affect (Thrift, 2004a, 2008; Anderson, 2006, 2009). Pre-eminent among approaches to the reconceptualizing of urban life is *Reimagining Cities* in which Ash Amin and Nigel Thrift (2002) sketch a new terrain for urban theory. Yet for all their dazzling insights, the authors confess that 'it is the cities of the North which we have had in mind while writing the book' while hoping that their 'new perspectives' can be subsequently 'explored by others' (ibid., 2002: 3). Similarly, in Hubbard's *City*, while capturing recent urban thinking about everyday life, the non-human, global mobility and creativity, the author concedes that the book's focus is on the contemporary West, before asserting that the key themes 'are relevant to the study of non-Western cities too' (Hubbard, 2006: 248). Though these problems of omission are troubling, this is balanced by numerous accounts of cities beyond 'the West' that have emerged in recent years.

For example, there been general accounts of 'non-Western' or Southern cities (Seabrook, 1996; Gugler, 2004; King, 2004) as well as publications that focus on particular cities; for instance Dubai (Elsheshtawy, 2009), Jakarta (Silver, 2007), Calcutta (Chattopadhyay, 2005), Jerusalem (Mayer and Mourad, 2008) and Bombay (Hansen, 2001) amongst others. Studies have compared one city with several (Nijman, 2007b; Davis, 2006; Neuwirth, 2005), and compared two cities, such as Mexico City and Sao Paulo (Sa, 2007) and Rio and New York City (Olivera, 1996). Others have adopted a 'regional' perspective, focusing on cities in Africa (Bryceson and Potts, 2006; Demissie, 2008: Locatelli and Nugent, 2009), Asia (Alexander et al., 2007; Dick and Rimmer, 1998; Dutt et al., 1994) and more specifically on Chinese (Zang, 2007: Wu, 2007) and Islamic cities (O'Meara, 2007). In addition, writers have sought to understand how specific processes or features recur or diverge across cities, for instance, in studies that spotlight demography (Montgomery et al., 2004), slums, informality and everyday life (Roy, 2005; Roy and Alsayyad, 2004; Rao, 2006), gentrification (Smith, 2002), women's writing (Lambrigh and Guerrero, 2007), water and infrastructure (Gandy, 2002b), traversing space (Langervang, 2008), social networks (Fawaz, 2008), religion (Valenca, 2006; De Witte, 2008; Henn, 2008), public space and gender (Jimenez, 2006; Alizadeh, 2007), economic development and regeneration (Fernandes and Negreiros, 2001; Bezmez, 2008) and sexuality (Visser, 2003).

Drawing on such key accounts of cities beyond 'the West', we now highlight how the chapters in this book build on and supplement this writing. However, we do not want to reify discussions about cities beyond 'the West', replacing old typologies with new ones. As such, the identification of key themes is intended to open up thinking about cities according to three key imperatives; first, to highlight some of the most interesting directions of study; second, to foreground key issues that emerge from the chapters in this volume; and third, to show how such work chimes with or diverges from recent thinking about how the city might be conceptualized, and in doing so we hence reveal opportunities for urban theory to 'travel' the world.

Urban modernity, globalization and neo-liberalism

As noted earlier in this introduction, developmentalist discourses have all too often positioned 'non-Western' cities as less 'modern' than those in North America and Europe. Lagging behind universal standards of development, such cities, it is claimed, should aim to attain the exemplary levels of modernity of Western cities. Echoing the critical interventions of Appadurai (1996) and Chakrabarty (2000), who regard expressions of modernity as multiple and decentred, Robinson (2006: 90) urges that 'all cities can be understood as both assembling and inventing diverse ways of being modern', thus highlighting changes, similarities and differences between cities.

Contributors to this book trace colonialist comparisons and discuss anti-colonial struggles that have resulted in national independence, with new elites frequently moving to assert their modernity by reshaping the city, trumpeting particular spaces and places as sites of sophisticated progress to rival the colonial metropolis. While urban streets were renamed and statues replaced following decolonization, cities were also venues for the importation of 'international style' architecture to assert that they were not 'behind' the West but were equally modern. As Sunil Khilnani (1997: 118) describes, Indian middle class elites 'aspired to the glistening fruits of modernity tantalizingly arranged before them – street lights, electric fans, tree-lined streets, clubs, gardens and parks', features broadcast by colonial rulers as symbolic of modern urban advancement. In Chapter 5, Swati Chattopadhyay shows how postcolonial states continue to instantiate modes of regulating urban space in accordance with the ordering principles of colonialism. However, these adopted expressions of urban modernity were, and are, contested and complemented by other modernist variants.

Such practices and processes are wonderfully exemplified in the contrast between the architecture of Le Corbusier, doyen of architectural modernism, and the sculptural works of Nek Chand in the Indian city of Chandigarh. Following independence, the modernizing elite captained by Nehru, articulating a philosophy that India needed to prove its modernity on the world stage, argued that this state capital of Punjab and Haryana would be 'unfettered by tradition' in the implementation of an audacious planning programme. Conceived as an ideal representative of this modernizing spirit, Le Corbusier was granted significant powers and the architect 'ensured that extensive visual controls were in place to nurture a homogenous aesthetic and visual standardisation' (Jackson, 2002: 52) in the production of a totalized, zonal structure, where signature modernist buildings, the State Assembly building, Secretariat and High Court were the highlights of a city divided into forty-six sectors, each including three-storey blocks of concrete apartments.

Contrasting with this utilitarian project is a more ambiguous, interstitial and fluid space. Following the ongoing creation of this modern city, and working against its logical and mathematical emergence, road supervisor and amateur sculptor Nek Chand secretively created a sculpture garden on informal scrubland. Built out of the debris caused by demolition of the 42 villages that surrounded the new city, a plethora of animal, human and mythical figures were constructed with pottery shards, broken glass bangles and numerous other scrap material, all bound together with concrete and complemented with unusually shaped rocks sourced from local riverbeds. Although threatened with demolition following discovery, the subsequent popularity of the garden inspired the local authority to provide Chand with 50 workers to develop a large-scale garden that is still being produced, despite the strict rules governing the planned city. In addition to the sculptures, there are non-utilitarian buildings and chambers, grottoes, waterfalls, bridges and balconies.

Jackson (2002: 52) contrasts the human figures in the Rock Garden with the idealized, universal abstraction of the 'Modulor Man', 'a six-foot human with his arms out-stretched' that served as a model to 'help architects adapt their designs to human requirements' through mathematical calculation. This notional citizen, whose ghostly presence pervades Chandigarh's planned city, was disrupted by the protean, multifarious human figures that populated the Rock Garden, devised according to 'circumstance and chance, as opposed to proportional precision, resulting in thousands of variations on several themes' (Jackson, 2002: 55). In creating an interstitial, indefinable realm, now a major tourist attraction, that confounds the rationality of Le Corbusier's Chandigarh, the Rock Garden expresses 'a popular and critical modernity, independent from revivalist or Modernist doctrines' that creatively recycles myth and past lives but certainly does not freeze the past (Jackson, 2002: 61). This specifically Indian modern space exists alongside Le Corbusier's emblematically modern city.

Waleed Hazbun (2010) critiques unreflexive notions of the diffusion of Western modernist progress across the globe as being blind to the local contexts in which ideas and practices travel. Hazbun argues that contemporary beach tourism in Morocco expresses a modern hybrid Moroccan identity that combines distinctive religious practices, familial gatherings and beach camping. Winter (2009) similarly calls for more Asian-centric histories, grounded studies and alternative theories in examining Asian tourisms.

All chapters in this volume thus implicitly engage with the distinctive ways in which modernity is expressed and grounded in settings around the world. In Chapter 10, Christina Jiménez shows how modernism from below in Morelia, Mexico, involves the initiative and agency of neighbourhoods themselves, actively participating in the construction of their own water supply and public squares, belying notions that modern change had to be enforced on a reluctant, 'backward' population by a coercive, modernizing state. By contrast, in Chapter 6 Yooil Bae highlights the strong role of the state in the modern development of East Asian cities while in Chapter 3 Choon-Piew Pow cautions that theorizing from an exceptionalist position with regard to China's urban development must not be at the expense of acknowledging global and regional agencies.

Globalization and cosmopolitanism

Contested notions of modernity have been extended through processes of globalization. Yet despite frequent assumptions that globalization is best located in those cities believed to be at the apex of a world hierarchy, notably New York, London, and Tokyo, this generalized Western gloss is rebuked by Anthony King's (1991: 8) assertion that:

> The first globally multi-racial, multi-cultural, multi-continental societies on any substantial scale were in the periphery not the core. They were constructed under the very specific economic, political, social and cultural conditions of colonialism and they were products of the specific social and spatial conditions of colonial cities. Only since the 1950s have such urban cultures existed in Europe.

Neglect of these earlier urban manifestations of global mixing has nonetheless persisted in theorizing the historical trajectories and geographies through which contemporary processes of globalization have emerged. For instance, cities have been situated as 'core primary' to 'semi-peripheral' and 'marginal' in hierarchical global city typologies that consolidate assertions of Western advanced urban modernity (Huyssen, 2008: 10). In particular, Connell singles out the ideas of Anthony Giddens and Ulrich Beck to critique theories that project universal notions of globalization despite their emergence out of the situated perspectives and pre-occupations of Western metropoles. Such theories neglect the colonial histories of 'non-Western' cultures, 'those at the sharp end of global social processes' (2007: 63). Furthermore, Connell draws attention to accounts that posit a vague 'local' as the other of the 'global', a hierarchical dualism within which 'local' and 'global' cultures combine to produce 'hybrid' cultures, or 'glocalization'. Frequently the local is reified as 'traditional' or 'indigenous' (and therefore static), exemplifying an urban cultural heterogeneity under threat from serial and homogeneous Western culture. Most globalization theory, Connell argues, relies on the production of archetypes, overarching phrases and 'panoramic gestures' to provide synedoches for a variegated, hugely complex process.

Such arguments draw attention to related critical notions of 'cosmopolitanism' where sophisticated, metropolitan urban elites possess attributes that enable them to 'appreciate' the multicultural difference in their midst whereas other urbanites dwell within the realm of parochialism (Binnie et al., 2006). Mamadou Diouf (2000) brilliantly exemplifies forms of cosmopolitanism that emerge in Senegal.

Diouf discusses the Murid brotherhood who set up extensive global trading networks which provide cheap commodities for residents of European, Asian and American cities, by circulating items between metropolitan and 'non-Western' settings and bringing back varied commodities for trade in Senegal. These traders have developed specific circuits of accumulation that do not conform to Western global capitalism but are innovative 'modalities of dealing with acquisition of wealth' that are produced out of continuously adapting their cultural traditions (Diouf, 2000: 680). Such located forms of global modernity are emphatically not 'ways of being' that represent 'incorporation into Western universality and the abandonment of one's own traditions in order to slip into new configurations uninfluenced by custom and religion' (ibid.: 683).

The Murid brotherhood is an Islamic group that originates from the political anti-colonialism of their founder, Amadou Bamba Mbacké, who erected a small mosque in what was the village of Touba. Now at almost a half a million people, with a huge and continually growing mosque, this city has become the centre of a network from which Murids begin their migrations and the place at which they aspire to be buried – Touba remains their spiritual home. The other important urban space for the Murid brotherhood in Senegal is the Sandaga market in Dakar where innumerable goods arrive from within the massive Murid global network. During their travels in Europe, the places where they temporarily reside are also symbolically called 'Touba', and the Murids practice their distinctive forms of ritual and worship, and possess key objects through which they are identified (clothes, bags, religious accoutrements and symbols). As Diouf (ibid.: 702) summarizes, instead of 'trying to demonstrate these modernities by the synthesis or the hybridization of the autochthonous and the global that current discourses on globalization seek to achieve, usually in an inept way', this story is able to foreground the 'creativity involved in the slow and shrewd deployment of the local in global space and time' (also see Kothari, 2008).

Pieterse (2010: 209) underlines how such pragmatic economic activities along with symbols of religion, fashion, body presentation, and musical taste are profoundly shaped by globalized imaginaries and that

> Given the profound presence of these translocal senses of place in the routinized activities of the urban poor, it is clear that we have hardly begun to scratch the surface in understanding what non-localism or translocalism means in terms of identities, interiorities, social practices, networks, intimacies, etc. in psychological, sociological and philosophical senses.

Similarly, Huyssen (2008: 4) draws attention to the plurality of globalized expressions, focusing upon 'spatial diffusion, translations, appropriations, transnational connections, and border crossings'. Emphasizing the multiple and contested effects of globalization in examining the varieties of cosmopolitanism in Bombay, Colin McFarlane (2008: 481) also contends that 'cosmopolitanism can be more or less inclusive or exclusive, and it can be predominantly global, national, or local'.

McFarlane refers to the modern cosmopolitanism of the open and tolerant city of postcolonial modernity championed by Nehru and characterized by planning and social justice (also see Prakash, 2008). This is contrasted to the exclusionary cosmopolitanism grounded in the lifestyle and consumption mobilized by the city's 'globe-hopping elite' (as often represented in Bollywood films) which many argue has, along with the perceived failures of modernist urban development, the communal riots and bombings of the early 1990s and the rise of the Shiv Sena Party, led to the decline of Bombay as a cosmopolitan city. However, McFarlane (2008) identifies a further 'critical cosmopolitanism' from within a collective of civil society groups that have emerged from the city's many informal settlements, namely Shack Dwellers International. This activist network connects similar groups across the world, involving translocal contact, travel and exchange in sharing knowledge and practice.

Ackbar Abbas (2002) maintains that historical and cultural context is crucial in identifying how different, often complex, forms of cosmopolitanism emerge. He grounds this claim in Shanghai, where a memory of a particular cosmopolitanism persists, one based on extraterritoriality, wherein concessions were granted to various European imperial powers to produce 'a polycentric, decentred city' subject to continual negotiations between the various power-holders' (Abbas, 2002: 214). This was a city characterized by great inequality and squalor but also glamour, decadence, diversity and the grotesque, productive of a 'necessary cultivation of indifference' as well as a medley of clashing architectural forms and new kinds of pleasurable public spaces such as cinemas, parks and racecourses that were subsequently utilized and adapted by Chinese locals (ibid.: 215). Subsequently quashed by communist rule, after a half-century hiatus a different, emergent cosmopolitanism is typified by corporate linkages to a global economy and rapid, massive building projects of concrete, steel and glass. Yet the state is also promoting the preservation of historic buildings, fostering a heritage that adds to the city's global cachet in referring back to this glamorous past. Interestingly, Abbas also discusses how the entrepreneurial class who patronize new, expensive restaurants fail to adapt to the protocol of switching off mobile phones to create a sophisticated 'cosmopolitan' atmosphere. In such rapidly transforming urban settings, people resist and adopt global conventions, as identities, mores and ways of being cosmopolitan clash.

One further way of articulating this global adaptation and absorption is to explore the conditions under which a currently pre-eminent neo-liberalism is worked out in different urban contexts. The Murid diaspora discussed above shows how a kind of 'non-Western' cosmopolitanism works within but is not absorbed by neo-liberalism. Our argument against Western-centric urban studies also contests contemporary universalistic assertions that the global expansion of neo-liberal economic and political strategies is inevitable and the sole option for urban development. Critical analyses can explore how such policies are unevenly enacted in particular situations, draw attention to their shortcomings, identify the cross-urban impacts and inequalities produced (such as gated communities and the spread of a transnational urban class), and proffer alternative strategies for development.

Crucially, such theoretical work can subsequently be exported back to Western metropolitan urban settings and be used to inform critiques of the assumptions that hold sway there.

There are, of course, many accounts which focus upon the production of widening inequalities, the diminished role of the local state, and the growth of exclusive private realms. For example, Teresa Caldeira (2008) discusses Sao Paulo as a site of neo-liberal structural adjustment, resulting in the erasure of modernist state programmes, the advent of privatization, rising inequality and crime, spatial segregation – together with fortified enclaves for the wealthy, and new ways of expressing personal and corporate status through architecture. Beatrice Sarlo (2008) shows that Buenos Aires has suffered similar urban processes, notably the tendency to reduce public space within which a range of everyday social practices are enacted and difference confronted. Nestor Garcia Canclini (2008) contrasts Mexico City's astonishing territorial and population expansion with the demise of the kind of state planning approach that conceived of the city as whole. Now, he argues, the city has devolved into fragments, exemplified by measures to create new centres of global service industry and consumption on the urban periphery in the production of an avowedly multinodal city, rather than one with a definable centre. Canclini further contends that increasingly commodification of public space is having negative impacts on urban understanding and experience. While proliferating billboards and media images saturate the everyday apprehension of the spectacle of Mexico City, he asserts that such images conflict with the historically and spatially tethered understandings of urban places produced by novels and other literature. An instantaneous, fragmented and present-tense experience of urban living is thus symbolized in such new audiovisual effusions which have replaced a slower, more gradual temporal experience.

Countering these pessimistic accounts, Arturo Escobar (2010) maintains that the playing out of neo-liberalism takes a very different form in Latin America, where disenchantment with the depredations wrought by structural adjustment policies has inspired a diverse array of social movements and progressive government programmes, notably in Ecuador, Venezuela and Bolivia, that have been particularly motivated by the knowledge and practice of marginalized ethnic groups. Crucially, Escobar maintains that policies of interculturality, pluri-nationality, decoloniality and *buen vivir* (good living for all) are shaped by a relational ontology that decentres the neo-liberal obsessions with individuality, nature-culture dichotomies, capitalized abstractions of space and the primacy of the market. Mixing indigenous, non-modern thought and modernizing reformism, such accounts and practices chime with the network and assemblage approaches of recent critical geographical thought discussed later. He suggests that such non-metropolitan, anti neo-liberal, communal, ecocentric and relational strategies represent innovative, grounded efforts to challenge current neo-liberal hegemonies. More than this, Escobar asserts that the provincializing of Euro-American orthodoxies is a progressive task that such movements fulfil for the whole world, and of course, academies of the 'West' can similarly learn from such thinking and practice.

In this book there are numerous examples of the ways in which neo-liberalism is variously implemented and resisted in particular contexts. Elsheshtawy explores how Dubai has become an epicentre of certain kinds of consumption-oriented urban development but that this urban change has produced shadow spaces for those who labour for this new metropolis but are largely disenfranchised and marginalized in the city. Iossifova investigates the state-led exclusionary gentrification processes in Shanghai and the ways in which these try to displace the activities of those who do not belong to such a vision but nevertheless strive to find their place within a reconfigured urban landscape; De Boeck depicts the grotesque state-led plans to build a giant and exclusive middle class enclave in Kinshasa; and Inzulza-Contardo reveals the specific form that gentrification takes in Santiago. Butcher and Jiraprasertkun highlight how people variously try to make sense of their place in the rapidly transforming cities of Delhi and Bangkok.

Urban mobilities and relational thinking

Many chapters in this book follow the assertion of Amin and Graham (1997: 418) that the city 'needs to be considered as a set of spaces where diverse ranges of relational webs coalesce, interconnect and fragment' and seek to contribute to new and exciting urban research agendas focused on mobilities and relationality/ territoriality. Robinson (2002: 544) considers how a focus on networks that connect different cities, and the overlapping networks that come together in particular cities to produce distinctive urban places allows a more creative 'consideration of multiple social networks of varying intensity, associated with many different kinds of economic and social processes, and with different kinds of locales, or places within the city'. Such overlapping networks of relations bring together people, resources, and ideas in diverse combinations, making cultural, political, design, planning, informal trading, religious, financial, institutional and intergovernmental connections. These complex geographies disavow simple binaries, hierarchies and categories.

For example, while assumptions often foreground key entanglements of cities as either 'Western' or 'non-Western', Percival and Waley (2010) contend that connections pertinent to the development of East Asian cities are largely driven by intra-regional mobilities, networks and transfers. More broadly though, accounts of flows, connections and complexities that have constituted the recent 'mobility turn' in sociology and human geography have tended to focus on the mobilities of elite groups. Here once more, an echo of colonial representations contrasts the immobile colonized with the globe-straddling endeavours of the colonizer. While it is crucial to acknowledge the inequalities of access and potential for being and becoming mobile, there is a danger that over-general assumptions about the relative mobilities of different cities reproduce the inference that 'non-Western' urban life is static, unchanging and inert. On the contrary, as Mbembe and Nuttall (2004: 351) point out, Africa 'has been and still is a space of flows, of flux, of translocation'. Indeed, Pieterse (2010: 208) asserts that the prevalence of forced and voluntary migration across the continent suggests that African cities 'are as much nodal points in multiple circuits of movement of goods, services, ideas and people, as they are anchor points for livelihood practices that are more settled, more locally embedded and oriented'.

In this context, Diouf (2003) importantly highlights how the collapse of systems of supervision and authority in many African cities has ensured that young people have become detached from their earlier roles as important actors in the construction of the postcolonial nation. Following economic crises, political corruption and failures of national development and state-building; unmoored from the duties and family structures and place and community based identities, young people are forging new, mobile allegiances. For example, their mobile practices can cross national and ethnic boundaries; can be inspired by global imaginaries that stimulate the pursuit of consumption of fashion, music, sexual adventure, or religious, criminal, political, or military activities. Criticized and feared, conceived as decadent and dangerous, young people, whether migrant workers, criminals, or artists, are now moving through and across 'the margins and the unoccupied areas in which emptiness and indetermination are dominant: places that are ready to be filled, conquered, and named' and where presence can be marked and broadcast and identities developed (Diouf, 2003: 5).

In Chapter 2, AbdouMaliq Simone concurs with these depictions of the demise of coherent and enduring locales constituted through stable social connections and more secure forms of state regulation. Simone shows how in the cities of the South people typically attempt to maximize the potential for future opportunities by contingently establishing connections with as many people, networks and scenes as possible. Often moving beyond their immediate locales, people disperse their social, commercial and leisure activities across urban spaces, constructing and sustaining social contacts and maintaining fragile reciprocities and obligations. Making use of the innumerable heterogeneous materials that flow into and through the city – stories, objects and ideas – people weave together fragile performances, offer commodities and services, and make ad hoc arrangements and collaborations in a fluid and ongoing series of affiliations (also see Simone, 2005). Simone posits a contingent relationality in which urban space is continuously remade through the forging of volatile and sometimes enduring connections, and he calls for state intervention that acknowledges and utilizes the potentialities that emerge out of such mobile interactions.

Such mobilities can also be discerned in formerly ferociously policed Johannesburg, where mobility and residence were strictly delimited. For example, Hilton Judin (2008) describes how post-apartheid conditions permits movement across urban space, and the colonizing of informal and public spaces for improvisational practices including dwelling, hawking, raising livestock, hairdressing and worship. Mbembe (2004: 369) concurs that this city's boundaries have recently become so 'geographically and socially permeable and stretched that the city seems to have no fixed parts, no completeness, and almost no discrete center', and is instead 'an amalgam of often disjointed circulatory processes . . . a place of intermingling and improvisation'. In a different take on the ongoing production of urban space by mobility and stability, Mehrotra (2008: 206) explains how the retreat of the state in Bombay has produced a 'new, bazaar like urbanism', created by those outside of the new elite, that 'slips under the laws of the city simply to survive' but without thought of cultural resistance. Key features of this urban scene are 'processions, weddings, festivals, hawkers, street vendors and slum dwellers' who create an 'ever-transforming streetscape', producing an urbanism that expresses the 'kinetic city', the city in motion, impermanent and composed out of recycled materials, continually modified and improvised upon, a 'temporal articulation and occupation of space' (ibid.). This 'flow, instability and indeterminacy' exists along-side the 'static city', the 'official, concrete and monumental, incorporating malls, gated communities, and architectures of commerce and power and the new high-ways, flyovers, airports, corporate hotels, convention centres, galleries, and museums that declare the city's integration with the global' (ibid.: 214). Yet the two realms often blend, for instance where informal bazaars occupy colonnades built to provide shadow and shelter for pedestrians.

Such inexhaustible urban mobility and relationality/territoriality is also recognized by Yingjin Zhang (2008: 239) who shows how recent cinematic portrayals of Beijing feature marginalized 'locals and translocals' who engage in 'drifting, a mode of operating within the city that privileges 'ambivalence, contradiction, contingence and improvisation'. However, it is vital not to simply celebrate urban mobilities that are often generated by a compelling need to escape endemic poverty but to acknowledge how people are often at the mercy of organized gangs who cream off profits and subject them to lies and violence as part of the exhausting, dispiriting process of moving across urban space to seek a livelihood (see Butcher, Elsheshtawy, Iossifova and De Boeck, this volume, for accounts of restrictions and divergent access to mobility).

While depictions of urban fluidity risk overlooking the disparities between the mobility of capital and labour, between the powerful and the weak and between relationality and territoriality, they nonetheless importantly point to a dynamic urbanism that seems qualitatively different to accounts of the mobilities of Western cities. For instance, in focusing on connections through which cities are continuously assembled, Colin McFarlane (2011) argues for relational/territorial thinking that is open to multiple spatial imaginaries and practices. He shows that while 'informal settlements' in São Paulo may be controlled and subject to iniquitous treatment, homemaking can nevertheless be conceived as an ongoing and incre-mental process through which shacks are gradually amended and supplemented with more enduring material through a process of tinkering and tweaking over time. These improvisations in dwelling are not merely impromptu but are the product of bringing together materials obtained through different connections with places and people as part of an ongoing assemblage, a process of practical adaptation that resonates with Simone's account. Accordingly, the 'slum' community that McFarlane examines, Paraisópolis, is continually being territorialized, deter-ritorialized, and reterritorialized, revealing that the idea of dwelling as a locally bound process rooted in the inertia of the poor should be replaced by a notion of continuous assembly, the gathering and dispersal of materials in transforming living quarters also fostering the sensual and practical apprehension of the city.

Regulation: modes of ordering/disordering

Issues of mobilities and relational and territorial thinking also resound through the modes and practices of urban ordering and disordering discussed by contributors throughout this book. Notions of urban regulation are a lasting legacy of colonialism, for the very structure of colonial cities was often amended, typically through the production of a 'dual city'. Sunil Khilnani (1997: 116) explains that the pre-colonial Indian city was a congeries of separate groups delineated into neighbourhoods by caste and religion, which did not appear as 'a single and cohesive space, which could be rationally administered, ordered and improved'. When confronted with what they regarded as chaos, Spurr (1993: 90) reveals that pre-colonial travellers first imagined an entire series of European institutions on the natural landscape and in urban terms this eventuated in the rigid division of the city into the 'native quarter' and the European area, or cantonment, a division still easily identified in contemporary post-colonial cities, as De Boeck describes in discussing the evolution of Kinshasa in Chapter 19.

Cantonments were produced in order to instantiate a perceived spatial order in contradistinction to the 'chaotic' 'native quarter', and was replete with references to European urban ideals – wide, tree-lined streets, parks and gardens, tennis courts and golf clubs, gothic style churches, domestic architecture that echoed suburban house and garden design, street names and statues of historic and colonial figures – and above all, in the case of the British, the key institute of the club, where national and communal rituals could be enacted through drinking, sport and civic participation. The residents of this highly regulated district employed servants, cooks and gardeners from the 'native' population but borders were carefully guarded to delimit the presence of unauthorized others. Crucially, however, for these colonizers, the 'native quarter was conceived as the urban antithesis to European civilized order' and was:

> the out-of-bounds city where the living and the dead intermingle ... the carnivalesque world of the bazaar city where nothing is delineated but everything exists in a chaotic state of intermingling: a carnival of night and a landscape of darkness, noise, offensive smells and obscenities.
>
> *(Parry, 1993: 245)*

Simultaneously feared and desired, the representation of these mythic realms emanated out of the suppressed fantasies of the colonizers and provided a 'malleable theatrical space' for 'self-dramatisation and differentness' (Kabbani, 1986: 11). Such representations continue to inform a Western imaginary that liberally applies chaotic characteristics to 'non-Western' cities, exemplified, for instance, in the representations of tourism and the responses of tourists to urban realms (Edensor, 1998) as we can see in the lure of Rio's favela tours discussed by Jaguaribe and Salmon in Chapter 15.

Such imaginaries have also informed the contemporary regulation of cities in certain 'Western' contexts. Richard Sennett (1994) has argued that such modes of regulation are reducing the potential for confronting and engaging with difference and that the urban fabric is becoming increasingly homogeneous, sterile and desensualized. Though such visions of cities in Europe and North America are contentious, imperatives and desires for order are a legacy that also persists in the regulation of post-colonial cities, reproducing expressions of power in urban space. Yet ordering policies coexist and intermingle with the informal, contingent appropriation of state utilities and urban space and the transgression of planning and transport rules. As Chattopadhyay shows in Chapter 5, theorists have yet to make sense of such processes and practices, and the conditions in such settings are all too often interpreted as representing disorder rather than considered as differently ordered. Accordingly, a key theme in this book is to look at the ways in which cities are organized, managed and regulated and how these processes are confounded and challenged. Such an exploration both identifies the specificities of particular urban cultures and critically interrogates the orthodoxies of 'Western' and 'non-Western' urban regulation.

In analysing urban regulation, Paasche and Sidaway reveal how neo-liberal imperatives are conditioning the ordering of Maputo, in the context of growing inequality, an intensified consumerism and privatization, and the development of gated residential enclaves, exclusive shopping areas and intensified security provision for a growing middle class. However, they describe a regime that is not systematically or centrally organized by the state but 'a complex patchwork of security enclaves' (Paasche and Sidaway 2010: 21) patrolled by a piecemeal collection of police-officers, private security guards and other, informal sentinels. Public order may be or may not be maintained by such interventions, which produce an unstable, contingent and fluid urban governance and a blurring of public and private space. This example highlights how the regulatory specificities of urban spaces should not be measured against some abstract and universal norm, for different cities mobilize various approaches to managing social life, diversity, the relationship between public and private, and the flows of goods, people, animals and traffic in very different ways. Indeed as Hooshmand Alizadeh illustrates in Chapter 9, Kurdish notions of ordering urban space according to concepts of privacy are at variance to Western concepts of public and private space, but also to the organization of other Islamic cities, revealing that ideals of spatial division may be grounded in local, everyday space-making. Similarly, in discussing political ecology in Chapter 18, Laura Shillington provides us with a grounded portrayal of the ways in which residents of limited means living in an informal settlement in Managua intensively manage the effects of urban nature, organizing domestic space and dealing with water quality and supply, sanitation, flooding and garbage disposal through inventive, adaptable, small-scale horticultural and conservation strategies in a context in which state regulation is unreliable or non-existent. At a different scale, in Chapter 7, Nora Libertun de Duren shows how a very distinct sense of European modern order influenced the evolution of San Isidro Jockey Club Park in Buenos Aires, in contradistinction to the valorization of native American nature in the design of New York's Central Park. Intended to seduce Argentina's rural elite into accepting the qualities of urban modernity, the park mimicked the geometrical designs of European cities in reinforcing notions that controlled nature represented the orderly modernity of the city.

In theoretical engagements with 'Western' cities, certain regulatory approaches are presented as taking hold which purify urban space, reduce meanings and practices and instantiate widely followed norms of bodily and social conduct. Urban elites further install a complex of surveillance, including CCTV, policing and security guards, increase the spread of privatized space, micro-manage flow and inaugurate rigorous sensory and aesthetic control. Practices of congregating, busking, loitering, begging, peddling, protesting, sleeping and forms of expressive behaviour are thus considered to be the subject to intensive surveillance and prohibited, with preferred modes of comportment limited to 'browsing' and 'grazing', as urban space becomes a function of rapid passage. Though interstitial urban realms remain, Richard Sennett (1994) laments the loss of diversity and the 'tactile sterility'.

In contrast, bazaar spaces of urban India are multifunctional, mixing together small businesses, shops, street vendors, public and private institutions and domestic housing (Edensor, 2000). Hotels co-exist alongside work places, schools, eating places, transport termini, bathing points, offices, administrative centres, places of worship and temporary and permanent dwellings. Bazaars are depicted as being 'centres of social life, of communication, of political and judicial activity, of cultural and religious events and places for the exchange of news, information and gossip' (Buie, 1996: 277), and so loitering, sitting and observing, meeting friends and carrying out domestic activities such as collecting water, washing clothes, cooking and childminding are part of daily experience. Moreover, mobile hawkers, beggars, musicians, magicians and political and religious speakers move across space and attract crowds, while sensual and performative practices of barter, according to Buie (ibid.: 227), are akin to an 'art', 'ritual' or 'dance of exchange'. The mixture of overlapping spaces and numerous micro-spaces also provides a labyrinthine physical urban structure that enables a flow of different bodies which crisscross the street in multi-directional patterns, with vehicles paying little heed to formal traffic rules as they jostle for position. Itinerant beggars and workers are rarely advised to 'move on', and animals are often left to share the streets. Walking cannot be a seamless, uninterrupted journey but must engage with interruptions and encounters, weaving a path that is accompanied by a variegated sensual experience of space. Visually, an unplanned bricolage of structures is infested with various signs, personalized embellishments and unkempt surfaces and facades, and an ever-shifting series of juxtapositions of diverse static and moving elements can provide surprising scenes. A particular haptic geography emerges from the continuous weaving amongst other bodies, the mix of diverse aromas produces intense 'olfactory geographies', and the combined noise of human activities, animals, forms of transport and music produces a changing soundscape.

Such 'non-Western' spaces are clearly regulated according to different conventions, with contingent and local planning practices and local power holders often exercising exclusion and control, perhaps on the basis of gender, ethnic and religious identity, but any systematic, over-arching regulation is replaced by low-level surveillance and tolerance of difference with no desire to limit the potentialities or ambiguities of place. The poor and all manner of human and non-human differences are confronted in a scenario that only appears disordered from a particular culturally located perspective (Chandhoke, 1993). Despite this Chakrabarty (1991: 16) expresses concern that increasingly, 'the thrills of the bazaar are traded in for the conveniences of the sterile supermarket'. Other accounts reveal how loose modes of regulation permit particular forms of resistance and expression in urban public space. For example, Victor Vich (2004: 61) shows how street comedians in Lima produce 'theatrical interpretations of social life' by using humour to address the dilemmas and difficulties of 'getting by' and the development of strategies for better livelihoods, developing a critical politics in response to harsh economic conditions and social inequities.

Lisa Weisenthal (2010) argues that these different forms of living in and moving through space are exemplified in state-subsidized housing in the South African township of eMjindini. State policies are concerned with fixing relationships, establishing permanency of residence and maintaining the static, legible, orderly appearance of buildings in this post-apartheid project. However, the fluid social networks across which inhabitants move and temporarily dwell according to contingencies of work, finance and relationships under conditions of uncertainty, violate these norms of propriety, fixed ownership and aesthetics, as temporary, traditional buildings are erected on household land and residents and owners come and go. In such a scenario, the state is unable to acknowledge the multiplicity and adaptability of spaces that are in constant negotiation and contestation. Similarly, in discussing dystopic representation of Lagos instead of understanding apparent disorder as connoting a lack of regulation, Hecker (2010: 263) suggests that we should focus on hybrid, fleeting assemblages that fill infrastructural gaps. For example, creativity involved in adapting and improvising in and around the traffic system, where road blockades are erected to funnel traffic slowly through small neighbourhoods, 'is a situation rigged most certainly to bring traffic and thus business to those areas' (ibid.). In identifying such contingent entrepreneurial strategies – and resonating with the earlier discussions about urban/global modernity, (dis)ordering and mobility – Lagos can be represented not as pathological but as a city at the forefront of globalizing modernity, emblematic of a more sophisticated, resilient and adaptable form of order, 'an emancipated city of intense flows, in which circulation is as important as any space or place; a landscape of mobile flaneurs crossing an interstitial terrain powered by leaded fuel' (ibid.). In the same register, Neema Kudva (2009) considers urban informality as a contingent manoeuvring within enmeshed networks of labour, employment and shelter, and between moments of entrepreneurialism, which relates to resistance against state repression, crime, religious fundamentalism and political violence (also see special issue of *African Studies Quarterly*, 2010 on urban informality). These conceptions and contestations are addressed in Chapter 12 by Iossifova who discusses adjoining spaces of informality and a regulated middle class enclave in Shanghai which exist in uneasy tension. She shows how state policies presently favour the city's increasing elite and have attempted to intensively regulate the mobility, trading and cultural practices of the residents of the informal area, minimizing their contact with, or intrusion into, the enclave. Nevertheless, she highlights how these impositions are being challenged and negotiations between residents of the two spaces are emerging to accommodate their difference. In a different context, in Chapter 19, De Boeck shows that the precarious informal social and economic – often agricultural – practices of residents of Kinshasa on land formerly controlled by the Belgian colonial rulers, are threatened by bureaucratic attempts to 'clean up' the city, a policy that also involves the demolition of roadside businesses and dwellings, the upgrading of transport, a giant elite housing project, and the severe regulation of street children. De Boeck argues that such modernist-utopian, neo-liberal visions seriously threaten the viability of Kinshasa's crucial informal economy.

However, while such work offers important insights into urban mixing, it is important not to over-romanticize such differently regulated realms. Indeed, Edgar Pieterse provides a sobering reminder of the impact of disorder and the absence of regulatory frameworks, of regular policing, employment and schooling. Cautioning against the optimistic interpretations of those who emphasize the innovative, entrepreneurial aptitudes of slum-dwellers and mobile youths, Pieterse (2010: 213) argues that the central experience of African urban life is 'first and foremost a story about terror; a narrative about multiple forms and patterns of abuse', characterized by the symbolic violence of poverty and deprivation and the very real violence of political coercion, extortion, prostitution, crime and corruption. This violence, 'that pumps energy through the spaces of daily survival and adaptation has a deep and expansive root network that it feeds off', originates in the mechanics of colonial rule (ibid.: 214). The depredations wrought by political parties, local gangs, and crime-lords are woven into the everyday experience of the poor, and while there are aspects of mutuality, this is a largely exploitative and brutal process that debases individuals. In calling for a more grounded, realistic understanding of contemporary African urban processes, Pieterse contends that developmental and state policies cannot circumvent scenarios in which 'politics and social life are irrevocably fragmented and reproduced by powerful vested interests that feed off dysfunction and low-intensity but routinized violence' (ibid.: 217). While Pieterse's arguments are timely, there is a danger that the desolate urban scenarios he portrays are taken as emblematic of all African and 'non-Western' cities.

Urban imaginaries

The final key theme which runs throughout the chapters in this book relates to Edward Said's (1993) notion of 'geographical imaginary' which refers to the attribution of meaning to a space or place so that it is perceived, represented and interpreted in a particular way, reproducing knowledge and facilitating particular courses of action. Said's conception primarily draws attention to the imaginative processes that fuelled discourses and representations of colonized spaces, and hence made colonialism viable. The colonial imaginary remains crucial in stimulating responses to the 'non-Western' city, influencing policies and ideals of post-colonial urban elites and having a continuing legacy in popular representations and contemporary developmentalist discourse.

In Chapter 16, Sofia Shwayri provides a chilling echo of such colonial imaginaries, revealing that American and Israeli military strategists imagine the city of the Middle East as a Muslim-Arab city, chaotic, lawless and underdeveloped, in contradistinction to the American city. Across the US and the world, as many as one hundred ersatz villages and towns have been constructed as sites for military training, pseudo-urban settings imagined as venues for war, replete with Arab actors dressed to simulate urban activity, lines of washing, mosques, taped calls to prayer, the olfactory imitation of corpses and sewage, and burnt-out cars. Though these imaginaries are informed by distorted conceptions of urban otherness, as Andreas Huyssen (2008) remarks, cities are sites of multiplicity, characterized by proliferating identities and perspectives, and a sense of place that is continuously produced by interacting with urban spaces at an everyday level. Emerging out of these engagements are manifold ways in which urbanites with different ethnic, class, gender and religion-based subjectivities imagine their own and other cities as 'the site of inspiring traditions and continuities ... the scene of histories of destruction, crime and conflicts of all kinds'. In short, an urban imaginary is 'the cognitive and somatic image which we carry with us of the places where we live, work and play' (Huyssen, 2008: 3). The urban imaginary is thus stimulated by the individually and collectively lived experience of the city, (re)produced by tourists, middle-class consumers, entrepreneurs, planners and architects, slum-dwellers and suburbanites, police officers, taxi drivers, street vendors and students as they work, dwell in and move across urban space. We have already encountered the urban imaginaries of Diouf's and Simone's African urban young people; the Murid traders of Senegal and the informal entrepreneurs in Lagos; Canclini's Mexico City inhabitants and the participants in the social movements depicted by Escobar; the inhabitants of the static and the kinetic city; and various Bombay cosmopolitans, as well as the fantasies of Le Corbusier and Nek Chand in Chandigarh. Our desire to multiply the readings of the (non-Western) city is further served by the numerous imaginaries depicted in this book.

For example, returning to Shanghai's mixture of nostalgia and futurity depicted by Abbas, the recycling of mediated images and ideas from 'old Shanghai' combine with extraordinary futuristic new architecture, proliferating electronic signage, sci-fi imagery and the saturation of information technologies in order to provide a complex urban imaginary that adds allure to its potential as a future corporate capital and global powerhouse (Lagerkvist, 2010). The fantasies of lifestyle, cosmopolitanism and modernity that pervade such attempts to market the glittering image of a 'global city', as Iossifova highlights in Chapter 12, are enticing a new middle-class elite who imagine that they will live and express their identities unhindered by others. However, such urban imaginaries do not necessarily involve a social and spatial segregation but rather an accommodation of the social and economic practices of poorer neighbours who impinge upon the newly emerging gentrified spaces. This process is also compounded by the numerous rural people drawn to Shanghai by this alluring cosmopolitan, opportunity-laden image. As such groups mingle and clash within the city, subjective imaginaries must adapt if they are to rub along together and negotiate shared urban space. Yet subjective imaginaries invariably provoke perceptions of other urban inhabitants and spaces. In Chapter 11, Butcher reveals some of the affective responses of people from different backgrounds as they move across the varied spaces of an increasingly gentrifying Delhi, where a sense of belonging or exclusion, often stimulated by sensory impressions and based on gendered, class and communal identities, highlights how the city may be read according to a host of contesting urban imaginaries, irrespective of official discourse.

Indeed, attempts to re-imagine cities are not only perpetrated by state and business elites. Mamadou Diouf (2008) describes how a Dakar riven with corruption and decadence has been re-imagined as progressive and vital by dynamic social movements, notably by the *set* campaign to clean up the city and the progressive and politically critical messages transmitted in the music of *mbalax*. In Chapter 17 Cuttaleeya Jiraprasertkun describes a melding of old and new urban imaginaries which informs the evolution of a sense of place in three very different areas of Bangkok, where ideals about the distinctively Thai qualities of place variously intersect and clash with the modernist imaginaries of planners and bureaucrats. Similarly, Alessandro Angelin (2010) describes how teenage boys in the favela of Pereirão in Rio de Janeiro built a scale model of their own community made up of a miniature world of houses, plastic figurines and toy cars. The boys have also developed theatrical, role-playing games that re-enact everyday practices and conflicts between gangs and the police on this miniature stage. In mimicking and critiquing state programmes of pacification in the favela, Angelin (2010) describes how such play is a 'technology of the imagination' through which young people make sense of the social order of the favela and their own roles within it.

Such imaginaries also offer a counter-balance to negative and essentialist presentations of the city beyond 'the West' by presenting it as an enigmatic object, described through mythic and fantastic accounts that recombine scraps of discourse and hearsay. For instance, writers critique 'Western' notions of urbanity by producing accounts of shadow or spectral cities (Appadurai, 2000; Simone, 2004; Neuwirth, 2005). Simone (2004: 93), for example, describes how residents of Douala,

Cameroon, express a belief that the city is haunted through 'a series of refractions among real life, artifice, imagination and actions whereby residents hedge their bets as to what events, relationships, resources and opportunities actually mean to their everyday navigation of the city'. Exemplary of such accounts is Chapter 19, in which Filip De Boeck critiques the modernist urban imaginaries, provoked by developments in Dubai and Shanghai, which have stimulated plans for an exclusive, gated community in Kinshasa, including office blocks and a conference centre, that is advertised on huge billboards across the city. He argues that this fantastic realm, which may or may not eventuate, has a corollary, a spectral vision of the intensification of poverty, favela-ization, and spatial division as the in-equalities of the city multiply under neo-liberal rule. Yet De Boeck carefully notes that such shining urban myths are not only dreamed of by the elite but can resonate with a widely shared imaginary of a utopian future.

Urban imaginaries are also importantly stimulated by images and narratives produced by film and television (not to mention literature, music, art and photo-graphy). The cinematic representations of Bombay in *Slumdog Millionaire* and of Rio de Janeiro in *City of God* have been particularly important in provoking an imagining of these particular cities but also by fuelling a sense of difference that adds allure to their consumption by Western audiences (as do, for instance, the urban spaces represented in the slew of American 'hood' films produced in the 1990s). McFarlane (2008: 483) emphasizes that 'film is a key repository of the urban imagination in Bombay, continually reproducing and contesting narratives and images of the city as variously cosmopolitan or divided, violent or hospitable, booming or in decline, collapsing or developing'. He argues that such filmic representations significantly impact upon urban discourse and imagination, pro-ducing the 'city of spectacle' in contradistinction to, but also sometimes imbricated with, the everyday, routine city of 'informal settlements, dense neighbourhoods, street hawkers, traffic congestion, construction debris and refuse' (ibid.). Illustrat-ing such impacts, in Chapter 15, Beatrice Jaguaribe and Scott Salmon show how the increasing popularity of favela tours have been stimulated by *City of God*, yet favelas are the subject of diverse imaginaries, serving as a site to broadcast the success of government social policies and policing, as realms of a diverse and hybrid creative popular culture, and as venues for vicariously experiencing the thrill of drug crime and potential violence, a set of competing and overlapping understandings that are embedded in tourist representations of the city.

In summary

Contributors to this volume have explicitly sought to respond to the challenges involved in advancing urban theory with reference to the themes and debates outlined in this introduction. In doing so each chapter responds in different ways to Ward's (2008) view that comparative urban research must be attuned to 'theorising back' in order to offer reflection on the geographically uneven founda-tions of contemporary urban scholarship and to engage with previous research.

Contributors explicitly acknowledge the complex and diverse ways in which urban spatiality is the product of distinctively situated social relations, practices and institutions whilst recognizing broader global processes. Throughout the pages of this book the city is theorised as mobile, fluid and relational, and territorial and fixed, being continuously contested, perceived and conceived according to a host of imaginaries. In pointing to the spatial and temporal openness of the city, chapters also offer important theoretical insights into multiple urban rhythms and experiences across time and space, in order to show how 'imprints from the past, the daily tracks of movement across, and links beyond the city' are key elements in understanding the multiplexity of urban life (Amin and Thrift, 2002: 8). The diversity of cities featured in this book reveal that urbanity beyond 'the West' cannot be classified in any simple way. Contributors thus tend to disavow the dualistic use of terminology – although often drawing on labels such as 'first, second and third worlds', 'developed' and 'developing', 'North' and 'South', 'majority' and 'minority' and so on as heuristic devices – in order to respond to the need 'to multiply the readings of the city' (Lefebvre, 1995: 159). Chapters in *Urban Theory Beyond the West: A World of Cities* not only seek to advance urban theory but do so by offering new and fruitful avenues for urbanists to make significant and innovative contributions to broader political, economic, social, cultural and spatial theoretical debates.

PART I
De-centring the City

2

NO LONGER THE SUBALTERN

Refiguring cities of the global south

AbdouMaliq Simone

Cities for those on their way somewhere

In Jia Zhang-ke's film *24 City*, Chengdu hovers like some unreachable object. The film concerns the demolition of an aeronautics factory that was a world unto itself. In its mixture of 'real' and 'fake' interviews with workers and children of workers it traces how a seemingly self-sufficient world moves to its own dissipation; how the city grows up around it, and thus how the sale of land enables the company to raise the money to modernize technologies that hadn't really changed for 50 years. But also how the factory complex – now to be the site of a massive luxury real estate development – persists as a kind of affective field; one of sacrifice and respect that enables subsequent generations to trace their way into the larger city. Districts such as these represent techniques of traceability – how the continuous struggles and innovations of urban everyday life trace their way across individuals, spaces and times; how they attempt to implicate themselves into other scenarios; how they aggregate and regroup.

Processes of urbanization are still ingrained conceptually as consolidations of specific kinds of places, sectors and scales (Brenner and Elden, 2009). Similarly, associational life tends to centre on participation in discrete organizations rather than on networks that associate different times, actors, actions and materials with different intensities, immediacies and access. Identifications of territories as embodiments of particular statuses and characteristics are truncated versions of the ways in which any location is a criss-crossing of bodies, materials and imaginaries passing through (Hillier and Vaughan, 2007). While such movement leaves traces that thicken into apparent structures through particular practices of investment, infrastructural design and human accommodation, every space is a product of extension and folding and is folded into various elsewheres in many different ways (Read, 2005).

People and materials that operate within any space actively or potentially can step in and out of different senses of what is required and possible, different performances and framing devices, different vertically layered strata of articulation, different ways of paying attention and of being implicated in what is going on. As Stephen Read (2005) states, places are always on their way somewhere, with different reach and possibilities, and always transformed by what people, materials, technical and discursive instruments do in the passing. This doesn't mean that people are always mobile, adaptable and flexible; it doesn't mean that some people are not incessantly cornered or enclaved. These processes, too, are aspects of urbanization, part of the trajectory of oscillating movement through which a wide range of economic mobilities are hedged through the cordoning off of others. So as physical and political infrastructures stratify movement into different interactional possibilities and steer people into specific densities and speeds, as well as open up weakly controlled reverberations, collective life makes itself known in various ways.

Still, when accumulation is privatized without regard to the ways in which access to critical resources are distributed, how urban residents are housed, fed and otherwise sustained becomes increasingly a matter of individual initiative, contestation and guile. Individuals, households, neighbourhoods and associations are set against each other, vociferous in their claims, cut-throat in their practices. Collective life and its institutions of governance and provision are recalibrated as instruments to remake everyday transactions into commodities. This is not the imposition of an overarching framework of rule where familiar modalities of deliberation and economic production are simply overcome or overwhelmed (Peck *et al.*, 2009).

Rather, turning social life into commodities stems from a series of cumulative impacts engineered through the enrolment of specific parts of cities, administrations and networks into inter-jurisdictional policy circuits. These circuits promise to enhance the authority and reach of actors and institutions preoccupied with often highly localized concerns about their legitimacy. That which perhaps at one time existed in order to take care of as many as possible or to ensure an equitable distribution of basic opportunities now takes care of itself by tending to others promising to bring development and modernity in return for a lion's share of available assets. Municipal and national governments, resource parastatals, customs and taxation agencies, development ministries and regulatory agencies that had found their viability in how well they orchestrated the competing claims or escalating demands of their constituencies – now increasingly act as 'fixers'. They become conduits of tricks, conditionalities, seductions and extortions that seem to deliver the compliance or at least the quiescence of those citizens they purportedly still represent but have left hanging.

The actions and regulation of majorities

In face of such urban conditions, what then are the majority of residents expected to do – from those dependent on subsidized consumption, income grants, theft and sporadic employment to those who manage to access some steady form of entrepreneurship or wage labour but who continuously fall behind one way or another? Even for residents who find ways to work almost around the clock, who can give and call on favours, and who can put small amounts of money away or make a succession of small investments, expenses must be smoothed over and shortcuts taken. For the millions upon millions of residents whose life in the city has been predicated upon 'doing the right thing' – saving, educating and toiling with dogged persistence, the pay-offs have often been few. Broad-based mobilizations of collective effort have been channelled into sustained demands for rights – to shelter, livelihood, political participation – and better access to water, power and sanitation. This effort has forced cities to accommodate a population pursuing more heterogeneous notions and practices about what urban life could be.

At the same time, cities became receptacles for individuals and entire communities displaced from towns and villages by the structural undermining of agricultural and craft economies. These people are largely maintained now in a constant state of suspension, with few bearings or prospects. As cities are opened up to increasingly vociferous demands for space, resources and opportunities, they become more differentiated, as residents live through the specificities of their daily circuits, occupations and conditions. It is often not clear who is poor and what criteria constitute the poor, or middle class, for there are prolific gradations of definition.

The capacities of capital to circulate across broader geographies and to leverage the specificities of place and time as advantage without making long-term commitments weaken the anchorage that residents might have had in neighbourhoods, factories and civil services (Roy, 2008). Residents must be prepared to constantly readjust to the vagaries of shifting money and territory – as businesses and public domains open and close. The mobility of capital and the quickly changing valorizations of what a city has to offer become ways of exerting power, forcing concessions, population controls and diversions of funds. The proliferating militarization of space, the miniaturizing of local administration and public services – all familiar strategies of colonial rule continue to be reworked and redeployed as a means to constrain internally generated urban productivity (Dick and Rimmer, 2009). Cities in the South do demonstrate insurgent planning practices based in various histories of anti-colonial struggles and urban social movements that have fought for citizenship rights, as well the destabilization of prevailing conceptions of citizenship (Miraftab, 2009).

At the same time, contemporary urban governance demonstrates an often effective ability to debilitate capacities for both grassroots and city-wide initiative and organization. This occurs largely in city government's unwieldy mixture of ineptness, indifference, incessant policy revisions, the partial and selective engagement of 'popular organizations' and the deployment of more proficient technologies of control (Coutard, 2008; Healy, 2007; Laquain, 2005). As municipal resources are dedicated to funding projects that might exert a gravitational pull on circulating capital, the capacities of the city to produce, to make things and to make viable lives diminish. As the gap between municipal government and the governed grows, despite the many procedures to decentralize authority and maximize public participation, the onus of social reproduction is increasingly placed on residents themselves. What then becomes possible in terms of mass organization, confrontation and political change? Even when authoritarian rule and direct repression is mitigated, the physical and psychological spaces for mobilization shrink. Opportunities for economic advancement for an urban majority may register sufficient discernible improvements to keep people from feeling that they have nothing to lose. But it is never enough to overcome a growing perception that most households are falling behind.

As such, most residents feel that there is no way they can go it alone. Even if as political and economic citizens they are accountable and eligible as individuals or households, bearing the responsibility of their choices, survival means collaborating with others in everyday life outside of the institutions of work, training, politics, or worship where the terms of such collaboration may be defined. As such, to not go it alone, to make a life with others is something that must be incessantly negotiated and a matter of speculation. Alliances and scenarios must be quickly adopted as well as discarded; long term trust and reciprocities can't get in the way of opportunistically finding a way to become a critical component of schemes happening elsewhere.

The tenuous outlines of collective life

What I want to emphasize here are the ways in which collective life entails polyphony of efforts, entailing cognitive, affective and technical efforts on the parts of residents whether or not effective political, labour, or religious organizations are available to concentrate sentiment and commitment. Acting in concert is something that takes place regardless of the forms we recognize as collective politics. And this acting in concert may indeed both support and undermine specific instantiations of politics at different times.

For example, in district after district that I have worked in, there is the following scenario. People show up, although they are not sure quite why; still with a sense of necessity – activists, city councillors, local thugs, entrepreneurs, fixers, religious figures, NGO workers and some concerned citizens. In the back banquet hall of an old restaurant food and drinks are served, and there is no real agenda. It is late, and no one knows quite what the outcome will be. But a deal will be hammered out; no one will like it very much; no one knows quite how it will be enforced or what the long-term implications will be. It is likely people will return soon, perhaps not here, but to some other fairly anonymous place that everyone knows. Still, it is an occasion where no particular expertise or authority prevails; there are openings to make things happen across a landscape of gridlock, big money and destitution.

These are gatherings that take place across many cities of the world; usually at off-hours and usually under somewhat vague pretences and aspirations. Nevertheless, it is an urban politics at work. It is a politics engaged in the arduous task of bringing some kind of articulation to increasingly divergent policy frameworks, administrative apparatuses, money streams and authority figures. All of these intertwine at abstract levels but their mechanics of interdependency are too often opaque within day-to-day routines of navigating and governing cities. The only way to work things out is often the awkward and tedious efforts of different actors trying to figure out what to do with each other when they find themselves in the same room.

Various forms of collective organization have demonstrated their importance in terms of enhancing secure tenure, of making visible the plurality of situations that constitute marginality and poverty, and of conduits of communication and coordination among these different situations and the people that live them. They have facilitated the consolidation of political and financial capital that attains concrete improvements in material and social conditions, as well as establishing mechanisms to equitably distribute the responsibilities of decision-making and fiduciary obligations. Enhancements of security, living conditions and political power may have been negotiated through community organizations and the networks forged amongst them.

Yet, the repercussions of these attainments manifested in more solidified platforms of access to and integration within the larger city, mean that individual residents – the components of these collective organizations – may use them in ways that are not easily subsumable to the agendas and coordinating mechanisms of those organizations. In other words, as people will use better material conditions and infrastructural inputs as a means of expanding their economic and social horizons, so they engage the larger urban space in ways in which existing collective organizations may face difficulties to engage, monitor and influence. It is here that the issues of security become tricky.

Collective organization here becomes the vehicle through which specific localities 'step into' the prevailing practices of municipal governance. Here, residents become accountable and manageable as fiscal subjects who have accrued indebtedness, not only in terms of the money they owe into the future, but also in terms of their obligations to particular modalities of self-presentation with its concomitant discourses of working in partnership with others, subsuming certain aspirations and ways of doing things in order to sustain those partnerships, and valorizing patience and incremental gains.

This is perhaps why some associations of the urban poor insist upon stepping 'outside' such corporate arrangements and take matters into their own hands. While of usually limited effect in terms of influencing policy, projects and the overall trajectories of urban development, such a position reiterates the city as a locus of intense conflict. It draws out, in an era of preoccupation with urban ecological security and its concomitant obsession with consent and partnership, the fact that impoverishment is a product of conflict, a particular assault on life, and that the self-valorization and nomination of the poor is their only real opportunity in that it restores urban space as a battlefield about what it is possible to do and say.

Additionally, in environments where people's vulnerability can be manipulated and traded for political advantage, where divisions within localities can be easily cultivated, and where individual assets and capacities never are sufficient to change much of anything, collective solidarities are important instruments of diligence, focus, and step-by-step concrete change. While residents will seek to defend their gains and protect the fruits of hard-won struggles, preservation alone is oriented toward the constant alert for possible threats and thus potentially limits what residents within a given locality do in relationship to the larger city, thus risking an atrophying of the very capacities that went into the collective struggle in the first place. New exteriors have to be plied, new intersections with the city, and these intersections often are exploratory, experimental – the locus of individual and small initiatives and not the outgrowth of collective decision-making; after all, the locality in its entirety can be 'tied up' with any one experiment.

Indeed, these individuated trajectories of engagement with the larger city are the 'ultimate actualizations' of that commonality that exists among residents. On the other hand, each of these individuated trajectories, built as they are on the platforms of improved conditions which are the product of collaborative efforts, contains within them varied aspects of that commonality – inclinations, capacities, and techniques – that are not used or made visible in these trajectories. This excess is the very material that goes into further developing the collective life of these residents – an excess that wouldn't exist without the tendencies of individuals to forge their own particular pathways out of the locality which is nominally the 'territory' of that collective (Virno, 2009).

Here, security is a process of extending the ways in which things are implicated in each other. Discrepant places, things, experiences are articulated, circulate through each other, not just as matters of speculation, but as a complex architecture of accumulating and dissipating energies and attentions. Usually we would assume that districts rely upon concepts such as 'property', 'neighbour', 'co-religionist', or 'co-worker' to help specify and regulate social distance and responsibilities among individuals. But for districts usually full of people coming and going, the sense of habitation does not rely upon plots, cadastral and social demarcations. Rather, residents imagine security and stability as located beyond what they can see and figure out – in dense entanglements of implication, witnessing and constant acknowledgements of other residents, whether physically present or not.

It is about who can do what with whom under what circumstances and what can ensue from the resultant actions; who can they reach, how can they be made known; how fungible are they and what kinds of other actions are impeded, compelled or opened up as a result. Nothing is completely ruled out or disciplined; as not everything is possible. This notion of security as acts of extension and conjuncture then involve practices of physical and cognitive movement – which I discuss later.

The discourses on rights and the practices of economies

Political rationalities that focus on activating citizenship on the part of individuals of various backgrounds and using citizenship as a means of equilibrating differences under the provision of rights and services, while obviously important to districts of any composition, do not really address the dynamics through which many urban residents attempt to secure their lives. Processes of urbanization do give rise to highly differentiated spaces of operation and make more provisional what could be identified as 'communities', 'localities' or 'districts'. These differentiations are not always clearly differences of advantage, authority and resources. What these differences mean in the intersections of labour, space, tools, time abilities and networks does not necessarily directly stem from the apparent visible and hierarchical distribution of capacities and privilege. Relationships can be simultaneously exploitative and excessively generous, competitive and collaborative. There are unpredictable oscillations of accumulation and loss which potentially introduces a dynamic egalitarianism over a long run that no one individually has the power to define or measure, even if in the present glaring disparities can be documented everywhere. This can be the case in situations where hierarchies appear quite obvious.

Let's take the case of Prumpung, a district in East Jakarta. Yule is the grandson of a Betawi small businessman who had left him some land in a *kampung* that was close to a market specializing in printing and the sale of carpets and tiles. The plot contained several houses which he reworked into a series of rooms to rent. With this inheritance and the money generated through rentals he received a loan from the bank to expand an ailing uncle's text book production business. Presently, a large proportion of the *kampung* is involved in one or more facets of this business. There are small production centres where ten to twenty young men assemble and glue sheets that have been photocopied from master copies that Yule secures from the agents representing different school districts across the country. There are hundreds of women who take these initially assembled books and work on the stitching in their homes. Many men in the *kampung* operate as sales agents, who attempt to drum up business across Jakarta and other cities.

All of these facets operate as small independent units, sometimes integrated and connected as a single production system. But they are more often dependent on securing work from different sources and competing with other units involved in the same kind of work. At the most basic level, a small businessman will retain the services of his immediate women neighbours, paying them per book stitched. The quantity of work will vary according to how much volume he is able to secure from the different networks that he is a part of. Some will be content to take whatever comes their way based on the character of the social relations they enjoy in the *kampung*. Others will consistently try to reposition themselves across different networks in order to secure favours and opportunities that will increase the quantity of orders. For the women, when work is plentiful, they will often indicate their ability to take on more work. They will then pass it on to friends and associates in other parts of the *kampung* who work for 'bosses' that either don't pay as well or who are having problems keeping up a steady volume of work.

In this way, investment in social relations across the *kampung* becomes a critical way of accessing opportunity and work at different levels. It operates as a way of smoothing out potentially difficult differentiations in opportunity and income accumulation within the *kampung*. But because almost all the different facets of this production system end up basically fending for themselves, it is difficult for managers, procurers, labour and salespersons to predict how much work and income they will have on a consistent basis. Salespersons, often the husbands of the women who take stitching work into their homes, are often dependent upon the specific seasonal patterns in which school textbooks are acquired. For other printed materials, such a bank ledgers, accounting logs, registers, or special editions of children's books or training materials, procurers may use different marketing agents. These agents outsource to different small companies. Some do the printing and reproduction, others do the finishing, and so forth, and so it is never clear with whom it is important to cultivate some kind of long-term relationship.

The disarticulations within such a production system introduce much flexibility. This keeps small scale operations going in face of economies of scale in these sectors, in that they minimize sunk costs, labour and marketing. The disarticulations enable a distribution of work and opportunities, but seldom at a scale that enables households in this *kampung* to really be able to save or exceed the status of the 'working poor'. Nurturing social relationships are important but do not change persistent inequalities amongst neighbours who are stratified in relationship to their history in the area, as well as their relationships to the offspring of the original landowners who retain their holdings and interests in many of these districts. These relationships are important in terms of securing work and mitigating glaring inequities. But while many neighbours are treated fairly by their 'bosses', who are also their neighbours, these relationships often preclude the articulation of specific demands. Neighbouring workers rarely group together to insist upon specific wage increases for fear of unsettling these relationships.

Securing a location to operate in requires the cultivation of and constant tending to specific relationships. This includes finding a job, selling things on streets or from store and house fronts, or offering services of any kind. These relationships are needed even when an individual attains a formal licence or registration. Of course the issuance of such formalities largely depends in the first place on the kinds of relationships an individual enjoys in a given neighbourhood or with a specific trade. At the same time, many urban residents know that they cannot afford to be 'captured' by too many obligations in order to operate somewhere. This is especially true for those who do not have the means, knowledge of the city, or commitment to invest in long-term relationships in a given neighbourhood. In other words, while social relationships remain thoroughly implicated in almost all attempts to generate livelihood, individuals know that they also must be able to get out of specific networks and arrangements if the cost, in terms of energy, time and resources, outweighs the advantages.

Even though generosity and reciprocity remain strong in most neighbourhoods, there is a certain volatility from below in terms of just how stable alliances and affiliations among neighbours may actually be. Then from the top, from the vantage point of local big men like Yule, these ambivalences are used as a tool for not only keeping his options flexible in terms of just who exactly performs a specific job for him at any specific time, but for manipulating the situation so that only he knows the entire 'game'. He will make sure that only he possesses the entire overview and map for the fabrication and sale of his product. By using the strong social relations that exist locally as a means to access committed and low-cost labour and other inputs, and by manipulating the ambivalence individuals feel about 'putting all their eggs in one basket' – investing in securing and maintaining a particular position in relationship to others – the big local men make the functioning of the overall system opaque to all but themselves. To a large extent then, their power derives from this opacity. If only they know how the game works, they can ward off challenges and efforts others might make to learn the system. At the same time, without sharing this knowledge, it becomes increasingly difficult to hold labour markets together, as many individuals, while maintaining residence in Prumpung, take their chances elsewhere.

There is much that is neither fair nor productive in such local economies. Much of their efficacy when they do work seems to derive from curtailing potential articulations rather than maximizing their possible synergies, and by playing people off each other. At the same time, it is not clear how such sectors – threatened as they are by new production methods – could absorb such a volume of labour in alternative arrangements. It is one thing to argue for rights based on enforcing certain accounts of equivalence and fairness but quite another to concretize them in a dynamic relationship with real urban economies that remain capable of sustaining, even if barely, heterogeneously composed residential districts close to the heart of the city.

The mobility of the senses

For many of us, we are constantly on the move because we sense we have no other choice. The same is also the case for residents from Jakarta to Dakar because livelihood has become increasingly uncertain. Through their own investments in mobility, many residents make inordinate efforts to expand the terrain they cover in the city. It could be a food cart dragged across miles of foreboding freeway to distant neighbourhoods or a car repair service installed at the periphery of a new shopping mall. Residents attempt to forge new perspectives by inserting their labour or limited money into small ventures away from the neighbourhoods they are accustomed to. They change jobs – not for an increase in wages or security – but to have access to new social networks. All of these efforts constitute working 'propositions' for how different districts of the city could be brought together. Through these linkages, neighbourhoods take on greater complexity and thickness that in turn engenders unforeseen opportunities.

Contrary to conventional wisdom that collective effort ensues from people who are linked by history, trust, or painstaking organization, many residents are taking risks precisely on those with whom they would have little recourse or adjudication if things go wrong. Still, you've got a little cash and someone else a truck, I've got a connection in the ministry, and someone else has tools or a small warehouse, or many favours owed. The idea is that opportunities come and go, and if we don't take the chance, than someone else will. What prepares individuals to take such risks, how do they practise and work them out with few available maps, expectations, or relevant authorities?

Part of the story here is that while the rigid administration and economic segregation of urban spaces may be severe, urban life largely remains a disparate collection and disconnection of fragments. Nigel Thrift (2004b) labels these fragments 'fugitive materials' – traditions, codes, linguistic bits, jettisoned and patchwork economies that are 'on the run', pirated technologies, bits and pieces of symbols floating around detached from the original places they may have come from. Not only does the city attract human migrants from elsewhere, but also all the bits and pieces of ways of doing things, long dissociated from their original uses, that 'wash up' on the shores of the city. Bits and pieces of discourses, things, signs and expressions are assembled into personal projects of survival, ways that people have of dealing with each other, of making deals. What is it in the way these materials are collected that supports life or acts against it? It is not clear. What in this practice supports the tendency of global capital to make people and local networks fend for themselves and what operates in ways that global capital will never get? It is not clear.

Clear or not, useful or useless, urban bodies are entangled with such collections (Colebrook, 2002). For many, inhabitation takes place in environments weakly insulated from the effects of producing life in conditions where the inputs have no consistent supply chain or vehicles of evacuation. Bodies are intimately entwined in scrap, fuel, rain, heat, waste, sweat, tin, fire, fume, noise, voices and odour, on the one hand, and multiple stories, generosities, violence, arguments, reciprocities and fantasies, on the other. There is no bringing all of this into account, into predictable means of calculating opportunity and reasonable future. Here, impetuousness coincides with cautious and seemingly endless deliberation. For many, the difference between gambling and planning, saving for years or spending whatever you have right away is non-existent.

Yet in situations where many forms of belonging and mediation have been lost, a capacity for discrimination emerges which turns what has been lost into an opportunity. If you look out onto the world and see few prospects, if doing the right thing doesn't get you anywhere any more, if everyone basically has the same ideas about how to get ahead and therefore there is a kind of traffic jam in front of any new initiative, and if you have got to use the people around you for things no one is ever going to be fully prepared to undertake, then everything stable in your life has to be looked at as if it were only one fleeting version of itself. The loss of mediation, of maps, of anchorage then is taken as an opportunity to reclaim various forms of paying attention to things and of being receptive to all that which circulates through the city as bits and pieces of different knowledge residents have brought from various elsewheres, times and circumstances. Often reduced to the status of being 'distorted traditional practices', 'magic', 'intuition', or 'street smarts', to name a few, these bits and pieces can be used as tools of inventing and implementing specific ways of thinking and feeling. Discrimination thus entails how one learns to pay attention to family influence, social affiliations, local and distant authorities of various kinds as if they were something else.

Usually individuals and households have particular ideas, norms and cultural rules about how people and things are to be considered: are they close to us, or are they far; should we take them seriously or just not pay attention to them? There are people and groups with whom one can exchange things, lend things, as well as forces and people that must be resisted. But here, discrimination is a way of paying attention to what one's neighbours or associates, co-workers, friends, or acquaintances are doing, not with the familiar conceptualization of what a neighbour is or should be, but through creative conceptualizations that enable a shift in the conventional patterns of how distance, proximity, reciprocity and resistance among people are orchestrated. Those that are familiar become something else – you are not quite sure what, and so you have to try different things on for size, or you have not care so much what they think or you hear them say the same old thing you've heard a million times before but now it makes a different kind of sense.

These manoeuvres become one way of anticipating what might happen if a person decides to take an unfamiliar course of action. When we act, we do so only if we have some sense about what is going to happen to us if we do something in a particular way; otherwise we won't do it. This is why we are hesitant to take risks or do something new. So what I am talking about here is the way that people invent probable outcomes for experimental actions in situations that no longer have a strong relationship to reliable institutions for interpreting what is going on. Thus what is proposed are ways of making connections among people and ways of doing things that don't seem to go together. This, then, opens up possibilities for individual residents to make new affiliations and collaboration, and take risks with them.

Here, households do not stand alone as discrete social and economic entities but are tied into various circuitries, positions and ways of being in the city. As these circuits scatter in all kinds of directions, households are not subsumed under a series of successive obligations – to which they must always defend themselves or be accountable. This does not mean that districts are free of conflict and manipulation. Claims are made and protected. Individuals have to cultivate relationships with patrons and institutions and are paid to the extent to which they become instruments of their will. But residents with different engagements in the larger city, with different paymasters, vested interests and loyalties, continue to pay attention to each other – not as a disciplinary manoeuvre – but with a real interest in what is going. Activists, gangsters, clergy, gamblers, government enforcers, marketers, municipal workers, school kids, women and men are often engaged in long animated discussions – in mosques, street corners, cafes and markets – about what they see out there and their theories about what is taking place in the city.

Economic practices and the horizons of policy

Of course, the complexity of these manoeuvres, these ways of seeing and doing, still have to have something to do with the logics and formats of the state and the corporation. And these relationships between the majority and the state are insufficiently or badly organized. Here, even the deployment of subsidies intended to foster greater inclusiveness – housing subsidies, basic grants and social funds – fail to relate to local livelihood practices and economies of accumulation. As mentioned previously, a politics that focuses on activating citizenship on the part of individuals of various backgrounds and using citizenship as a means of equilibrating differences under the provision of rights and services, while obviously important to districts of any composition, does not really address the dynamics of many urban districts as specific kinds of economic spaces.

Here, building lines, plot size, distribution points, service reticulations are systematically violated – sometimes in the interest of greed – but more often as mechanisms to keep different kinds of residents in close proximity to each other. Equations that link training to skill, skill to occupation, occupation to set modalities of entrepreneurship, and entrepreneurship to specific forms of spatial organization and use can be thoroughly mixed up. Districts known for furniture production, auto repair and parts, printing, floral decoration, textiles, or ceramics usually contain a wide range of plant sizes, technologies, specializations and degrees of formal and informal organization. There are as many venues and instances of collaboration and clustering, as there are differentiated approaches and competition. Just because there is a medium-sized factory that uses a hundred workers to make jeans doesn't rule out that just right next door are two small workshops, each with two to three workers, making the same kind of jeans for the same market.

Even under the rubric of a common sector, many economic activities are difficult to organize as associations, chambers, or unions – subsumed to a formalized set of business practices and representations. For example, in the car parts and repair business, the persons who have the facilities, the contacts, the skills, the money and the capacity to organize workers, shipment and sales are seldom the same set of people over time, as they move in and out of different facets of the business. Shops, homes, streets, offices take on many different simultaneous or alternating functions. Services and parts are bundled in different ways and quantities from many different sources depending upon the situation and consumer. Interdependencies and calibrations among discrete facets of the production and commercial systems of textiles, furniture, foodstuffs, household items and electronics are so intertwined that conventional tools of regeneration, agglomeration and regulation would likely disrupt their functioning altogether.

At the same time, how they function is, if not necessarily opaque, difficult to grasp in any semblance of their entirety. While it possible to trace the intersections of actors, materials, spaces and times, those that are involved in these economies have great difficulties explaining how they work – after all, they have to start a narrative somewhere and if accounts require a certain linearity to be translatable across situations and actors, simultaneities and complex synchronizations will be inevitably left out. Here, involvement seems sufficient. In other words, however inclusiveness is attained, it doesn't matter so much the circumstances or mechanics involved – for participants believe that their inclusion gives them access, not to just one position or vantage point but many, and so the constraints of any particular job or relationship can be worked out by virtue of leveraging other potential jobs and relationships that are accessible without great difficulty or negotiation.

For those whose inclusiveness then is provisional, the absence of explication – of an ability to account for how the game of local economies operates – becomes deeply troubling. Many residents go to inordinate efforts to get out of where they are and go somewhere far away; some save and plan for years; others make spur of the moment decisions and are on their way in no time. Many residents are indeed stuck, unable to go anywhere, make any changes. Some are buried under a pile of obligations, mourning, depleted confidence or just too many expectations. Some run interference for others, and here sacrifices and obligations mount, and much time is wasted calibrating at the level of meters and grams just what constitutes a fair settlement of indebtedness sufficient to ward off threats. Basic feelings of trust and belonging are leveraged as the very means to keep them operational. But without unspoken confidence in the endurance of critical family and local ties as a means of framing reliable interpretations of people's actions, individuals become uncertain about how their words and behaviours might be construed by both known and unknown others. An exaggerated transparency then often ensues, as individuals display their incapacity, their wounds or their good intentions. The point here is not to represent 'things as they are or should be' but rather how to perform in order to be seen; how to make oneself visible in a crowded field of attention, how to know, as Danny Hoffman (2011) says in his work on fighters in Sierra Leone, 'what the camera wants'.

Increasingly, supernatural forces are seen as having overflowed the categories and domains by which they could be either domesticated or kept at bay. If such forces are increasingly diffuse, and where distinct categorizations – for example the differences between the devil and the holy ghost – are viewed as complicit, extraordinary measures have to be taken in order to circumvent the constant sense of crisis that this generalization of the supernatural portends. The city is then seen not as a terrain of specific politics, dispositions of capital flows or ecologies of socio-technical systems, but of spiritual warfare that civil processes are unable to mediate and that require different forms of accommodation to powers that render visible identities and places something other than they are. Exclusion here becomes then a privileged position for avoiding a diffused sense of immanent catastrophe (Sanchez, 2008; Tonda, 2005).

States will have to be more proficient in including these economies – especially once new accords on energy use are ratified – which will fundamentally alter the processes of economic growth. But inclusiveness does not mean simply drawing the majority of urban residents into new modalities of work, household formation and governance. Inclusiveness will also have to entail reconciling or at least co-ordinating very different logics of what really constitutes value, worth and efficacy. We often forget to ask questions that are at the very heart of city life. What happens when people operate in very heterogeneous mixtures of history, contacts, skills and interests with others also operating in such conditions. The cumulative effect of what takes place, of how this collection of actions impacts on larger spaces and larger groups of people is difficult to predict or order. Orders imposed upon this heterogeneity, if recent past evidence is any indication, tend to constrain the capacities of residents to interweave their time, money, aspirations, comparative advantages and occupations (Corbridge *et al.*, 2005).

Yes, a kind of ordering, a kind of governance must take place, one that substitutes working with specific representatives of localities – no matter how they are chosen – with distributed engagements and different forms of resource allocation. In other words, formulas that attempt to match taxation with income level, economic sector, or property holdings, that match commercial regulations and subsidies with the size of firms and work force, don't always work. Municipal authorities still must regulate, still must tax, still must issue subventions, exemptions, and they still must fund. But instead of demanding that individuals – individual persons, enterprises, households, associations, and so forth – be addressed and made responsible as individuals, the formal entities of governance need to better understand the clustered relationships that link schooling, marketing, producing, transporting and residing within specific spaces. These relationships sometimes entail the interlinking of skills, labour, responsibilities and money so that what appears on the surface to be discrete functions and settings are tied together.

Municipal governments are under increasing pressure to find ways in which existing resources can attain wider and more diverse use and therefore diversify potential revenue streams and costs (Harrison, 2008; Robins *et al.*, 2008). Such a process needs to be built on the existing ways residents flexibly use different networks of connections – categorized in different ways – to access resources that are, in turn, distributed in quantities that acquire particular value by again flexibly using categories that point to different kinds of social relationships and responsibilities. As municipal institutions always have to figure out what things cost, as well assess the value of the materials they have to work with, a critical aspect of municipal governance is to flexibly take the infrastructure, services and other assets it nominally controls and make sure that different uses of it come into play, and where those different uses have different costed values that can be set in relationship to one another. The question here is how municipalities can use their ownership of assets, spaces and the instruments of governance to build up more viable urban economies by supporting and extending the densities of transactions that already articulate many of the critical dimensions of everyday urban life.

It is increasingly clear across ideological spectrums and geographical locations that definitions of what works in terms of governance and the methods entailed in getting things to work have more to do with the design of relationships between spaces, institutions and populations than it does with effective discourses – the right words, the right authority, the right participation. Making things work is not so much a matter of bringing the 'informal into the formal' or to 'upscale effective local practices' or to 'coordinate scales'. Nor is it about integrated practice, intersectoral coordination or the appropriate sequencing of policy interventions.

Given that any domain within cities is now intensely networked across various scales, locations and processes, and that the differentiated positions of domains within a city are critical facets of the capacity of urban systems to manage some form of coherence, governance requires both a language of articulation that enables these domains to recognize themselves as part of the 'same city' and, at the same time, allows singular forms of decision-making and administration. In the latter, as local economies often rely upon the complementary relationships of different actors, capacities and ways of life, the discreet components of these relationships need flexibility in order to figure new ways of responding to the changes brought about by their conjoint actions. Small scale producers, service providers, fixers, brokers, repairers, drivers, freelance labourers, supervisors, financers, retailers and artisans always have to recalibrate their relationships to each other as markets come and go, as consumption patterns change, as commodity chains and regulations shift. As particular trades, production and marketing systems are spatialized, the kinds of articulations among infrastructure, rules, financing and zoning can't be usefully generalized across the city as a whole.

At the same time, different residents – with different backgrounds and livelihood – have to have recourse to solidifying their positions without having to constantly negotiate. There must be an equilibration of power that recognizes the value of all residents as the starting point for all subsequent development. This is the functionality of notions of citizenship – where residents have access to an easily graspable language that engenders a sense of connection among all that exists within the city; a language that connects the heterogeneity of lives and as a connector makes no judgment about that which is connected. This language may include the specification of certain rights and obligations and the ability of residents to have access to the institutions which implement specific demands, as well as the right to shape these demands.

But in terms of how resources are mobilized and used, how services are conceptualized and delivered, how local economies are managed, or how specific principles of equity or compensation are concretized, these are not matters of deciding between the public and the private, the centralized or decentralized, the city and the region, or the entrepreneurial and the social. Rather these are matters of critical design. In other words, these are matters whereby the fragments of practices and institutions – seemingly formal, informal, private, public, social, political, or cultural – are interwoven to maximize the inclusion of all that takes place within a given locality, neighbourhood, or domain and then to enable the capacity of these components to both continuously reshape their own activities and territories as well as enter into various kinds of relationships with others – some exploratory, some contractual.

Therefore, education, environmental management, industrial regulation, financial mobilization, development investment, conflict mediation, social services and residential management may take variously designed forms. Such a diversity of governance might be understandably perceived as a potential nightmare, especially for cities already struggling to provide adequate infrastructure, transportation and basic resources. But such singular arrangements also necessitate ways of specifying articulations to the larger urban system – for the relative autonomy deployed to consolidate particular local economic and political practices will not do anything unless it finds ways of connecting to the city as a whole. Placing part of the onus on individuated domains and territories within the city to come up with ways to connect to the city as a whole may perhaps prove to be even more productive than specifying vectors of connection from above. As for urban theory then, understanding the systematicity of the city entails thinking of cities as many different cities at the same time, not as a plurality of fractals, but as the designs and struggles of many attempting to recognize each other as one, always imposing themselves on the other, as well as finding ways to leave each other alone.

3

CHINA EXCEPTIONALISM?

Unbounding narratives on urban China

Choon-Piew Pow

Introduction

> Any consideration of geography in the fullest sense of the word must face up to the theoretical problem of the analysis of the unique. In one sense the very thing that we study is variation: each place is unique.
>
> *(Massey, 1984: 8)*

As a theoretical rejoinder to dominant 'Western' urban theories and discourses, scholars working in the context of the global south have often underscored the local uniqueness and particularities of their respective cities as a way to 'speak back' to Western theorization. A typical argument advanced in some of this work lays claim that conventional urban paradigms are grossly inadequate for studying 'non-Western' cities, given the unique historical as well as political–economic conditions that have inflected local urban development. Drawing on some of the geographical work on urban China as a starting point, this chapter critically examines the claim that China's urbanization model is unique, exceptional and hence defies conventional (Western) theorizing. Specifically, the use of the term 'China exceptionalism' as it is applied in the context of China's urban development refers to the assertion that political and economic events peculiar to China have rendered the Chinese urbanization trajectory *more different* than similar from Anglo-American cities. The Chinese state, in particular, is seen to respond to and/or create conditions and institutions that render urban China's experience unique and exceptional. The chapter will first take stock of some of these arguments and then proceed to examine some of the theoretical pitfalls inherent in such thinking. Specifically, the chapter addresses the ways in which the thesis of exceptionalism have been sustained and, more importantly, what is lost or ignored by framing geographical inquiry and knowledge within such exceptionalist discourses.

The notion of China as a 'unique' and 'exceptional' country/regional case study has been put forth, explicitly or otherwise, by various China urban scholars (see, for example, Lin, 1994; Ma, 2002; Friedman, 2005; Logan, 2002, 2007; Sit, 2010). In Woo's (2001) paper, for instance, he critically evaluates the claims of China's 'economic exceptionalism' with its new institutional forms (dual track pricing, fiscal contracts) that are considered to be necessary for China's *particular* urban economic circumstances. Writing about China's urban transition, Friedman (2005: 54–5) also poses the question of whether China's urban development with its unique features such as 'in situ urbanization' is indeed a case of 'Chinese exceptionalism'. In similar veins, geographers such as Lawrence Ma and George Lin have observed that conventional Western development and urban paradigms are grossly inadequate for studying China's urban development, given the *unique* historical as well as political-economic conditions that have inflected Chinese development.

Specifically, the use of the term 'China exceptionalism' as it has been applied in the context of China's urban development refers to the assertion that political and economic events peculiar to China have rendered the Chinese urbanization trajectory *more different* than similar from other capitalist and even socialist countries. At the broader level, exceptionalism may be defined as the study of the idiographic and the incomparable. The idiographic approach stems from chorographic studies in traditional regional geography which attempt to establish and explain the differences between particular places (Schaefer, 1953). The concept of 'exceptionalism' and 'particularism' thus contrasts with 'nomothetic' and 'universal' explanations which strive to find similarities between divergent phenomena and to formulate general laws that account for a broad range of cases.[1] As Entrikin (1991: 21) further points out, 'the uniqueness arguments are associated with the view that place and region matter. Social and economic forces shape places and in turn are shaped by places'.

The bulk of this chapter examines recent works on China's urbanization, highlighting the different aspects in China's urban trajectory that may be seen to be different and/or similar and considers the theoretical and methodological implications of framing geographical inquiry and knowledge within such exceptionalist discourses. At the outset, it is important to state that this chapter does not make an explicit case for or against exceptionalism. Neither does the chapter claim that the authors cited here are necessarily advocating for or against an exceptionalist position. Rather, the purpose is to show how distinctions about China's unique and exceptional urban conditions have been made in the urban literatures, with the aim of contributing to the nascent theorization of China's urban trajectory. In particular, the paper points out that arguments on China exceptionalism are often underpinned by the omnipotent state that is seen to either respond to and/or create conditions and institutions that render China's urban experience unique or exceptional. In this conception, cities are often treated as staging platforms where national urban policies are articulated and enacted. Even localized urban development 'from below' is often seen as the (in)direct result of centralized state plans and visions. For scholars attempting to read and interpret Chinese urban development, cities invariably bear the imprint of the power and influence of the state. As will be further elaborated, such state-centric thinking runs the danger of leading to a form of 'methodological nationalism' (Wimmer and Glick-Schiller, 2002) that takes the nation/state-society as the natural social and political unit of analysis (see also Agnew, 1994; Taylor, 1996).

The Chinese road to urbanization

At the risk of oversimplification, scholarly works on China's urbanization may be divided into three main phases: the pre-modern traditional Chinese urban centres, socialist urbanization (1949–78), and post-reform urbanization (post-1978), with the main bulk of contemporary research focusing on the latter. In the case of pre-modern China, work has included writing by historical geographers such as the late Paul Wheatley and Ronald Knapp as well as other specialists such as Jonathan Spencer, Sen Dou Chang, and Lawrence Ma. Paul Wheatley's (1971) magisterial work *The Pivot of the Four Quarters*, for example, looked at how the functional unity of ancient Chinese city is underpinned by its symbolic role. In particular, Wheatley pays great attention to the cosmo-magical symbolism of the Chinese city, and the parallel between the macro-cosmos (the universe) and the micro-cosmos (the city). Ma's (1971) book on commercial development in Sung China (960–1279) examines the close interrelationship between commercialization and urbanization in traditional Chinese cities and the extent to which Chinese urbanism differs from European counterparts. Research on the socialist city is mainly concerned with how socialist ideologies of equality and egalitarianism and state-directed programmes have made their imprints on the distinctive urban development in China from the Great Leap Forward (1958–60), Cultural Revolution (1967–76), and the Third Front Construction.

Since the implementation of the economic reform in the late 1980s, scholars have begun to focus their attention on the impact of marketization and of the opening up of the command economy on socialist urban patterns. This can be seen in the voluminous work on rural–urban migration, uneven regional development policies, the changing *hukou* system, housing reform, etc. In particular, the roles of the state and market forces have been central on the research agendas of scholars. To this extent, a political economy approach to studying urbanization has generally been favoured by researchers. As Ma (2002: 1,545) highlights, 'Chinese economy is inherently political. Political economy should be fore grounded to enrich our understanding of the complexity of China's economic and urban transformation'. However, if Chinese urbanization has been considered to be a product of state policies, such state-centric analysis also draws on the idea that a unique set of political and economic conditions in China which have given rise to the emergence of a powerful, if authoritarian party. In these terms the state functions as the ultimate decision-maker, regulator, and participant in the urban economy, notwithstanding pressures from globalization and the decentralization of administrative and fiscal power from the central to local levels.

Sources of China exceptionalism

Judging from recent debates in the social sciences, more specifically in the burgeoning globalization literature, the role of the state has been a contentious issue. On the one hand, hyperglobalists have argued forcefully that the 'death of the nation-state' is imminent due to the breaking-down of national borders as a result of technological innovations. Sceptics have, however, pointed to the resilience of the nation-state and its continued viability in the global economy. On the whole, while China scholars have not been far removed from such globalization debates, there seems to be a general consensus that one unique trait in China is that the state remains a deeply entrenched political and economic institution strongly embedded in the social fabric of the society. This is not to claim that strong states do not exist elsewhere, but as Ebanks and Cheng (1990: 48) note, 'where China's experience of urbanization is unique rests on the Chinese Government's uncompromising command over human resources . . .'.

The idea of a dominant Chinese state is obviously not a novel one, with some of the earlier interest in the area focusing on the dynastic (or 'despotic') history of China. For example, Paul Wittfogel's (1957) controversial work argued that the historical emergence of 'hydraulic states' and civilizations in 'Oriental' societies were quite different from those of the West. Pointing to the example of China (amongst others such as ancient Egypt, Mesopotamia and India), Wittfogel believed that the mass organization for irrigation and agricultural development required the centralized control of the state, and government representatives were able to monopolize political power to dominate the economy, resulting in the formation of an absolutist managerial state. More recently, the dominant role of the 'developmental state' in East Asia has also captured the attention of political economic analysts, some of whom argue that China's governance has displayed attributes of state developmentalism including the prioritization of economic development in national policy and the establishment of economic bureaucracies to 'guide' the market as well as the close alliance between business elites, party officials and the state bureaucracy (Johnson, 1982; Oi, 1992; Duckett, 1998; Wu, 2002). China geographers too have pointed out that the state has played a distinctive and pivotal role in the shaping of urban development. As Ma (2002: 1547) notes, the invisible as well as the visible hands of the state, sometimes wearing gloves to conceal its true identity, are everywhere, including in the nebulous area of property rights reforms (also see Zhu, 2002). Similarly, Fan (2002, 1995) emphasizes that the enduring and decisive power of the state and its leaders are salient in areas ranging from macro level regional development policies to influencing and shaping the 'institution-based opportunity structure' of individual migrants.

The state and urbanization in the early Socialist period

In the early Socialist period, the imprints of state policies on the urban landscape were clearly evident. For example, during the First Five Year Plan (1953–7) and the Great Leap Forward (1958–60) that emphasized non-agricultural economic activities, hundreds of thousands of peasants flooded the cities and suburbs to work in poorly equipped factories, leading to a period of 'over-urbanization' or 'spurious urbanization' (Zhou and Ma, 2000). Such massive mobilization of the peasants was driven by an ultra-leftist ideology to spur industrialization and economic growth especially in steel and iron production. As a result, the gross value of industrial output index rose from 100 to 535.7 between 1952 to 1960, but the growth of agricultural output dropped to a drastically low level at only 5 per cent annually, leading to the disastrous famine which plagued the nation in 1970s (Ebanks and Cheng, 1990: 33).

The failure of the Great Leap Forward movement drew a roundabout turn in official policies, leading to the strict application of measures designed to remedy the over-urbanization of the 1950s and to maintain China's aggregate urban population at a sustainable level. During the Cultural Revolution (1967–76), 26 million migrants were sent back to the countryside with some 17 million urban young people being forced to go 'up to the mountains and down to the countryside' to be 'rusticated' and 're-educated' by the peasants. The Cultural Revolution was inspired and launched by Mao who looked to the peasants as source of revolutionary creativity and a reliable base for China's permanent revolution. At about the same time, the state-sponsored migration to the interior under the 'Third Front' Construction (*sanxian*) programme from 1965 to 1971 led to a concerted shift of Han Chinese population to the western interior. For reasons of national security, large numbers of factories and industrial workers were moved from the eastern part of China to widely dispersed mountainous regions in the western interior (Fan, 1997). At the same time, a large number of peasants were recruited to work in urban factories, giving rise to the mass mobilization of people and a peculiar and massive interchange of urban and rural populations. On the whole, urban population was artificially depressed, such that in the 1960s and 1970s, the level of urbanization never exceeded 25 per cent (Lin, 1994: 13). During the entire period 1961–76, there was an overall decline in the urban population's share of the total population from 19.7 per cent at the end of 1960 to 12 per cent 16 years later (Ebanks and Cheng, 1990: 33).

The mass mobilization strategies of the Chinese socialist state during the 1960s and 1970s were extraordinary by any standard, and many scholars believed that their impact had shaped Chinese urban development in peculiar ways. In particular, it has often been stressed that no other urban population in the world has undergone such dramatic fluctuations in such as a short span of time. Ma (2002: 1550), for instance, comments that 'the Great Leap Forward was a unique, ideologically ultra-leftist and economically chaotic period the like of which has never been seen in human history'. What is even more remarkable and unique about Chinese socialist urbanization is that there was no large-scale urbanization despite intensifying industrialization. According to Lin (1994), contrary to the Western academic ideas of a 'similar path' or repetitive urbanization process in 'third world cities' suggested by Bauer, Rostow and Smailes, the growth of urban population in China remained surprisingly stable and low for decades while industrialization had increased substantially. This was made possible by strict control of urban–rural migration through the state imposition of the *hukou* or household registration system that effectively barred people from the countryside from migrating en masse to cities (this will be returned to later). Between 1952 to 1982 output value of industrial production increased by 21 times while urban population grew by only 2.3 per cent per annum in the 1960s and 1970s. As pointed out by various scholars, this anomaly could be attributed to the distinctive anti–urban and pro–rural attitudes of the Chinese communist party-state which adopted drastic de-urbanization policies and rustification campaign such as during the Cultural Revolution period (Ebanks and Cheng, 1990).[2]

Scholars have thus observed that 'those distinct fluctuations in the growth of urban populations show a remarkable coincidence with the cycles of political movements in China (e.g. First Five Year Plan, Great Leap Forward, Cultural Revolution, etc.) that were interposed by the Chinese Government' (Ebanks and Cheng, 1990: 33). During the socialist period, the peculiar circumstances and state policies had rendered Chinese urban development as unique and different from many Western cities. Understanding of a 'unique' Chinese urbanization was also tempered with a strong dose of rationalism and scientism. As Lin (2002: 1820) observes:

> The theory that the Chinese socialist state had adopted a strategy to minimize urbanization costs so as to maximize industrialization was a clear testimony of the belief in the power of reasoning and rationality as well as of the existence of logic and order even in Communist China. On the surface the discourse seemed to be in the arena of exceptionalism. The fundamental belief, however, remained under the shadow of the Enlightenment school of thought . . .

Another unique feature of China's urbanization which differs markedly from 'Western' cities is in the nature of everyday urban life. Socialist policies and the commune system meant that urbanization has not led to the sort of metropolitan life commonly expected in big cities, as noted by Western scholars such as Simmel and Wirth. In fact, during the pre-reform period, urbanism in Chinese cities was less pronounced with lower population density, greater homogeneity, and relatively less urban inequality. Here, the impact of unique state institutions such as the work-unit (*danwei*) needs to be elaborated. As various scholars have pointed out, during socialist China (and even in the contemporary post-reform period), the work-unit is a highly integral element in the urban structure. In fact, the work-unit has been described as the 'basic spatial and social cells' of Chinese cities; and an extension of the state apparatus that undertakes the important function of labour reproduction (in particular housing provision), social organization, and control (Wu, 1996; Chan, 1994). Thus, the function of the work-unit in a planned economy is far beyond simply that of the organization of production.

Spatially, each work-unit has clearly defined boundaries that are marked by walls or gates. For example, factories and universities *danwei* are separated into workspace and residential space, where within each work-unit, social cohesion is strong and inequality low. However, while spatial segregation exists in Chinese cities division is not underpinned by factors commonly expected in Western cities such as household income or social class. Rather the pattern of spatial segregation is linked to a differentiation between land-use zones and different work units with a considerable degree of social mix within work units. Even though not all work units are the same (certain types of work units, especially those with government employees, are better paid and had higher status) this system produced unique and different urban patterns from those in the West.

The hukou *system*

Another distinctive state institution that warrants our attention is China's urban household registration system or the *hukou* system that prohibits non-urban *hukou* residents from migrating to the city without official approval. While not a unique system by itself (similar cases can be found elsewhere such as the Soviet *propiska* internal passport), what is remarkable about the Chinese *hukou* system is the extent to which it has been fully integrated into the broader social and economic structure (e.g. entitlement to food vouchers, urban jobs, health care) and successfully filtered into social life to determine life chances (see Chan and Zhang, 1999; Fan, 2002). In this respect, the Chinese state clearly possesses what Mann (1984) calls 'infrastructural power' with the capacity to penetrate civil society, and to implement logistical and political decisions throughout the realm, including extracting taxes, access to personal information/data; state laws; influence on economic activity (employment; pensions; family allowances, medical expenses; welfare; public services and so on).

In Chan and Zhang's (1999) detailed study they argued that the *hukou* system was not designed mainly as a system to block rural–urban migration, as commonly portrayed in the Western literature. Instead, it was part of a larger economic and political system set up to serve multiple state interests to secure social and political order and other related objectives. In the early years of the system, it served largely as a monitoring, not a control, mechanism of population migration and movements. However, as influxes of peasants into cities escalated and began to be a serious burden, the central government tried various measures to stem the 'blind flows' of rural labour. Consequently, this led to the promulgation of China's first set of *hukou* legislation by the National People's Congress in 1958. This new legislation established a fully-fledged *hukou* institution and granted state agencies greater powers in controlling citizens' geographical mobility through a system of migration permits and recruitment and enrolment certificates.

The *hukou* system essentially comprises a dual classification by residential location and socio-economic eligibility. Anyone seeking officially sanctioned or 'formal' rural–urban migration first has to complete a dual approval process, changing the place of regular *hukou* registration, and second, convert the *hukou* status from agricultural to non-agricultural (*nongzhuanfei*). In essence, the *hukou* system in the pre-reform era acted very much like a domestic passport system, dividing the population into two social groups, one (the non-agricultural population) economically and politically superior to the other (the agricultural population), with vastly different opportunities, obligations and socio-economic statuses (Chan and Zhang, 1999: 830).

During the pre-reform era, the *hukou* system was thus able to keep in check non-state sanctioned movement of people from the rural to urban areas, thereby contributing to the peculiarly low level of urbanization (though the system was not absolutely fool-proof). The effectiveness of the *hukou* system was in part due to the fact that during the early socialist period, economic activities were strictly controlled by the bureaucratic system with the state monopolizing the distribution of important goods and services, few of which were available in the market at affordable prices. People's daily lives were thus closely connected to and monitored by the twin state apparatuses of the work-unit and the *hukou* system. For instance, urban recruitment and job transfers were strictly controlled by the government and there were few chances for urban employment outside the state channels. It was hard for people to survive outside their *hukou* registration place without proper permits. In addition, their daily lives were tightly bound to the work units (*danwei*) and watched by the police and the residential organizations (street residents' committees in the city and village committees in the countryside). As a result, the *hukou* system was greatly enforced and any violation could be easily detected. Overall, the unique *hukou* system worked with other social and political mechanisms to form 'a *multi-layered web of control*, with each institution administering one or more aspects of rural-urban migration' (Chan and Zhang, 1999: 830).

Ideological justifications in the reform era

Before proceeding further, it is necessary to examine some of the ideological justifications that have been marshalled by the Chinese government in order to garner support for its economic reform policies. Here we encounter yet another argument for China's exceptionalism. Using Su and Feng's (1979) theory of 'primary stage of socialism', the Chinese government has argued that, owing to China's unique historical condition as a 'semicolonial' and 'semi-feudal' country when it adopted socialism, the country has *not yet* been able to reach the socialist stage as defined by Marx and Lenin. In particular, Deng has suggested that classical Marxism was developed in relation to mature socialist societies, but not for nations at a 'primary stage of socialism' and with low economic development and that China needed to develop a distinct, 'Socialism with Chinese characteristics, that is the application of socialism with respect to China's unique situation' (quoted in Fan, 1997: 622). During the late 1980s, when party leaders desperately sought legitimization for the reform movement and the non-state economy, Su and Feng's idea became popular and was adopted by state officials. If China's development is underscored by its unique historical background, various scholars have further contended that Chinese urban and economic transformations are different from the 'path dependent' model of the West in that the situation in China is not one of pure market transition (Nee, 1989) but of a mixed socialist/market economy based on partial reform (Bian and Logan, 1996). Yet other scholars have clearly dispensed with any pretension of socialism to proclaim that the unique form of capitalism in China is 'red' (Lin, 1997); local (Smart, 1995) and 'Chinese' (Dirlik, 1997). Essentially, economic reforms which begun in 1979 have generated a distinctive set of conditions that profoundly changed the previous dynamics of urbanism.

Urban policies in the post-Mao era

If socialist urban development is seen to be peculiar to China's unique conditions in the 1960s and 1970s, reform-era urbanization under the official rhetoric of 'socialism with Chinese characteristics' could be interpreted as being equally exceptional given the often uneasy blend of socialist market economy along with the often unpredictable changes that arise from China's political-economy. For example, scholars have argued that China's economic transformation away from state socialism would not represent a direct copy of a Western-style capitalist model. Ma (2002: 1,546) for one deliberately avoids using the term 'transition' to describe China's reform-era changes as the term 'assumes a process of change toward a preconceived and fixed target' which is not entirely appropriate for China. Instead, Ma opts for the concept of 'transformation' which avoids the implication of 'inevitability'. More specifically, Ma develops a 'state-centered political economy' perspective in order to understand China's economic restructuring that is *different* from Western social science conception (ibid.: 1546). Pannell (2002: 1572) however outlines the notion of 'urban transition' with reference to 'the movement of population from rural to urban locations, the shift of their work activities to things normally associated with urban locations, such as from manufacturing and various other services and the changing nature of their lifestyles in cities and towns'. Yet for other authors, the two terms are used quite interchangeably, in order to describe the transitory aspects as well as more radical transformative dynamics of urban change in contemporary urban China (also see Friedmann, 2005; Logan, 2007). Other scholars have advocated examining Chinese urban reform with a new set of analytical lenses that pays particular attention to China's 'special blend of policies' and its unique socialist legacies (Parish 1990). For Logan (2002), contemporary urban China research thus has an unusual kind of 'frontier character' which cannot be readily subsumed under conventional theorizing. The exceptional quality of China's rapid urbanization is further underscored by the fact that 'there has never been such a case, a Third World country propelled so quickly toward first rank of world power, at the same time as making a transition from a centrally planned to a market society' (Logan, 2002: 21).

Unlike eastern European countries, China's reform has not pursued massive privatization of state property. Rather, it has adopted partial and gradual reforms which mainly involved decentralizing the right of control and management of state property (including housing and land). The partial nature of the reform process is reflected in the delay in reforming state-owned enterprises and the state's (more precisely various state work-units') involvement in housing provision and land development. Nevertheless, with the increasing decentralization and marketization of the Chinese economy, a distinctive set of new conditions has emerged, transforming the urban process from *above, below, outside* (Ma and Fan, 1994; Fan, 1995b). One unique occurrence in China is the spontaneous development of county level cities and certified towns where joint efforts of local state, migrant workers, and foreign and domestic investors have given rise to the localization and expansion of collectively owned small-scale private industries. In addition, as the state decentralizes decision-making to local governments and introduces free market forces to transform the economy, the previous rigid urban hierarchy organized by vertical linkages and political functionality is disintegrating and being replaced by a new system of cities shaped primarily by horizontal connections and economic exchange. Oi (1992; 1995) for example, describes this as a new 'local state corporatism' which represents a qualitatively new variety of developmental state and not merely a modified Leninist system (1995: 1133). In this distinctive model, local officials act as the equivalent of a board of directors and sometimes more directly as chief executive officers in order to push for local developmental goals. Such urban growth impetus comes mainly from local and foreign sources without direct financial assistance from the central state; hence the term 'urbanization from below' is used to describe such processes. For example, in Lin's (1999) study on economic reform in southern provinces such as Guangzhou and Fujian, he argues that spatial restructuring in China since the economic reforms has been essentially a result of state disarticulation rather than increased state intervention. In particular, Lin notes that capital allocation from the central state has substantially dropped since 1979 and the balance has been made up by local governments through fundraising from various local and foreign channels. The bulk of 'state' capital is therefore being provided and handled by local government rather than by the central state.

However, it should be highlighted that the role of the state has not been totally written off in the euphoria of the reform era. As Ma (2002: 1,546) argues, one of the distinctive characteristics of China's development is that 'despite the forces of globalization, liberalization, decentralization, the state that has powerfully shaped China's urban transformation is omnipresent'. In fact, one may argue that the opening up of China would not have occurred without the central government's reform initiatives. For example, the spontaneous expansion of production spaces in small towns and the countryside, particularly in Southern China, has been facilitated by the relaxation of state control over local developmental affairs. The regulatory role played by the central state through legislative and fiscal measures has remained significant, and under special circumstances crucial, as evidenced by the state's determination to maintain the currency value of the Chinese yuan.

During the early reform period, the state also played an active and dominant role in designating and developing the 'three economic belts' by directing specific provinces, cities and zones to enjoy preferential policies and receive large state funding for infrastructure development as well as various financial and legal inducements for attracting foreign investments. According to early government policies, the particularities of China's regional differences have necessitated a regional division of labour following a ladder-step model which refers to the declining levels of factor endowments and economic development from the east to the west of China (Fan, 1995a). Specifically, selected areas in the eastern region were designated as prime 'growth zones' that will spearhead China's economic growth and its further integration with the external economy. These areas include the special economic zones (SEZs) of Shenzhen, Zhuhai, Shantou, Xiamen (1979) and Hainan (1988); 14 open coastal cities (OCCs) (1984); free trade zones in coastal cities (1993); open economic zones such as the Pearl River, Minnan, and Yangtze Delta; and other open cities and counties. These open zones constitute what is known as a 'golden coastline' (*huangjin haian*) in the eastern region (Fan, 1997). Not only was the eastern region set aside for export-led industrialization, but it was also designated the first region to experience speedy economic growth. Bias toward the eastern region is also manifested through a series of 'preferential policies' (*qingxie zhengce*) that heavily favour the eastern region at the expense of inland China (Fan, 1995a).

A final example of China's uniqueness may be gleaned from recent works on suburbanization. In particular, it has been pointed out while suburbanization in China has displayed some similar characteristics with the North American scenario; there exist distinct differences, owing to China's unique condition (see Plate 3.1).

PLATE 3.1 Privately constructed suburban homes at the fringe of Kunming city, provincial capital of Yunnan in south-western China
Source: Choon-Piew Pow

Zhou and Ma (2000), for example, document that population in the old city core declined in the 1980s and 1990s while suburban population rose. In particular, they argue that suburbanization in Beijing, Shanghai, Shenyang and Dalian is essentially the result of economic restructuring arising from reform-era initiatives where the state and foreign investors, private entrepreneurs and various domestic danweis have played pivotal roles. Unlike the current metropolitan landscape in the US where the suburban growth has given rise to a polycentric spatial structure, suburbanization in China is still in its incipient stage (barely 20 years) with suburbs dominated by cities, both administratively and functionally. The state in China has a more direct and powerful role in gearing up the suburbanization programme (albeit in concert with other actors). In addition, for most Chinese, the suburbs are not seen as better places than the city core to live or work, unlike the 'suburban flight' in North America. In fact, most urban residents tend to avoid moving to the suburbs if they have a choice, as Chinese suburbs are not yet fully developed places with adequate basic amenities and shopping, schools, and social activities are still lagging behind those in the city. It is also observed that unlike the United States where the notion of core-periphery has lost much of its significance in urban spatial structure because the urban landscape is politically fragmented and spatially and economically polycentric, the Chinese urban scene is still decidedly dominated by the city in all aspects of city–suburb relationship (Zhou and Ma, 2000).

Overall, China scholars are generally in agreement that urbanization in China is indeed exceptional, given the often '*unique* patterns of urbanization and urban growth as compared with both developing and developed countries; the *unique* measures and policies taken by the Chinese state to restrain urban growth and the *unique* future goals of national urban policies' (Ebanks and Cheng, 1990: 30, emphasis added). Such an exceptionalist position invariably raises further questions and poses several pertinent methodological and theoretical concerns. For one, the merits of China's exceptional urban conditions need to be further studied through more rigorous comparative analysis. Furthermore, by over-stating the 'exceptional' qualities of Chinese urbanization, we may end up overlooking or ignoring some common urbanization experiences that can usefully inform comparative urban analysis. The following section examines these issues further.

China the incomparable?

In arguing for the uniqueness of China's urbanization, one must bear in mind that claims of uniqueness and divergence need to be balanced with areas of convergences. One area of convergence, for example, can be found in the literature on 'new regionalism' and global city-regions (Scott *et al.*, 2002). According to these authors, all across the world, the increasing globalization of capital and transnational flow of resources are spurring the development of new global city-centric capitalisms marked by inter-city cooperation and the mutual building of regional economic and political competencies. There are now more than three hundred city-regions around the world with populations greater than one million, ranging from metropolitan agglomerations dominated by a strongly developed core such as London region or Mexico City, to more polycentric geographic units as in the cases of urban networks of the Randstad or Emilia-Romagna. In China, several urban regions can be identified including Beijing-Tianjin-Tangshan and Qingdao-Ji'nan-Weihai in the north, Shanghai-Suzhou-Nanjing-Hangzhou in the lower Yangtze Delta in the east, and Guangzhou-Shenzhen in the Pearl River Delta in the south. While there are differing criteria for defining what constitutes a global city-region, it has generally been contended that China's phenomenal growth had come less from its unique economic policies than its 'convergence to a prototype WTO economy' (Sachs and Woo, 2000).

In seeking to theorise this convergence, Wu (1997) paper has provided momentum for comparative analysis of urban processes between contemporary Chinese and Western cities. In post-reform China, Wu argues that in order to understand the built environment, it is imperative to critically examine *general* processes of production and reproduction. Specifically, Wu reviews four theoretical approaches commonly applied to Western urban analysis – Harvey's (1978) 'capital switch thesis', Ball's (1986) 'structure of building provision' and Smith's (1979) and Clark's (1995) 'rent gap hypothesis' and theories on 'bundle' property rights – and draws on their relevance and usefulness to examining post-reform urban changes in China. Researchers have also began to be interested in the emergence of a new 'state entrepreneurialism' in China, compared with more established market-model notably in the United Kingdom and the United States (Duckett, 1998; Wu, 2002, 2006; Hall and Hubbard, 1996, 1998). Other scholars have looked at urban landscape transformation in major Chinese cities, debating whether the new cityscapes are becoming 'more Western', as evidenced by the proliferation of high-rise office buildings, 'designer skyscrapers', shopping malls and suburban housing as well as the role of global business elites in shaping them (Lo, 1994; King, 1996; Olds, 2001). In an edited volume by Wu (2006) on globalization and the Chinese city, contributors examine urban development and local practices as part of globalization processes in an attempt to overcome the dichotomy between the East versus West. In this sense, globalization is treated as a 'neutral' term and seen as a trans-local process embedded in particular places, giving rise to a new urbanism through 'multiple materialization of globalization' in diverse Chinese cities (see Plate 3.2).

In contrast, however, recent work has challenged the universalizing assumptions in 'neo-liberalism' that have played out in the Chinese context, drawing out various parallels with 'Western' urban experiences but also highlighting interesting departures (see Lin, 2009 on 'neoliberal' land politics in China). Zhang and Ong (2008), for example, show that while privatization in China embraces Western neoliberal principles of private accumulation, entrepreneurialism and self-interest, there are multiple meanings, localizations and entanglements of privatization in contemporary China. The particular configuration of neoliberalism in China spawns new kinds of 'socialism from afar' in which privatizing norms and practices proliferate in symbiosis with maintenance of authoritarian rule (Zhang and Ong,

PLATE 3.2 Skyscrapers and shopping malls now dominate the urban landscapes of many Chinese cities such as Shanghai (seen here)
Source: Choon-Piew Pow

2008: 4). At the micro level of privatized urban forms, Pow (2009) also demonstrates that 'gated communities' in Shanghai cannot be adequately understood as the *inevitable* diffusion of an global/American-style urbanism into Chinese cities. While not refuting the influence of globalization (see Wu, 2004), gated communities in Shanghai need to be examined in terms of the local complexities of urban housing reform policies and the politics of middle-class place-making. Fundamentally, a major objective of such research is to link local changes occurring in contemporary China process operating at the global level.

Still, as Doreen Massey (1984: 8–9) cautions, 'in the search after general laws, the intellectual dominance of certain forms of 'top-down' structuralism, the (quite correct) desire to relate the individual occurrence to the general cause', researchers should also not lose sight of the local uniqueness of places (see locality debates). For Massey: 'local uniqueness matters' and it is just as important to understand the underlying general processes as it is at the same time to recognise and appreciate the importance of the specific and the unique (Massey, 1984: 299–300). Such views notwithstanding, too often has a solution been sought through an uneasy and sometimes untenable juxtaposition of two kinds of explanation; on the one hand, the 'general', whether it is in the form of immanent tendencies or empirically identified wider processes, is treated in deterministic fashion. On the other hand, since such infinite variety of reality does not readily conform to this logic, additional factors are added on, in an ad hoc and descriptive fashion, in order to explain away 'deviations'; or researchers simply conclude that the cases in question are too unique and exceptional to be generalized. At yet another extreme, geographers of various postmodern/post-colonial persuasions have seriously taken on board Lyotard's (1984) dictum of an 'incredulity toward meta-narratives' and adopted a deeply sceptical view of 'grand theories' which tend to over-generalize entire regions and the world. Consequently, the general intellectual climate has been to repudiate or 'deconstruct' grand intellectual statements of 'truth' ('reality' is seen to be too complex and the cases too individually unique to be reduced to a singular theory); while encouraging the celebration of 'polyvocality' (many voices) and differences. To some extent, this is evident in the recent writings of some China scholars. For example, Ma (2002: 1551) comments that:

> It must be cautioned, however, that grand theories or metanarratives suitable to all transitional economies, or even to a large country, may never be possible, and we might have to be content with a number of country-specific and even region-specific mini-theories suitable for different places in the diverse transitional systems in the world.

Belatedly, such a 'deconstructivist turn', when taken too far, has also taken its toll on 'unique' disciplines such as area studies (unique in the sense that area specialists study the attributes of particular places or region, for example China, African or East Asian specialist). As Ludden (1998: 24) claims:

> The production of area-specific knowledge about the world has no firm theoretical foundation. It seems to be an intellectual by-product of modern state territorialism and of those state-supported institutions of area studies which became prominent during the heyday of the nation state, in the decades between 1945 and 1990.

Given this tension between the unique/particular and the general/universal, the fundamental methodological question seems to be how to keep a grip on the generality of events, the wider processes lying behind them, without losing sight of the individuality of the form of their occurrence (and *vice versa*). Pointing to general processes does not adequately explain what is happening at particular moments or in particular places. Yet any explanation must include such general processes.

To raise another methodological problem, a major consequence of the exceptionalism argument, when taken to its logical extreme, is to imply that lesson-drawing is difficult since no comparable cases are available; conversely, the Chinese experience would also provide little value for other countries to learn from since the particular conditions in China, by definition, cannot be replicated elsewhere. In other words, exceptional countries can neither draw lessons from other countries nor can other countries draw lessons from them.

More fundamentally such exceptionalist positions in China geography scholar-ship with its 'embedded statism' (Taylor, 1996) invariably leads to a form of 'methodological nationalism' that takes the nation/state-society as the 'natural unit' of social and political analysis. Wimmer and Glick-Schiller (2002: 306) further point out that:

> The naturalization of the nation-state is in part a result of the compart-mentalization of social science project into different national academic fields and a process strongly influenced not only by nationalist thinking itself, but also by institutions of the nation-state organizing and channeling social science thinking in universities, research institutions and government think tanks . . . To the extent that social science is deployed to provide pragmatic solution for "national problems" in economy, politics and the social services.
>
> *(Wimmer and Glick-Schiller, 2002: 306)*

The national framing of social science research and discourses is especially pro-nounced in the Chinese context where academic research mostly comes under the auspices (and regulation) of state-funded institutions such as the powerful Chinese Academy of Social Sciences (CASS) and nationalized university system. As the CASS mission statement unequivocally states, the role of the institution is undoubtedly 'to promote research and to undertake and fulfill key state research projects in light of China's national conditions'. In these terms, the 'territorializa-tion of social science imaginary' and the reduction of the analytical focus to the boundaries of the territorial state container not only reifies the social-political construct of the nation-state but it has also led social scientists to become obsessed with describing and emphasizing particularistic characteristics and processes oc-curring within nation-state boundaries as opposed to those more general processes outside (Wimmer and Glick-Schiller, 2002: 307).

Critical reflections: unbounding China geography?

By surveying work by China scholars, this paper interrogates the very idea of 'China exceptionalism' as it is applied in the context of China's urbanization experience. In particular, I have shown that there is now a growing awareness in urban China academic circles that the patterns of urban development in China appear to be more sophisticated than the Western norm and that the conventional model of urban transition simply cannot give an adequate explanation of the Chinese experience. As this paper has pointed out, this supposedly exceptional quality has be attributed in part to the dominant role of the Chinese party-state and its unique institutions (such as the work-unit and *hukou* systems) as well as various historical and contemporary events that have rendered the Chinese trajec-tory of urbanization more different than similar from other urban experiences in the world. To say that Chinese urbanization is unique and exceptional also means that it does not follow the course of established knowledge derived from (Western) cases. This position, as I have argued throughout this chapter, raises several important theoretical and methodological problems, in particular regarding the role of idiographic and nomothetic explanations in geographical analysis. Fundamentally, this argument also raises a wider question: how is it possible to reconcile both the universalizing and particularizing discourses that have characterized the study of cities and regions? (Entrikin, 1991).

To this end, this chapter has considered the extent to which the exceptionalist position leads to a form of methodological nationalism that downplays and ignores alternative scales and spatialities of urban processes and dynamic interscalar relations. There is hence a need to go beyond a state-centric perspective and move away from the view of 'China less as a self-contained "geobody" than as part of a broader set of global and regional processes; from the *outside-in*' (Duara, 2009). As Cartier (2001: 25) notes: 'the lesson to be drawn is that insisting on beginning from any fixed scale is deeply antithetical to understanding dynamic processes of human realities'. Yet in attempting to 'unbound' area studies (China geography) from its national framing and challenge its territorial presuppositions, it is equally important to acknowledge the importance and value of 'local knowledge' on the ground. Writing about the tensions between localist perspectives and the universality of science, Shapin (1998) points out that the obstacle faced by a 'geographical sensibility' towards science has been the *view from nowhere* claim that localized knowledge that is geographically embedded cannot be authentically true. However since the mid-1980s, the geographical sensibility towards science became philosophically deeper with the realization that 'locality and spatial situations needed to be attended to in order to understand how scientific knowledge is made' and there is 'no alternative to being there, being where knowledge was made' (Shapin, 1998: 5–6).

And while many scholars have rightly cautioned against circumscribing social science analysis to the territorial boundaries of the state/society container, Bunnell and Thompson (1998: 1,518) usefully point out that: 'There is a difference between cautioning against falling into the 'territorial trap' (Agnew, 1994) on the one hand, and suggesting that the nation-state is no longer a useful scale of analysis in social science, on the other.' In fact, as opposed to the Anglo-American context where most geographical scholarship on rescaling has emerged, state power and national scale framing may retain or even resume greater importance in Asia. It is inconceivable to think that the role of the state in China could wither or become irrelevant anytime soon. For Bunnell and Thompson (1998: 1517), the critical task then is to 'avoid the tendency either to hold "area" (territorial state-container) or "flow" studies apart'. Indeed, what needs to be further theorised in China urban research is how the general and specific come together in multifarious urban processes, without necessarily collapsing or reducing one to the other. For (China) geographers, this is an important theoretical endeavour for as Massey (1984: 8) observed over two decades ago: 'Any consideration of geography in the fullest sense of the word must face up to the theoretical problem of the analysis of the unique. In one sense the very thing that we study is variation: each place is unique'.

Notes

1 While the meanings of the terms 'idiographic'/'nomothetic' and 'universal'/'particular' tend to overlap, it is prudent to note that they are, strictly speaking, directed at somewhat different, though not wholly unrelated, issues. Idiographic/nomothetic distinctions are used primarily in the discussion of scientific concept formation whereas particularism/universalism debates are generated in relation to ethical and moral theories. Nonetheless, these terms have often been used interchangeably but readers should be aware of their differences.
2 While ideological conviction of equity and egalitarianism certainly held sway among the ranks and file in socialist China at that time, other scholars such as Kirkby has stressed that the real motivation behind the controlled urbanization were more pragmatic considerations of 'urban manageability' and 'military preparedness' (also see, Fan, 1997; Lin, 1994). As Kirkby (1985: 14) wryly puts it: it is the industrialization imperative that has shaped China's urbanization, not abstract notions like anti-urbanism.

4

URBAN THEORY BEYOND
THE 'EAST/WEST DIVIDE'?

Cities and urban research in postsocialist Europe

Slavomíra Ferenčuhová

While the differences and similarities between 'capitalist Western cities' and cities in 'socialist' Central and Eastern Europe were discussed well before the fall of the Berlin Wall, since that time postsocialist urbanity has been considered by increasing numbers of local and international researchers. In the Czech Republic and Slovakia, for example, the institutional background of urban studies disciplines have undergone important shifts and attention has turned to defining the boundaries of urban theoretical and methods agendas. In this context, *urban change* has become one of the leading topics. Concepts from Anglo-American urban theory have tended to be studied, reviewed and applied in order to understand urban change observed since the early 1990s and to propose hypotheses about the future development of postsocialist cities. In particular, discussion has focused on defining the character of the 'postsocialist city' as specific and different from other cities, as well as postsocialist transition(s) and transformation(s) as the main theoretical framework for studying cities in Central and Eastern Europe.

This chapter reflects on this body of knowledge and the theoretical approaches stemming from recent research focused on cities in former socialist European countries. The main goal is to discuss the relationship of this research to contemporary international urban theory. I begin with a short overview of urban research since the 1990s in the Czech Republic and Slovakia as well as referring to work undertaken in other postsocialist countries. The second half of the chapter then discusses theorization of 'postsocialist' urban development, comparing conceptualizations of transformation, transition and postsocialism in relation to broader urban debates. In particular, I discuss the role of academic knowledge production in promoting a vision of postsocialism as an unfinished (and probably endless) project of 'catching up with the West' – in terms of both urban development and urban research. In doing so, the view of postsocialism as a 'transitory' and/or 'corrective' period is argued to have a parallel in the reproduction of the image of local urban studies as 'lagging behind' international research, being the 'object' rather than subject of studies, as well as lacking the ambition to contribute to urban theoretical debates.

Urban theory after 1989: key themes and strategies

In the Czech Republic and Slovakia, urban theory since the 1990s has tended to focus on the development of towns and cities through comparison with the (state socialist) past, and/or with references to 'Western' urban experiences. Topics for consideration included socio-spatial inequalities, housing and residential change, commercialization and gentrification or urban transformation in general (Stein-führerová, 2003; Sýkora, 2009a, 2009b; Růžička, 2010; Ouředníček and Temelová, 2009; Falt'an, 2009; Matlovič, 2004; Mulíček, 2009; Musil, 2003). In these cases writers have tended to draw on terminology and theory developed in the study of 'Western' cities with reference to issues such as suburbanization, the emergence of gated communities, and inner city revitalization (e.g. Ouředníček, 2003; Brabec and Sýkora, 2009; Sýkora, 2003, 2007; Sýkora and Posová, 2007; Sýkora and Ouředníček, 2007; Mulíček, 2002, 2009; Temelová, 2007, 2009; Mulíček, 2009; Burjanek, 2009). For example, Brabec and Sýkora (2009: 84) adopt characterizations of the 'major features of gated communities' from Blakely and Sinder's book *Fortress America: Gated Residential Communities in the United States* in order to consider new residential development in Prague. In a similar vein, Jana Temelová (2007: 2–5) considers urban revitalization with references to 'North American and European cities since the 1970s', and describes the relatively weak position of local authorities in Prague, 'learning how to deal with urban problems and challenges'. Sýkora and Posová (2007) also draw attention to the specificity of postsocialist suburbanization by adopting methodological tools commonly used by 'Western' scholars. Such examples highlight how consideration of the differences and similarities between urban development in the 'Western' and 'postsocialist' European cities has been undertaken, often to inform policy formation.

In such cases 'Western' cities are often cited specifically with reference to the period of chaos that is associated with the early 1990s, characterized by national reforms (e.g. administration reform, privatization of state-owned housing) but also little (financial or administrative) support for local municipalities (Sunega, 2005; Surazska, 1996; Sýkora, 2002). An aversion to planning as a practice associated with the rule of the communist party (Halás, 2007; Sýkora, 2002, 2006) has also been described in association with a strong 'neo-liberal drive', and laissez-faire attitude (Valentová, 2005; Musil, 2005). Sometimes researchers have described a progressive evolution of urban planning and policies as a process of 'learning' to use new policy tools adopted as best practice from the urban policies in Western European cities (Sýkora, 2002, 2006).

Ethnographic research has also been particularly important to understanding urban change in Czech and Slovak cities in a number of different ways. For example, studies have considered changing attitudes to housing with reference to personal biographies or memories (Barvíková, 2010). Everyday life, discursive practices and materiality have also been analysed in relation to food provisioning and consumption (Smith and Jehlička, 2007; Smith, 2002; Smith and Rochovská, 2007), urban identity re/formulation and attempts to 'deal with' socialist history (Ferenčuhová, 2009; Bitušíková, 2002; Young and Kaczmarek, 2008; Cochrane and Jonas, 1999; Eckardt, 2005). In such work, authors have sought to map symbols of the 'socialist past' in relation to the 'postsocialist present' (see Light, 2004; Czepczyński, 2008) as well as considering the placement of symbols and monuments associated with ethnic and/or minority/majority social identities with reference to 'cultures of remembering' (Szaló and Hamar, 2006; Ferencová, 2008).

Hörschelmann and Stenning (2008) argue, however, that while such work has been important in outlining the complexity of everyday life as a fruitful avenue for academic attention, researchers have nonetheless tended to oversimplify an understanding of postsocialism as a set of political and economic reforms and to present a narratives of a straightforward 'transition to capitalism' (Smith, 2002; Smith and Stenning, 2006; Round and Williams, 2010; Smith and Jehlička, 2007). For example, studies of the privatization of housing have described the emergence of new types of social roles (owners, non-owners) and relationships (landlords and tenants) (Smith and Jehlička, 2007; Smith and Rochovská, 2007; Stenning *et al.*, 2010; Shevchenko, 2002; Bodnár, 2001; Caldwell, 2004; Keller, 2005; Oushakine, 2000; Patico, 2009). However, as Hann *et al.* (2002) note in seeking to theorize such work on postsocialism point out that practices (such as self-provisioning, alternative consumption patterns, mutual help, and the importance of social networks) are either presented as representing the 'socialist' period or indicating the 'already capitalist' – or non-socialist – era. Such insights thus lead Alison Stenning (2005a: 124) to describe postsocialism as 'partial and hybrid', pointing out the combination of, on the one hand, the past and present, and, on the other hand, local and global social forms and practices, as typical of postsocialist everyday life (Shevchenko, 2002; Caldwell, 2004; Smith and Jehlička, 2007; Keller, 2005). Such criticism notwithstanding, perspectives on everyday life in postsocialist urban contexts have been developed in relation to manifold practices, habits, places, relationships, institutions, and so on (Hörschelmann and Stenning, 2008; Patico, 2009). Since the 1990s, then, research into postsocialist cities has focused on topics and issues such as experiences of poverty and neo-liberal reforms (Smith and Rochovská, 2007), routine and extraordinary consumption activities, and everyday uses of public space as key examples of 'postsocialist urban life'.

Theoretical debates: perspectives on (post)socialism

Although individual scholars in Czech and Slovak urban studies have drawn on a range of different theoretical traditions to consider what they consider postsocialism, post-communism, post-totalitarian period, transformation and post-transformation, there is a general agreement around a number of principal assumptions. There are two approaches to conceiving the difference between the socialist and postsocialist city. The first underlines similarities between modernity and urbanization in the West and East as a point of departure, while the second simply understands the 'socialist city' as radically different from the capitalist city. The latter approach stresses how principles of investment and accumulation have been the basis for the development of the socialist and capitalist city, as the mutually exclusive urban forms (Růžička, 2010), or simply seeks to explain patterns of urbanization as typical of the socialist or capitalist cities (Matlovič, 2004). The former, and larger, body of work refer mainly to Enyedi's (1996) and Szelényi's (1996) conceptualization of socialist urbanization as parallel and similar to capitalist development (Mulíček, 2009; Musil, 2003; Sýkora, 2009a; Gajdoš, 2009).

For example, Jiří Musil (2003: 138) considers countries with centrally planned economies and those with market economies prior to the 1989 as 'two variants of modern, or modernizing, societies', suggesting that some of the macro-social processes were 'common or at least analogical'. Commonality of the modernization process in both East and West cities was stressed by György Enyedi in what he observed as basically the same 'model of urbanization'. The specificity of the Eastern case was nonetheless considered as caused by 'a delayed economic and urban modernization and, second, the social political system' (Enyedi, 1996: 102). The parallel development of cities in the West and East is thus explained by their shared experience of industrialization, as well as by the influence of planning policy, with differences seen as the – intended and unintended – effects of state socialism (Sýkora, 2009a: 390).

The specificities of (post)socialist urban development is also described by Iván Szelényi's 'under-urbanization' thesis (first proposed in 1971, see Szelényi, 1996). The concept was used to designate discrepancies between high levels of industrialization in socialist Hungary (and other countries in the region) and the slow growth of the 'permanent urban population' (Szelényi, 1996: 294). Although in 1996, Szelényi himself pointed out that his concept was meant to be 'value-neutral' and that he was not suggesting that the capitalist urban system represented 'normal' development from which the socialist cities diverged (Szelényi, 1996: 294–5), it is this interpretation that has prevailed in most conceptions of 'urban socialism' in the Czech and Slovak scholarly work. Both the idea that cities were 'delayed' during the period of state socialism and suggestion that those cities are now returning to 'normal' urban development (a conception which was not fully accepted either by Szelényi or by Enyedi) are often pronounced without discussion. This approach has thus lead to the understanding that the transformation period is a process of 'correction', a 'return from deviation', a 'return to the normal', or a time of 'catching up', and is clearly seen as 'evolution', or in Ondřej Mulíček's (2009: 159) words:

> [i]t is especially Szelenyi's view of the development of the socialist cities that allows us to perceive the transformation in the postsocialist city as a combination of 'corrective' processes putting the postsocialist city back to the natural path of evolution of the West-European cities and of a whole range of general global processes and changes, which express the shift of the European cities from the fordist period to the postfordist phase of development.

Similarly, Luděk Sýkora (2009b: 285) understands the revolution at the end of the 1980s as an 'outcome of *evolution*' [my emphasis] defined as 'wide process of societal change towards more advanced, organized and complex society' where the socialist period is conceived as a 'path with a dead end', or a 'blind alley'. The revolution therefore had one main goal – correcting an 'unviable evolutionary path and to put society on the track that promised a future with higher freedom and more wealth' (Sýkora, 2009b: 286). A parallel conception of transformation as a process of 'rectification', or return to the 'normal evolutionary trajectory', also appears in the work of Peter Gajdoš (2009) and the writing of L'ubomír Falt'an (2009) where the term *evolution* is replaced by *civilization*.

Taking a different approach, studies of postsocialist cultural landscapes have described the past socialist period as a phase of 'liminality' (Czepczyński, 2008: 109). The concept of liminality is taken from the vocabulary of anthropology and refers to a rite of passage. Such an approach turns the idea of correcting the wrong path of evolution into a true caricature. It indirectly proposes a picture of cities as individual organisms and represents the transformation period as a phase in the process of acquiring new (higher) roles, such as passing from childhood to adulthood. It is not surprising that Czepczyński (2008: 113) points to liminality as 'an ambiguous period characterized by humility, seclusion, testing and haziness' – childhood in the form of the mistaken project of state socialism being left behind, with the capitalist adulthood yet to be reached.

In attempting to theorize postsocialist urban change as an aspect of evolution, such writing agrees on one crucial point; that the current situation in cities in both the East and West are nonetheless influenced by globalization. Postsocialist cities are thus described as undergoing 'double/dual transformation' (Sýkora, 2009a, 2009b; Mulíček, 2009; Matlovič, 2004; Gajdoš, 2009; Temelová, 2007, 2009; Ouředníček *et al.*, 2009). This process is understood as a combination of locally specific changes (especially with regard to the shift to market capitalism and the above mentioned process of 'returning to normal'), which differ slightly from one country to another, and of the parallel changes induced by the effects of globalization (Musil, 2003: 159). Such globalization processes include the shift to post-fordist (capitalist) economy, 'internationalization' or 'post-modernization' (Falt'an, 2009; Matlovič, 2004; Sýkora, 2009a; Musil, 2003) and that:

> The specificity of the post-socialist transition is in its nature of double transition. [...] It includes the local transition to market economy and the global transitions conditioned by economic globalization and its influence on local political, economic, social and cultural restructuring. The post-socialist transition is not only about local change from socialism to capitalism, but also about the integration of former socialist states into global capitalist order.
>
> *(Sýkora, 2009b: 286)*

Such explanation thus paints a picture of juxtaposed (but not necessarily inter-acting) changes: those that are locally and historically specific and those that are shared with the rest of the world. The specificity of postsocialist cities in this perspective emerges from the transitory and revolutionary character of local development.

As such, when it comes to *transition* and *transformation*, a conceptual inconsistency appears in recent theoretical work on the postsocialist city. While the words are often used in an interchangeable manner, Sýkora (2009b: 284) makes a distinction between the two – transition is understood as a 'broad, complex and lengthy process of societal change', which consists of 'multiple transformation processes' concerning various areas in society or related to different time periods. This conception allows the author to further differentiate between the *types* of transformations including an initial, fast 'intentional transformations controlled by the government', and a longer-lasting 'spontaneous transformation' or 'evolutionary adaptation of existing social as well as regional and urban systems to the new societal rules' (Sýkora, 2009b: 285). The importance ascribed to the first induced transformations is highlighted by Sýkora as an adaptation of a 'path dependency and path-shaping' approach to postsocialism (see Beyer and Wielgohs, 2001). The expression that presents the revolution of 1989 as a way to 'put society on the track that promised a future with higher freedom and more wealth' (Sýkora, 2009b: 286) is now more easily understood – the revolution represents the first moment of *controllable* change. The rest is left to the 'natural evolutionary development' leading to a future that is quite indefinite, but generally identical for the cities in the West and in the East of Europe.

Moreover, as Beyer and Wielgohs (2001: 362) observe the 'path dependency and path-shaping' perspective suggests that the transition process is, in general terms, uncontrollable by political means. A 'natural evolution' can hardly be perceived as a project that has an alternative, or allows (or requires) substantial control or intentional shaping. Some scholars note that Czech and Slovak cities have, due to the 'delay', a possibility to 'learn' from the example of Western cities in order to cope with the development and to avoid the greatest risks that the Western cities encountered years ago. However, this conceptualization of transition proposes a rather antithetic picture – those who are affected by the transformation processes are in the position to observe the inevitable.

In defining a (double) transformation, such academic work has thus sought to theorize the *postsocialist city* as a transitory phenomenon, and postsocialism as a temporary (and dynamic or turbulent) period (Sýkora, 2009a, 2009b; Matlovič, 2004; Mulíček, 2009; Sýkora, 2002; Sýkora and Ouředníček, 2007). When it comes to the obvious question that arises from such a perspective – when does the postsocialist city as a temporary phenomenon end? – responses are often vague and unsatisfactory, along the lines that:

> The transition from the socialist to another type of city proceeds in the time-period between change in the societal rules of the game and the completion of corresponding transformations in built and social environments. When such transformations are accomplished former socialist cities

will become one of the many variants of urban places under capitalism. Of course, they will reflect their past. And they will be subjected to other transformations stemming from new impulses coming from the capitalist society. The evolutionary adaptation to revolutionary change of the principles of social regulation will be over.

(Sýkora, 2009b: 288)

Stenning and Hörschelmann (2008: 319–21) identify such notions of a 'return to normalcy', the 'catching up' imperative and descriptions of the postsocialist world as lagging 'behind', as characteristic of the 'dominant discourse of transition'. While the 'transition narrative' has been criticized (Brandtstädter, 2007; Stenning, 2005a, 2005b), such critique rarely appears in the writing of Czech and Slovak scholars and dominant accounts of postsocialism as transition are seldom challenged. Still, the arguments proposed in the works which do question the 'narrative of transition', as Stenning and Hörschelmann (2008) highlight, opposition between 'postsocialism' and the 'Western capitalism', to which the transition is supposed to lead, as well as the search for the *final point* of postsocialism is unconvincing. Moreover, based on inspiration taken from post-colonial studies, Stenning and Hörschelmann (2008) highlight how the West is also postsocialist, also without an end point, and in doing so make obvious the mutual discursive construction of the relationship between East and West, in which the 'other' serves as positive or negative reference points in defining the self.

In the case of Czech and Slovak urban theory the prevailing acceptance of the explanation of postsocialist urban development as 'delayed' and undergoing the process of 'transition' is mirrored by depictions of urban studies as similarly 'catching up'. The representation of Central/Eastern Europe as undeveloped, and not yet civilized, has been depicted in the political discourses in both the 'West' and the 'East' (see Domański, 2004; Kuus, 2004) and in urban studies, a parallel picture of the *theorization of cities* as getting back to normal and catching up is replicated in the representation of local urban studies as not (yet) reaching 'international quality'. For example, Hampl *et al.* (2007) describe how particular research agendas are missing from Czech geography, and that empirical rather than theoretical work has tended to dominate. However, authors have commented that this focus has probably had 'rather positive outcomes (especially in the sphere of policy application)', and thus praise the theoretical modesty of local researchers (ibid., 2007: 486). Taking a different perspective, Judit Timár (2004: 371) reflects on the tendency of Hungarian urban studies to unquestioningly adopt international academic research agendas and that a current 'unevenness of the convergence' of postsocialist and Western geography is ensuring that academics from the postsocialist countries tend to take a subordinate position in urban debates.

In these terms, the idea of postsocialism as a transition to the (normative) standard, and the (self)perception of the local urban studies as being under-developed and lacking in the capacity to propose valuable theoretical insights are clearly interconnected. However, Stenning and Hörschelmann propose an overview of critiques Western theory for overlooking what is being produced elsewhere in the world. On the other hand, the tendency to accept uncritically 'western knowledge, *because it is western*' was noted in the postsocialist academic world in the early 1990s (Timár cited by Stenning and Hörschelmann, 2008: 316). The overview of recent Czech and Slovak urban studies presented in this chapter similarly indicates that the image of research and academic knowledge production as lagging behind Western urban theory nonetheless stubbornly persists in the works of contemporary writers. In these terms relations between academic worlds negotiated and reproduced in encounters between the 'Western' and 'Eastern' scholars, *and* within the local academic communities and writings, continues often to be based on misrepresentations and stereotypes of the postsocialist city (Hörschelmann and Stenning, 2008: 350–1).

Conclusion

In concluding this chapter I wish to offer a few autobiographical accounts of my own experiences of the East/West divide in urban theory. The first relates to a conversation with a colleague from a postsocialist country. When discussing our plans to submit abstracts for consideration to an international conference, he confessed with a sigh that he was annoyed, and even ashamed, to still be writing about postsocialism. He perceived that simply by using the term he was admitting his provinciality and incapacity to follow recent academic trends and concepts, observing that postsocialism was not *à la mode* any more in the Western science. Indeed, Stenning and Hörschelmann (2008: 329) note that the concept of postsocialism is used less and less by people in everyday life in postsocialist countries, precisely because it is generally associated with backwardness. In academia, too, the term is also accepted 'reluctantly' by researchers (Stenning and Hörschelmann, 2008). At the same time, though, in this story the feeling of backwardness was strengthened by the idea that the topic (postsocialism as an 'object' of study) is now outdated. Hence, it is not just that there are negative connotations associated with the term postsocialism itself but that the relevance of research into postsocialism itself that is being questioned via the self-referring of local scholars to international theory. Studying postsocialism thus appears as inappropriate, *because it seems to be out of fashion in the 'West'*.

The second story also points to the complicated discursive construction of the relationship between the 'West' and the 'East' in urban theory. In a recent article, Smith and Timár revisit an important argument previously proposed by Smith and Stenning: that theory should 'flow' between the West and the rest of the world in order to 'decentr[e] the role of the West' (Smith and Stenning, 2005, quoted in Smith and Timár 2010: 122). Such a cosmopolitan approach to theory is thus presented as being a 'shared interest of researchers studying the spaces of postsocialism, *irrespective of where they live*' (Smith and Timár, 2010: 122, my emphasis). However, on the very same page, an excerpt from an article by Pickles (Pickles and Smith, 2007) is used to illustrate the relevance of the past in understanding postsocialism today. In this excerpt, the following sentence appears, 'perhaps what first strikes you *when you travel into [Central and Eastern Europe]* ... is the incredible mélange of practices, rhythms, and identities that flow through particular places ...' (Pickles, 2007, in Smith and Timár 2010: 122; my emphasis). In this reflective and thoughtful article, where statements about the 'shared interest' of researchers 'irrespective of where they live' appear, the researcher studying postsocialism nonetheless *travels*, or at least addresses his/her reader as 'you who travels to Central and Eastern Europe'. Reading this article as a theorist of postsocialist cities based in Central Europe, I do not find myself in either of the roles. I do not travel to postsocialist Central Europe; I live here.

Both examples thus illustrate a simple but very important point – the meanings of 'postsocialism' in urban theory, are constantly being re/created in writing, conversations, and perceptions of the other, of the self, and of the relationship between the two. These perceptions, representations, and propositions, while often been incongruous, still appear together. Such contradictions therefore form part of what it means to 'be postsocialist'. However, at the same time they foreground the question of how (and where, by whom, and for whom) urban theory is produced, perceived, used, developed, or rejected, and it is useful to turn to the work of Pierre Bourdieu (1994) to explain this situation. Bourdieu describes 'microcosms' (such as arts or science) as *social space*, i.e. as fields of artistic, literary or scientific production, defined by the positions and social relations between the artists/scientists and the other figures intervening in the process of production/ consumption. In these terms, 'light' and 'serious' genres, recognized and refused oeuvres, are defined and explained by their mutual and differentiated positions. According to Bourdieu (1994), positions and relations defining 'the field' have an influence on the production of 'the art' itself and it is through the effects on social space that, for example, political or social changes are reflected in the resulting work. Such conceptions can, of course, be easily and fruitfully transposed to the sphere of academic work and used as a useful analytical tool to explain the actual shape and outcomes of urban theory. Mapping the positions of topics or genres, for example, of representations as postsocialism as being of 'marginal' interest in contemporary academic knowledge production, thus allows the role of theorists, as well as institutions such as universities, research agencies and the 'consumers of scientific knowledge' and the 'geography of knowledge production' (Smith and Timár, 2010: 121), to be interrogated.

The need to reflect on the differences in work conditions and contexts experienced by non-Western academics appears in reflections on the position of postsocialist (or, in general, non-Western) science to be an important goal (Stenning and Hörschelmann, 2008; Timár, 2004; Hörschelmann and Stenning, 2008). Together with Smith and Timár (2010) I believe that in order to understand 'postsocialist urban studies' – be it studies *on* postsocialist cities, *in* postsocialist cities, or by authors, who themselves claim to be 'postsocialist' (by institutional affiliation to the local academic bodies, by the experience of socialism in their biographies, because they live in the countries termed postsocialist, or because they simply accept the idea that we are all postsocialist) – serious consideration of the everyday character of academic production is vital if we are to understanding the discursive practices involved in defining postsocialism, postsocialist urban development, and postsocialist urban studies, and how such catergorization and theoretical work can add to urban theory in general and specific ways.

5

URBANISM, COLONIALISM AND SUBALTERNITY

Swati Chattopadhyay[1]

What consists of urban theory beyond the 'West'? Nothing. And everything.

The 'West' remains the subject of urban theory, even when the putative object of discussion is a city 'beyond the West'. The un-nameable region carries in its more typically used designation as 'non-West' the burden of difference, of being the 'reverse' of the nameable subject; the prefix 'non' stands for an 'absence', a 'lack' of all that is 'important' and 'worthwhile' (*Webster's Dictionary*, 2001). Our insistence on redressing this scholarly disdain should not obscure the long history of theoretical reflections on urban processes in the non-West. In the nineteenth century it took the shape of concerns about colonial governance, the city considered as a tool for expropriation, an experimental field, a problem of tropical disease and sanitation. Sociologists and urban planners in the early twentieth century theorized and modelled new towns all across the colonial world. In the wake of decolonization in the twentieth century attention turned to these cities and urban regions as sites of development,[2] renewing and recasting the older conception of colonial cities as laboratories and nodes in the capitalist market. In its most recent incarnation, urbanism beyond the west has come to be re-viewed in the light of a planetary ecological crisis and global political insecurity.

What is remarkable in the last instance is that a critique of neo–liberalism feels compelled to rely on Dickensian metaphors to describe contemporary urban processes in the non-West that are structurally different from that of nineteenth-century Europe (Davis, 2006). The descriptive vocabulary cannot seem to keep pace with political critique or changes in urban morphology, unable to see through the metaphors of death, disease and toxicity the contours of creativity and resistance that give depth to these 'abominable' geographies. The impasse suggests a theoretical lag and an inability to move beyond the limitations of a colonialist urban history that is preoccupied by the 'West'.

We face a paradox in the phrase 'beyond the West': the 'beyond' suggests an 'outside', and yet the outside remains tethered to the 'West' as the point of reference. Relaxing this attachment would require a new conceptual frame. By focusing on the idea of 'colonial urbanism' and the concepts used to describe and theorize cities in the nexus of capitalism, colonialism and modernity, this essay works toward a problem statement that might help articulate such a frame.[3] I am interested in the theoretical consequences that spring from interrogating the universality of the laws of political economy shaping the form and culture of cities.

The culture of capitalism

In theorizing urban processes under capitalism, specifically the relation between culture and economy, scholars have assumed that Western cities have worked out in their triumphs and trials the historical development of cities in the world.[4] Crucibles of capitalism, they have *theoretically* foretold possible outcomes of culture's relation to capital. This is directly tied to the way we think of the global political economy: state agents and corporate actors in the West set the rules of the game that the rest of the world follow.

We are familiar with the broad contours of this story. Cities are instruments of capitalist accumulation, and following the Marxian understanding of base and superstructure, the capitalist economy generates the framework of social and cultural possibilities. Initially, however, the relation appears to be quite the opposite: capitalism seeks and must find cultural capacities that it can build upon. Capitalism in Western Europe originated and flourished – took hold – because the urban conditions enabled it to do so (Braudel, 1992). The already existing urban network supported the 'appropriation, mobilization, and geographical concentration' of surplus capital and labour power that had been freed up through primitive accumulation (Harvey, 1989: 23). The classic case of primitive accumulation in the form of rural enclosures occurred in Britain, uprooting the peasantry and turning them into wage labourers, ultimately creating the rural–urban split that marks urban capitalist development. If 'spatial fix' in the form of urbanization is a necessary moment of capital (Harvey, 1989), the grounds for it has to be ready in the form of tangible and intangible assets, namely the built environment, technological innovation, and social infrastructures to control civil society. In the process of maturation, capital makes of these same social and institutional capacities 'slaves of accumulation' (Harvey, 1989: 24).

Where does one locate colonialism amid these mechanisms of capitalist accumulation? From Europe capital reached out into the world beyond to colonize territories, giving rise to urban regions 'dependent' upon metropolitan Europe.[5] Modern colonialism is traced back to the conquest of the new world to service mercantilist competition between emergent nation-states. Capitalism in the eighteenth and nineteenth centuries inherited this credo of conquest from mercantilism; if it did not *need* colonialism, or if its logic was indeed contrary, the latter would have disappeared. As Edward Said (1993) noted, colonialism was maintained at 'great cost' to the European nations, only less than what it cost the colonized, of course. It took much convincing 'at home' to promote the merits of empire to entire segments of society who did not profit from it directly, but who served as fodder for imperial ambitions in battlefields across the world.

Colonialism is the strong arm of capitalism, gathering raw materials and labour, creating markets and sites for investment with profit directed towards the metropole. There are two readings of this global reach of capital: one is that colonialism, as a tool of capitalism, extends capital's reach. Liberal histories of the British Empire, for example, have portrayed the colonial state as an 'organic extension of the bourgeois metropolitan state and colonialism as an adaptation, if not quite a replication, of the classical bourgeois culture of the West in English rendering', and positive proofs of the 'universalizing tendency of capital' (Guha, 1997: 4). Colonial cities are the end nodes of the network of accumulation centred in Europe. They are thus physically and economically 'peripheral' while being ensnared in the universalist logic of capital. The rules of capital apply to these cities just as they apply to those in the metropole, and yet they may not share the same outcomes, at least initially. The visible differences in the colonial and ex-colonial terrain emerge as so many deficiencies. A historicist understanding of the progress of capitalism, in which nation-states are arrayed in a temporal scale of development, explains these colonial outcomes as embryonic stages of fully-fledged capitalist development; third-world cities are 'premature'. This argument could be used to claim that third-world cities today represent what first-world cities were in the nineteenth century; the former are going through the growth pangs that the latter experienced decades and centuries ago. If the most advanced stage of capitalism harbours within it traces of earlier phases, third-world urbanism has nothing original to contribute to urban theory per se. It is theoretically redundant. To address third-world cities is to define and solve them as 'problems' according to already available formal and sociological models. This is the project of development: to modernize peoples and institutions – to correct cultural incapacity – so they may serve capital well or, alternatively, gain the capacity to wage an internal struggle with it (Appadurai, 2004).

The other reading is that the colonial and ex-colonial contexts would continue to show deficiencies, because the grounds for harbouring a fully capitalist urbanism did not and will not occur in these contexts beyond the West. They are constitutively different, and thus the principles of bourgeois liberalism are not applicable in the colonies. Much colonialist urban policy was shaped by a view that the territorial and cultural conditions in the colonies are inherently different on account of race, climate and history – producing entire fields of 'tropical' study – and this justified the application of rules and provisions in the colonies different from the ones used in the metropole[6] (King, 1976; Wright, 1991; Çelik, 1997; Myers, 2003; Chattopadhyay, 2005; Legg, 2007; Kidambi, 2007). The liberal political view of empire (the first reading) could even overlap with this interpretation, because the violence of colonialism and its role as a mechanism for making distinctions along racial lines, between Europeans and non-Europeans, colonizer and colonized, had to be explained away (Chatterjee, 1993; Mehta, 1999). This colonialist view at least has the advantage of stating explicitly that liberalism obeys a different set of rules in the colonial terrain. The seemingly universal laws of capitalism have to be bent to serve the needs of colonialism. Colonialism is not merely an extension of capitalism; it has other functions that do not fall within capital's economic ambit proper. It is a material manifestation of authority, a means of clarifying relations of domination and subordination. Colonialism and capitalism, though allied, are functionally and structurally distinct.

A variant of this suggests that colonialism/imperialism in the nineteenth century indeed prevented capitalism from fulfilling its potential: the territorial logic of imperialism 'inhibited' the capitalist logic (Arrighi, 1994; Harvey, 2003). According to this view, the capitalist logic resides in 'opening up' spaces (it must do so continuously to ensure further accumulation) and generating economic 'flows' across territorial boundaries. Even though expansive, imperialism with its focus on territorial acquisition and control works contrary to the logic of capital. Colonialist policy refused to the colonies equivalence in terms of social and physical infrastructures, and diminished the promise of capitalism significantly beyond the stage of primitive accumulation (Harvey, 2003). Since the primary story is of development of cities under capitalism, urbanism in the *not*-West is theoretically peripheral; as aberration, it only delineates the true trajectory of capitalism's relation to the culture of cities in the West.

In his earlier and most important work Harvey (1989: 25) provides the classic reading of patterns of accumulation to explain the difference between first-world and third-world urbanization:

> Primitive accumulation, and other processes of appropriation do not guarantee ... that the surpluses can be assembled in time and space in exactly the right proportions for strong capital accumulation to proceed. In eighteenth-century Britain, for example, the strong capital surpluses more than matched the surpluses of labor power. Wages rose, and much of the surplus was absorbed in consumption projects. In contrast much of contemporary Africa, Asia, and Latin America is faced with the situation in which immense quantities of labor power has to be dispossessed to release very little capital, creating massive and chronic surpluses of labor power in a context of serious capital shortage. . . . This is, for example, the hallmark of much of contemporary Third World urbanization.

There are two notable points about this explanation. It treats Britain's eighteenth-century development as an internal dynamics, uncomplicated by external movements of capital. The crucial role that colonialism played in generating Britain's surplus is not key to the equation – it is parenthetical. Neither is colonialism's role in inflecting how much or how little capital can be squeezed out of primitive accumulation in the third world given any theoretical weight. At the same time, the statement subsumes cities in Europe and the third world under the same laws of economics, a macro theory making room for comparative assessment (though the latter is not the focus of his work), without asking why capital makes a distinction between the metropole and the third world, between white and black.

When Harvey launches his argument about imperialism's spatial and economic imperatives in *The New Imperialism* (2003) to explain contemporary imperialist sabre-rattling to defend capitalist supremacy, he finds scant need to engage with the substantial post-colonial literature. The result is that the meditation is not so much about imperialism as it is about imperial powers: the European and American imperialists are the only meaningful agents in his story, and he ends up proposing a kinder, more accommodative American imperialism (at least in the short run) as well as the need to accept limited forms of 'accumulation by dispossession' (Harvey, 2003: 209–10).[8] Making a distinction between dispossession for expanded reproduction and dispossession purely as a solution to over-accumulation of capital, he argues that in the interest of 'macro' and 'long-term' impact, resistance movements must abandon 'nostalgia for that which has been lost' and discriminate between 'progressive and regressive forms of dispossession', to 'guide the former towards a generalized political goal that has more universal valency than the many local movements, which often refuse to abandon their particularity' (Harvey, 2003: 178–9). The fragmentary nature of 'local' movements against dispossession waged in different parts of the world, their feeble focus on expanded reproduction, and most importantly their 'failure' to obtain global-universal transcendence are at issue here. His discomfort is with the refusal of many resistance movements to take 'command of the state apparatus' and the 'non-progressive' mentalities that will not conveniently align with the call for expanded reproduction. This is allied with the anxiety that 'the reversion to older patterns of social relations and systems of production will be posited as a solution in a world that has *moved on*' (ibid.: 177; emphasis mine). The temporal structure of his argument has great difficulty disengaging from the historicist tradition in Marxist thought even when he acknowledges its failure in sustaining resistance to capital. It is at this point the new imperialism emerges as a universalizing possibility that might, in the short term, bridge over the many fragmentations, presumably by pushing the cause of expanded reproduction.

Harvey's hope is that the Western powers would learn to act as true hegemons and establish leadership in investing in a global New Deal to arrest the aggressive pattern of 'accumulation by dispossession'. He does not ask what in the long historical experience has given the rest of the world any reason to *consent* to this new European imperial hegemony. More importantly, in trying to 'update' the concept of imperialism, Harvey misapprehends the very principles that power imperialism and offer it resistance. This has much to do with the diffidence in delving into colonialism as a phenomenon integral to capitalism (one he shares with many fellow Marxists).

Colonialism and imperialism do not proceed from the consent of the governed, but from outright domination. Coercion and violence, rather than persuasion and

consent, are the better part of imperial calculus. Harvey borrows Antonio
Gramsci's notion of hegemony (via Arrighi) and by applying it to competing
nation-states (the pre-Gramscian use) rather than to social classes within states,
inserts an idealism that is not realizable or historically supported (ibid.: 36–9).[7]
As Ranajit Guha has examined at great length, the misreading of dominance as
hegemony proceeds from the presumption that capital 'succeeds in overcoming
the obstacles to its self-expansion' and subjugates 'all precapitalist relations in
material and spiritual life well enough to enable the bourgeoisie to speak for all
of that society' (Guha, 1997: 19). It can only read dominance as consensual by
ignoring the production of subalternity, and the corresponding forms of resistance
to colonialism and imperialism.[8]

It is possible to derive one of two inferences from Harvey's account of capitalist urbanization and the new imperialism: that, indeed, colonialism is not a
significant factor in explaining the relation between capital and urbanization, or,
that we need a different formulation to understand this relationship. Let me
elaborate with a proposition.

Structuring absence

Colonialism is, to borrow a term from Jacques Lacan, a *structuring absence* in
urban theory. Urban theory's epistemological domination is coeval with modern
colonialism and capitalism, and while the latter is a mainstay of theory, the
former – colonialism – remains theoretically peripheral. In its peripheral, 'invisible'
role it allows capitalist accumulation to make global sense and enables the hoist-
ing of capitalism as the primary figure of modern urbanism: colonialism is alluded
to, shows up at the edges and crevices of argument, but is rarely addressed as an
explicit analytic. Two decades ago, Anthony King complained that given its large
role in shaping the world economy, colonial urbanism has been conspicuously
neglected as a subject of research (King, 1990: 2). A summary glance at some of
the most imaginative works on modern urban culture would bear this out.

Consider Raymond Williams' *The Country and the City* (1973). He began and
concluded with references to colonialism and imperialism as a plea for comparative study, although again, it is not an important analytic in his understanding
of the culture of capitalism. While he clearly signalled the mutability of social
relations and the landscape, he made the argument that the English experience
held the explanatory substance for understanding the landscape of capitalism in
the rest of the world:

> We are touching, and know we are touching, the forms of a general crisis.
> Looking back . . . on the English history, and especially on its culmination
> in imperialism, I see in this process of altering relations of country and city
> the driving force of a mode of production which has indeed transformed
> the world . . . this mode of production began, specifically, in the English
> rural economy, and produced, there, many characteristic effects . . . which

have since been seen, in many extending forms, in cities and colonies and in an international system as a whole.

(Williams, 1973: 292–3)

Within his larger argument that 'capitalism, as a mode of production is the basic process of what we know of the history of country and city' (ibid.: 302), colonialism is indexed in two ways: first, implicitly, as a process of the city colonizing the countryside,[9] and second, explicitly, as Europe's domination on the rest of the world. Both of these readings are in line with Marx's understanding of the relation between colonialism and capitalism.

Williams did not see colonialism as structurally different from capitalism (he uses colonialism and imperialism interchangeably), although he differed from his peers about the inevitability of its consequences. Pointing out the forms of resistance to capital in the twentieth century that had taken a different route in China and Cuba, he critiqued the 'abstract chauvinism' in the 'old imperialist countries' of the assumption that 'what has happened to them is what was happening or would happen to everyone'. His statement did not contravene Marx's understanding of capital's unfolding (see Guha, 1997: 13–16), but the focus of his critique was the ambivalence in Marxist theory towards the countryside: the desire to create a balanced relation between city and country (under communism) and the parallel characterization of the peasantry as 'backwards', in need of the progressive thrust of industrialization/urbanization to be turned into a revolutionary force, along with the backhanded compliment paid to the bourgeoisie for rescuing 'a considerable part of the population from the idiocy of rural life'. Williams argued for a different vision of city and country, one that would reverse the model of domination by the city over an 'available' countryside, precisely because of the 'dialectical twist' that inflicted long-term damage to the metropole itself (Williams, 1973: 303). Yet he did not see colonialism substantially inflecting the social relations and landscape of eighteenth- and nineteenth-century England, and therefore the theorization of the city–country relation.

Indeed, the view of colonialism as an extension of capitalism resides deep in the recesses of classical political economy, one that also offers explanation of why capitalism manifests itself differently in the West and the not-West. Following Marx and Engels, the process through which the world beyond Europe succumbed to the industrial power of the bourgeoisie despite the immense cost to peoples and cultures has been viewed by theorists of modernity as an essentially progressive phenomenon.[10] Materially, it enables Europe's industrial revolution. Politically (and theoretically), it purges these Western nation-states of social and historical contradictions.

In reviewing political economy's complicity with colonialism and imperialism, David Blaney and Naaem Inayatullah argue that classical political economy produces 'an implicit or explicit theodicy – an explanation and rationalization of persistent malevolence in a world inexorably destined … to produce the human good' (2010: 4). For Hegel, a racialized colonialism was the only answer to the 'internal disintegration' of European civil society and a 'crucial and necessary part of his attempt to preserve the harmony and universality of the modern symbolic order' (ibid.: 116). Europe could export its impoverished 'rabble' to the colonies 'returning them to the family principle in a new country' and at the same time provide impetus to new industrial activity (ibid.: 139). Colonialism came to be justified by shifting Europe's internal disintegration to other societies, while seeming to perform the noble task of bringing modern civilization to these peoples and regions that now, by a circular logic, are deemed 'historically moribund'. In Hegel's hand, political economy's premise of systematic death – 'necro-economics' – becomes a 'necro-philosophy': '(t)he production of poverty is merely displaced onto the colonial world, and to pre-figure Marx and theories of dependency and the world system, reappears as a global wound' (ibid.: 140).

Capitalism's dispersal is also a displacement, the securing of a space 'outside' of itself (Luxemburg, 2003). This outside is structurally necessary for capital, one that it must create continuously.[11] Colonialism and imperialism have served and continue to serve in creating this outside, not just as a site of primitive accumulation, but much more. This dispersal/displacement also allows the developed West to appear as the sovereign subject of theory. The non-West disappears in a death glow. Colonialism vanishes as a signifier, only to surreptitiously show up as the 'modernism of underdevelopment', where first world movements are echoed as so many painful distortions (Berman, 1989).

The spatial logic of coloniality

'Coloniality', Walter Mignolo notes, is the 'hidden face of modernity and its very condition of possibility' (2002: 158). The key concept in this 'colonial matrix of power' is *'race* (in the sense of racism), and not class (in the sense of classism)' (Mignolo, 2005: 383). The 'logic of colonial difference', to use Partha Chatterjee's (1993) phrase, rests on this distinction, and throughout the colonial period, European administrators, historians and philosophers refused to admit that the universal laws they upheld in Europe applied to the colonies in matters of law and government. Consider a key factor in capitalist urbanization in the West: private property. The process through which land was confiscated from the peasantry in the West to become the private property of a new group of proprietors was premised on the idea of land as freely alienable. The very character of land was transformed by being subjected to the system of monetary exchange as a commodity. Its profitability was magnified by investment (improvement) and realized through a system of wage labour in the form of the dispossessed peasantry (which further necessitated the idea of juridical equality), giving rise to the familiar class relations. In the long run, the dispossessed peasantry-turned-proletariat would be reincorporated into the state in their role as citizens through various means (civic, educational, military), a process that could not be accommodated within the logic of the colonial system (Lloyd, 2005). The pattern of primitive accumulation that dispossessed the indigenous peasantry in the colonies was fundamentally different.

Property relation was among the first issues tackled by colonial authorities in their effort to transfer land rights and profits to the colonial state. These processes were integral to the economic logic of capitalism. The French and the British adopted different strategies, and the outcomes depended upon the political context and existing contingencies of property relations. Where it was not outright theft, it was justified by a patina of legality.

In eighteenth-century Bengal, the British introduced a notion of private property that amounted to a quasi-feudal system, one that managed the double trick of 'freeing up' land and retaining a feudal system of production. As part of revenue reform under the East India Company, the older mode of landholding was dismantled; land was auctioned off to a new class of landowners in return for a fixed revenue due to the colonial state. If the revenue was not delivered by dusk on a set date, the property could be sold to the highest bidder. The new landlords did not come to 'own' the property, but were bestowed the right to rent the land and collect revenue on behalf of the colonial government. At the same time the customary relation between the peasantry and the landlords, including that of forced labour and the many 'coercive privileges' of landlordism, were left largely undisturbed. The revenue system, called the Permanent Settlement of Bengal (1793), produced a class of landowners who had no incentive to 'improve' the land, and lived off the rent in cities far removed from their rural holdings. In addition it allowed the colonial state to avail of forced labour for government work whenever needed. And there was no fear that private property would introduce a principle of equality and thereby induce competition with the colonizers: private property was designed to make Indian landlords not just 'subjects' but 'slaves' (Guha, 1996: 32). The domination of the peasants (or the landlords, for that matter) did not require hegemony, because unlike the European peasantry, they were not on their way to becoming citizens within the colonial state (Lloyd, 2005). The system produced a city–country relation far removed from the phenomenon described by Williams (1973). It produced subalterns who were identified by their 'difference' from the colonial state and the indigenous elite, and remained largely beyond the purview of the state (Guha, 1983). Unlike Gramsci's subalterns, they were not on their way to proletarization (Lloyd, 2005).

David Lloyd has noted that Rosa Luxemburg made an important point about the process of primitive accumulation in the colonies. There the assault on the colonial peasantry, she noted, was economic as well as cultural:

> Since the primitive association of the natives is the strongest protection for their social organizations and for their material basis of existence, capital must begin by planning for the systematic destruction and annihilation of all the non-capitalist social units which obstruct development.
>
> *(Lloyd, 2005)*

Economic penetration required cultural dismantling. Nineteenth-century history of colonial India is replete with examples of such cultural assault. The physical community – entire villages – were burned and destroyed, and villagers transported to plantations, precisely because as long as the socio-spatial integrity of the indigenous community was unharmed, the colonial state stood powerless, despite possessing superior arms and economic leverage. The annals of the Santal Rebellion in the mid-nineteenth century are a testament to this face off between superior state technology and community solidarity.[12] The broad lineaments of this process has occurred before and after in the colonial world. The economic logic of capital crucially *depends* on the cultural and spatial logic of colonialism and imperialism and not vice versa.

Imperialism has two territorial logics: one is the occupation of the interior of a country for which military might must be supplemented by cultural assault, and the other is patrolling and expanding the borders of an imperial territory by the use of arms. In India and elsewhere in the colonial world imperialism did not inhibit the economic logic of capital, as long as that capital was looking to benefit Europe.

Discussing the dual worlds of the colonizer and colonized in the white town and native quarters, Fanon (2004: 5) made a pithy comment on the perversity of the economic logic that drives coloniality: 'In the colonies, the economic infrastructure is also the superstructure. The cause is effect. You are rich because you are white, and white because you are rich,' adding that 'Marxist analysis should always be stretched when it comes to addressing the colonial issue'. The perpetuation of this closed system required a spatial strategy that would address the problems of property, access to resources, and governance as a unified set of concerns. This is where capitalism and colonialism reveal their functional and structural differences. Colonialism's function went well beyond ensuring economic surplus to generate self-referential modes of situating the colonizer and the colonized in the world. It fell to theories of colonial urbanism to work out this unified schema. The biggest contribution of scholars working on colonial cities is to demonstrate the workings of the 'logic of colonial difference' in the most tangible and permanent shaping of the physical spaces of habitation in which the economic and cultural logics were designed to coincide in an explicit and outward expression of colonial authority.

Not all colonial cities were founded by the colonizers, and neither were all of them commercial hubs. In most, the colonial logic was grafted onto a pre-existing urban fabric, thereby articulating and even highlighting the level of physical transformation deemed necessary to accommodate an extractive economy oriented towards Europe. The well-studied examples of the refashioning of an existing urban fabric as a colonial matrix of power include Rabat, Algiers, Delhi, Lucknow and Lahore. While researchers have focused on the large port cities that served both commercial and administrative function of colonialism, most were smaller provincial towns that carried the sinews of the colonial economy from the port cities inland. This required attention to settlements as well as the infrastructure of roads, railways, canals and bridges to ensure supply lines between city and country. Because colonial authority exerted itself as a network of governance, the

connection between provincial and larger colonial towns was even more important than in Europe, where towns tended to retain a degree of independence.

Scholarly examination of colonial cities as a system began in the wake of decolonization.[13] As instances of colonial domination, these cities came to be viewed
as controlled windows into colonial society that provided a 'privileged insight'
into the 'essence of colonial life' (Ross and Telkamp, 1985). From the very beginning colonial cities were placed in a comparative global field, and seen as a 'type'
whose formal logic could be understood by studying the relations of domination
and subordination characteristic of European colonialism (Redfield and Singer,
1954). Importantly, however, they were seen as metropolitan transplantations in
the tropics, and in *essence* European. Here one could look closely at the spatial
lineaments of European social and economic power writ large in the themes of
dual cities, segregation, *cordon sanitaires* and apartheid (King, 1976; Abu-Lughod,
1980; Oldenburg, 1984; Metcalf, 1989; Wright, 1991; AlSayyad, 1992). The
colonial 'mentality' expressed itself in its obsession with order, elaborated in grids,
neo-baroque schemes and garden city plans, and extended across the spectrum of
colonial spaces: plantations, mines, bungalows, camps, courts, barracks, gymkhanas,
cantonments and civil lines. Pioneering research sought to understand the role
and location of colonial urban formations within global economic and cultural
processes (King, 1985; 1990). Focused on colonialist interests they tended to emphasize metropolitan designs and the outcome of metropolitan institutions super-
ficially grafted upon colonial soil. Since metropolitan institutions were thought
to 'ossify' in the colonies, King argued that it is important to 'relate institutions
and behaviour to contemporary developments "back home"' (King, 1985: 13).
The combination of looking 'back home' and a world systems theory that viewed
these colonial sites as determined by the movement of European capital ensured
a reification of the point of view of the colonizer. Ostensibly about cities in
Africa and Asia, these studies embraced Europe as the subject (Kusno, 2000: 6).
As such their theoretical contribution was not so much about urbanism in Africa
or Asia, as it was about Europe beyond itself.

As this research matured it offered important avenues for rethinking the West
itself. It showed that to understand the social, scientific and urban developments
in the West one must recognize the role that colonial experience played in the
construction of metropolitan modernity, precisely because social and urban experiments could be conducted more ruthlessly in the colonies than in the metropole. Metropolitan developments 'owed' something to colonial developments: the
'outré-mer' was 'a terrain for working out solutions to some of the political, social,
and aesthetic problems which plagued France' (Wright, 1991). Paul Rabinow's
study of French modernism sought to understand the 'practices of reason' that
were mobilized to bring together modern fields of knowledge with modernist
urban forms to produce new spaces (1989: 9), and argued that in concept, method
and style, the fashioning of a French identity crucially depended on colonial
experiments in Africa. While Rabinow's work was firmly centred on questions
of French social norms, Janet Abu-Lughod (1980), Wright (1991) and Zeynep

Çelik (1997) foregrounded the conflictual basis of colonial urbanism and the complicity of architects and urban planners in fostering racialized landscapes. The legal infrastructure had to be manipulated to enable dispossession and building codes devised to produce a vision of 'order' and 'beauty' in a blatant effort to induce colonial investment in real estate. Abu-Lughod (1980: 151) explained the process by which 'segregation on economic and cultural grounds was buttressed by law' in Rabat-Salé. This involved the automatic removal of lands from the jurisdiction of Muslim courts, expropriation of private property of native residents for 'public good', and differential building codes for different parts of the city, processes that were designed to ensure a 'systematic transfer of Moroccan resources on to French colonists and to their new and elegant urban quarters' (ibid.: 147). The modernist urban forms of these cities are a record of massive inequity. The histories of modern urbanism and colonialism are so deeply interlinked that to extricate one from the other (for example, to discuss Le Corbusier's urban vision without referring to his 'contribution' to French colonialism) provides a one-dimensional and severely distorted vision of modernity and modernism.

That the tracing of the colonial-metropolitan logic of these cities often began with studies of world fairs is significant. In world fairs the nationalist/imperialist ideology was given a controlled and full meaning – they were microcosms of the colonial universe – unadulterated by the contingencies that colonialism encountered in the colonies (Mitchell, 1991; Wright, 1991; Çelik, 1992).[14] Among the many dubious contributions of world fairs is that they celebrated and taught metropolitan citizens a way of seeing colonial–spatial relations and their own location in it. Imperial spatial imagination was brought home and domesticated through a variety of popular cultural artifacts. A contrapuntal reading of metropolitan texts and spaces (Said, 1993) has thus been integral to the study of colonial urbanism, and continues to provide valuable insights into the pervasive coloniality in the heart of the new empire and 'post-colonial' spaces of the nation. Colonial practices have become foundational to the workings of the post-colonial state. The principles of coloniality have been internalized in the methods of the state and desires of the bourgeoisie to such an extent that in countries as varied as Indonesia, India, Mexico, UK, France, Australia, or South Africa these show up in everyday and exceptional violence against the disenfranchised (Jacobs, 1996; Kusno, 2000).

Recent research has moved away from focusing solely on colonial domination and the Westernization paradigm to restore the power of imagination to the colonized and ex-colonized. There is a willingness to see in the historical record alternative voices and visions, often incommensurate with the authority of the colonizer (Yeoh, 1996; Kusno, 2000; Hosagrahar, 2005; Chattopadhyay, 2005; Glover, 2007; Legg, 2007; Kidambi, 2007; Chopra, 2011). For example, the 'colonial spatial imagination' (Glover, 2007: 29–31) of British India was borrowed from and built upon Indian building and urban traditions as much as it was imported. Further, it was liable to change because it was not operating in a socio-spatial vacuum: it was forced to negotiate a thick social field. The colonized population countered, replaced, modified and bypassed colonial intentions in innumerable ways, in the role of patrons and architects, builders and masons, residents and politicians, conferring entirely new sets of meanings upon colonial built forms and their own habitations. Whether it was protest against sanitation measures (cholera and plague in particular), the right to representation in municipal politics, or the anti-colonial nationalist campaign that was determined to make the colonial city over in its own image, organized and not-so-organized social movements demonstrated the possibility of inaugurating a new structure of meaning in both ordinary and exceptional spaces in the city. In so doing they reconfigured the social parameters of colonialist urbanism. The battle was about the power of representation and ways of seeing the landscape. The focus on indigenous actors in recent studies of colonial urbanism has thus made a convincing case for understanding these alternate urban visions.

Let me summarize the theoretical movement within studies of colonial urbanism to suggest a point about such ways of seeing. The 'first generation' of historians of British and French colonial urbanism were concerned with colonial governance and the forms of dominance that city planning was designed to express; these were in a larger sense a critique of the knowledge structures that informed planning decisions. And yet, in their success in tracking ideology they also reiterated the role of colonial authorities and *planned* decisions in shaping cities.

The focus on planned outcomes throws in relief the dominant view of cities, and the language of critique tends to take on a dominant idiom. The status of elite representations – government reports, statistical accounts, urban plans – and their limited capacity to describe the complex negotiations of power remain largely unaddressed. This is, of course, valid for all areas of urban research, and not limited to studies of colonial urbanism. The 'second generation' of researchers have not been able to entirely break this habit, despite their success and willingness to read dominant representations 'against the grain' to produce a more complex rendering of cultural and spatial contestation. The prevalent focus on accounts of elite actors, colonial or indigenous, produce stories that tend to privilege some forms of urban visuality over others, leaving outside its hegemonic orbit most of the built environment.

Attention to indigenous actors, elite or subaltern, by itself does not do much more than produce a 'fuller' picture. More problematically, it could resort to a reversal of the colonialist logic, where the latter is replaced by an elite nationalist logic without any change in conceptual parameters.[15] It does not address the problem of representation that lies at the heart of the colonialist paradigm. For example, one could view cities within the same analytic structure, with a focus on the architecture of authority, planned developments and grand geometries, privileging urban forms that are more easily comprehended by modern techniques of representation and modernist planning discourse. From within such a frame, the everyday culture of third world cities would continue to appear merely 'chaotic', 'uncouth', 'inchoate' (Berman, 1989), bywords for describing cultural forms whose logic is errant. The outstanding difficulty with rectifying this viewpoint is that subalternity, by definition, presents us with a problem of representation.

Outsides?

Colonial cities changed and mutated over the period of colonial rule, and the success in aligning the economic and cultural logic varied, giving rise to different degrees of colonial control. These variations make explicit the conditions under which colonialism succeeds or fails in achieving its intended goals, or rather the multiple ways that colonialism's legacy contributes to the shaping of urban culture in the colony and postcolony. There are two primary strands of argument that might help link urban forms and conditions of subalternity. One is that the distinction between the economic, political and cultural is all but dissolved in the colonies, a phenomenon Fanon described as the flattening of infrastructure and superstructure. The other is the conceptualization of historical difference *vis-à-vis* the logic of capital.

In the first, the interchangeability of superstructure and infrastructure (not a dialectic) ensues from the role of governance as *commandement* – 'the right to demand, to force, to ban, to compel, to authorize, to punish, to reward, to be obeyed', and the blurring of distinctions between private and public property, that Achille Mbembe has remarked upon eloquently (2001: 34). Such forms of governance, Mbembe notes, do not rest on a covenant, and the resultant political culture is inadequately apprehended by the concept of civil society and an elitist notion of democracy. Following independence from colonial powers, the new welfare state, and newly instituted system of universal franchise must still contend with this colonial legacy. When the infrastructure is the superstructure, the 'politics of the governed' takes on a contour all its own.

Partha Chatterjee argues that the effects of the enduring structure of the state and the relation between the elite and the popular assume two configurations in postcolonial states. The first may be described as a 'variant of the colonial strategy of indirect rule':

> This involves a suspension of the modernization project, walling in the protected zones of bourgeois civil society and dispensing the governmental functions of law and order and welfare through the 'natural leaders' of the governed population. The strategy seeks to preserve the civic virtues of bourgeois life from the potential excesses of electoral democracy.
>
> *(Chatterjee, 2004: 50)*

Its urban impact is manifest in the proliferation of gated communities in the suburbs, privatization of public institutions, and efforts to control the visual 'disorder' of cities in the postcolony.[16]

The other effect is the 'less cynical' effort to steer the project of modernization, via the welfare functions of the modern state, through a wide terrain of beliefs and practices that are administered by authorities other than the state and over which the legal arm of the state has only limited access. This is the terrain of the most historically significant encounters between the state, elites and subalterns.

Chatterjee has used the term 'political society', as opposed to civil society, to describe this nebulous zone of representation and assertion of claims. Chatterjee points out that in *Prison Notebooks* Gramsci 'begins by equating political society with the state, but soon slides into a whole range of social and cultural interventions that must take place well beyond the domain of the state.' In attempting to make the colonial subaltern into a national citizen the modernization project must operate in this domain outside the state. It is also here that it encounters 'resistances that are facilitated by the activities of political society' (Chatterjee, 2004: 51).

In political society the debates about urban resources revolve not around rights, but around force and entitlement. Marginalized groups are compelled to secure 'devious' means to access power; they have to 'pick their way' through an uncertain terrain 'by making a large array of connections outside the group – with other groups in similar situations, with more privileged and influential groups, with government functionaries, perhaps with political parties and leaders, often making 'instrumental use of the fact that they can vote in elections' (Chatterjee, 2004: 40–1).

Chatterjee cites the example of vendors in Calcutta who illegally set up shop on sidewalks along the major arteries in the southern part of the city, carrying on a vibrant informal economy that provides livelihood, goods and services to a vast number of people. Repeated attempts have been made by the ruling parties and the municipality to rid the sidewalks of vendors. For example, in 1996, facing criticism from a certain section of the city's elite residents and in their desire to attract foreign investment, the communist-party-led state government launched 'Operation Sunshine' that destroyed the illegal structures on sidewalks. However, after a few months of absence and negotiations the occupants managed to return. The negotiations are conducted not on the basis of 'rights' but on the basis of 'entitlement' and are highly contingent on the ability of the squatters to create lines of communication with middle-class residents and government functionaries, and mobilize popular support.

Political society breaks down the spatial relations that were supposed to remain inviolate within bourgeois state formation, namely the distinction between public sphere and private sphere and their corresponding correlation between public space and private space. This in turn shapes urban form and experience, often visibly indexed in the 'disorderly' – multiple and contradictory – uses of public space that pay scant respect to its bourgeois conception (Chattopadhyay, 2009). Apart from the ubiquitous scene of hawkers appropriating streets and sidewalks in Latin American, Asian and African cities, this includes squatter settlements of various degrees of permanency on public land, jerry-rigged public transportation and the pirating of public utilities. The overlap of spaces and uses and their illegal or quasi-legal status obscure the systematicity and creativity of these practices (Neuwirth, 2005; Holston, 2008; Larkin, 2008).

The post-colonial state, having inherited the structure of colonial governance and the project of modernization, has also inherited its concomitant view of spatial and temporal order, manifested in the desire for exactitude, certainty, permanence, objects carrying fixed meaning, discrete bounded events that could be placed in a narrative of cause and effect. These principles of spatial and temporal ordering are at odds with the politics of contingency of political society. Not necessarily a refusal, it constitutes the inability of the state to comprehend the spatial dynamics of the marginalized within its own logic, and is the founding crisis of 'insurgent citizenship' (Holston, 2008).

But what sources do these marginalized groups draw upon to inhabit urban space in such a contrary manner? Here we need to look to the second strand of the argument. Even when of the most brutal sort, colonial dominance was not able to shut out all contingencies that opposed the colonial logic; rather the nature of colonial dominance ensured that there remained life worlds that were outside the purview of the colonial state and the logic of colonial capital (Chakrabarty, 2000).

Dipesh Chakrabarty discusses two ways of conceptualizing capital's encounter with historical difference. One way is as 'something external to its own structure' that in the course of struggle it manages to wipe out to reproduce its own logic. This is 'capital's antecedent posited by itself' and lends to one kind of historical reading. Characterizing it as History 1, he notes: 'this logic is ultimately seen not only as single and homogenous but also one that unfolds over (historical) time, so that one can indeed produce a narrative of a putatively single capitalism in the familiar "history-of" genre' (Chakrabarty, 2000: 48). The other mode in which capital encounters difference is 'not as antecedents established by itself, not as forms of its own life-process', but as that which resides outside its logic and does not lend itself to its reproduction. The latter are 'not pasts separate from capital; they inhere in capital and interrupt and punctuate the run of capital's logic' (ibid.: 64). These social formations that reside in a contrary relation to capital are not, as David Lloyd clarifies, 'merely awaiting development or destruction, but elements produced in differential relation to capital'. They are 'the difference that is produced at the interface between the modern and the non-modern, in and from the encounter between capitalist colonialism and the social formations of the colonized' (Lloyd, 2005: 431). The latter lends itself to History 2 in Chakrabarty's characterization. Considered together, History 1 and History 2:

> destroy the topological distinction of the outside and the inside that marks debates about whether or not the whole world can be properly said to have fallen under the sway of capital. Difference, in this account, is not something that is external to capital. Nor is it something subsumed into capital. It lives in intimate and plural relationships to capital, ranging from opposition to neutrality.
>
> *(Chakrabarty, 2000: 65–6)*

Such social formations, residing in intimate and yet contrary relation to capital, provide an alternate account of subalternity, thus creating several openings for a new framework of urban history and theory. They suggest that we read *subalternization* – the process through which the colonized peasantry is made into a subaltern figure – as something that ensues from their encounter with colonialism, a process that might enable the survival of modes of creativity and resistance not fully theorized by capitalist histories of urbanization.

This formulation is in marked difference to Antonio Gramsci's notion of sub-alternity. In Gramsci's classic formulation, the subaltern does not have access to means of representation and therefore to forms of resistance that will enable it to appropriate the state. Their history is 'fragmentary' and 'episodic' (Gramsci, 1971: 54–5). Their cultural world has to be remade to mobilize them as a proletariat, a corrective response to their 'fragmentary' history. When Gramsci's concept of the subaltern crossed the 'colonial difference' in the work of South Asianist historians, the problem of racialization entered debates (Mignolo, 2005). The term came to refer to peasants, tribals and those groups who are not easily assimilated within the logic of the state and capital. Ranajit Guha used the word to refer to a differential in power relationship: the elite made the subaltern 'aware of his place in society as a measure of a distance from themselves – a distance expressed in differentials of wealth, status, and culture' (Guha, 1983: 18). Gayatri Spivak (2005) has argued that the term be used to refer to those who are removed from all lines of social mobility – 'subalternity is where social lines of mobility, being elsewhere, do not permit the formation of a recognizable basis of social action'. David Lloyd has summarized these positions succinctly: subalternity is 'that which the state does not interpellate, and therefore what lies outside it; it lacks a subject, what lies beyond the capacity of the historian to represent'. It can be a subject 'merely contingently' (Lloyd, 2005: 422).

Three points emerge from these understandings of the colonized subaltern: 1) subalternity is a space – *location* – within/outside the structure of domination and subordination; 2) there is a difficulty of *recognition* (recognizing subaltern social action as such), and therefore representation; and 3) any possibility of representation, building upon the fragmentary and episodic, is *contingent*. But this contingency or problem of recognition need not be disabling. Rather, it is 'the most characteristic and suggestive element of subalternity', carrying 'traces of an alternate social logic' (Lloyd, 2005: 428).

This social logic, ubiquitous in cities across the world, yet theoretically opaque – evident merely as chaos – cannot be grasped in terms of planning frameworks, their temporal horizons and formal preoccupations. It will require a method quite contrary in terms of temporality, duration and spatiality to engage in a sustained exploration of the *principles of contingency* that shape these spaces. Urban theory 'beyond the West' must learn to recognize these contingencies, if it is to de-limit the overextended claims of dominant theories of capitalist urbanization.

Notes

1 I want to thank Jeremy White for his helpful comments and suggestions on the text.
2 See Amin (1976). Jonathan Crush (1995) notes the nineteenth-century colonial legacy of 'development'. Also see in this context Michael Cowan and Robert Shenton (1995).

3 For a more elaborate articulation of this conceptual frame, see Chattopadhyay, *Unlearning the City* (Minnesota, forthcoming).
4 See, for example, Peter Hall's (1998), *Cities in Civilization* that narrates the genius of Western civilization in shaping cities. Yet not surprisingly, Japan, as an example of advanced capitalism finds a place in it. Hall noted that history of other cultures and cities would be written 'in due course'.
5 Here I am restricting myself to colonialism that accompanies the industrial revolution.
6 Both tropical medicine and tropical architecture had immense impact in shaping colonial urban and rural environments.
7 At one point Gramsci notes that hegemony as an 'educational' relationship occurs in the 'national' as well as 'international and world-wide fields' (Gramsci, 1971: 350), but this is not a substantive point in his conception of hegemony.
8 Ranajit Guha's critique of neo-colonialist historiography of empire is relevant here. See Guha, 1997, 85–9.
9 This is similar to one that William Cronon (1991) adapted in explaining the emergence of Chicago as the gateway to the American West in *Nature's Metropolis*.
10 *The Communist Manifesto* is most often cited as the source of this view of capitalist progress.
11 See David Harvey's (2003) elaboration on Luxembourg's use of primitive accumulation as a continuing process in capitalism.
12 See Banerjee (2006) and Guha (1983).
13 Although historical monographs on colonial cities predate mid-twentieth century decolonization, the first study to address the phenomenon of colonial city is Redfield and Singer (1954).
14 See Kusno's study that begins the same way to articulate the colonialist practices of the post-colonial state.
15 This is the critique of the elite perspective in Indian historiography and nationalism.
16 In India this has taken the form of the prohibition of political wall writing in urban centres and changes in laws that govern the size and number of commercial hoardings in cities such as Bangalore in favour of massive digital billboards, rather than multiple small hoardings that characterize the small shops of merchants.

PART II
Order/disorder

6

GOVERNING CITIES WITHOUT STATES?

Rethinking urban political theories in Asia

Yooil Bae

Introduction

Over the last two decades, political analysis of urban growth politics in Western democracies has been dominated by growth machine and urban regime theories. At the same time, the ascendancy of these theories has made them the subject of critical scrutiny. In the context of transitional societies undergoing rapid economic growth, urbanization and democratization, one might expect that growth-oriented local businesses and their coalitions would emerge as dominant stakeholders in the formulation of urban policies. This expectation might be consistent with how urban political theories have driven the development of Western democracies, but recent studies of urban regimes and growth machine theories beyond Western settings such as cities in Asian countries show somewhat different patterns of urban political economy. This paper argues that the missing link in Western experience-based urban theories is the role of the 'state'. I argue that the role of the state should be clearly conceived to provide the basis for an understanding of the causal mechanisms underlying both political-economic and social change in Asian cities. The rapid development of big cities such as Tokyo, Seoul and Singapore and of small cities as well requires a more comprehensive understanding of the role played by a strong state in framing the urban political agenda as it nurtures the growing power of business and social forces at the sub-national level. At the same time, the globalizing economy also carries with it an ambiguous set of implications for politics in these cities. Through theoretical and empirical sketches of Asian city politics, this paper attempts to reformulate urban theories from a comparative perspective.

The predominant form of economic development and modernization around the world for several decades has been one of rapid industrialization and urbanization. The process of modernization in advanced economies in Western Europe and North America clearly showed that interconnectedness of industrial revolution and spatial patterns of urban concentration. After the nineteenth century in particular, the explosion in manufacturing and the growth of factories in selected urban areas brought about the migration of job seekers from rural areas to most urban areas. In the meantime, urban capitalists including merchants and landowners who accumulated their wealth during the industrialization period became urban elites and formed coalitions for further economic development in their cities. Despite the emergence of several types of urban problems including class divisions, environmental degradation, housing shortages and so forth, this pattern of capitalist development became prevalent and remains so today.

This paradigm of urban economic development is therefore deeply rooted in city politics in advanced democracies. To explain this process of rapid industrialization that resulted in a massive wave of migration to urban centres and drove urban elite-led development, a longstanding stream of urban literature has focused on how 'place-based' local interests work together and maintain coalitions for local prosperity in a rapidly changing environment. Given that globalization now provides additional contextual and structural elements supporting further development, localized interests have started to seek opportunities and organize themselves across borders. These local economic interest-based explanations have dominated the American urban studies literature (Cox and Mair, 1989; Logan and Molotch, 1987). Some urban sociologists have even utilized the word 'growth' as an appropriate term to describe the essence of the city political economy in Western societies (Molotch, 1976).

This pattern of development and urban growth in advanced countries has been followed by newly industrializing countries since around the 1960s. In particular, the rapid industrialization and urbanization of Japan and the four East Asian Tigers has gained scholarly attention due to their strong resemblance to the pattern of wealth accumulation and growth seen in the western world. For most of the twentieth century, East Asia was among the least urbanized parts of the world, but thereafter cities grew rapidly. For instance, East Asian cities such as Tokyo, Seoul, Singapore, Hong Kong and many Chinese cities have become command-and-control centres of global capital and now occupy an important position in the world economy. The global city rankings of South Asian cities such as Bangkok and Kuala Lumpur, capitals that articulate their highly successful newly industrializing economies despite economic crises (Douglass, 2000; Newman and Thornley, 2005). This developmental pattern among Asian cities represents a trans-regional process of industrialization, urbanization and globalization that has provided somewhat favourable conditions for the type of urban growth politics that proliferated in Western democracies.

A longstanding strand of US-based urban literature highlights several important elements of urban growth politics that provide a suitable comparative framework for Asian city politics. On the surface, economic interests that have emerged at the local level occupy crucial positions in local economic development and set highly favourable conditions for further development. For instance, Zhang (2002) tested urban regime theory in the context of Shanghai and highlighted the importance of business organizations and private investors.

However, unlike many Western democracies, where business and landed elites took the initiative to expand the local and national economy, countries in the East Asian region, with occasional exceptions, fall into a highly distinctive developmental trajectory: one of 'state-led growth' (Bae and Sellers, 2007; Hill and Kim, 2000). In Asian countries like Japan, Singapore and South Korea, the state actors – government agencies, bureaucrats and public officials – exert leverage over the private sector through various industrial policies, regulations and highly controlled markets. All political activities in the age of industrialization focused on ensuring national economic growth. Although rapid growth of the economy has favoured the emergence of new landed elite groups and increased business power, social actors were successfully incorporated into the developmental coalition for the single national goal of rapid industrialization. Even in the subsequent era characterized by a more internationalized economy, the legacy of strong state power continues although greater openness to global markets has constrained the state's possible range of options.

Since the 1990s, there has been an increasing trend of comparative studies in urban politics (DiGaetano and Strom, 2003). While growth machine and urban regime theories have played a pre-eminent part in major urban political debates since the 1980s, much of the growing comparative research on urban political economy has started to point to the 'atheoretical' nature of these two bodies of literature, which are in many ways an abstraction of the experiences of US cities (Cox and Mair, 1989; Davies, 2003; Pierre, 2005). Sellers (2005) contends that a far-reaching multilevel approach to urban growth politics is necessary to fully capture similarities and differences among countries. By taking account of these comparative strategies, this chapter examines the missing link common to studies of Asian city politics and development – the role of the 'state'.

Urban growth and politics: a theoretical overview

Studies on how community debates raised urban political power in American cities became popular in the 1950s. Yet urban elitism and pluralism – the major theories put forward – have been replaced or supplemented by political-economic approaches that effectively capture non-political influences on city policymaking such as fiscal austerity, local business, and property-related interests. Among these political economy approaches, the growth machine and urban regime theories have emerged to form the dominant paradigm for the study of local politics. These two bodies of literature provide a relatively stable and clear explanation for increasing local competition, the dominance of politics in local economic development, public–private partnerships, and economic interests in the urban decision-making process, questions that cannot be simply answered by community power debates. Politics have also played an overarching role in urban economic development elsewhere in the world, yet the theoretical frameworks used for analysis were extracted from US experiences (Wood, 1996).

A key question addressed in these community power debates was 'who governs?', an enquiry that concerned the advantages and disadvantages of certain groups in the urban political arena. The growth machine theory which received widespread recognition in the 1980s and 1990s instead focused on the influence of specific groups (property-related) in the process of physically restructuring cities (Logan and Molotch, 1987). Many urban sociologists were dissatisfied with community power studies based on paradigms such as urban elitism and pluralism because they failed to link everyday political activities to specific land-use patterns and the distribution of urban resources within a city (Jonas and Wilson, 1999). Based on the experience of US cities, Logan and Molotch (1987) argued that the city population is highly dependent upon private investments and profits for employment opportunities and that the politics of local economic development are largely driven by local growth elites representing an alliance among the rentier class, property owners, developers, politicians, the media, utility companies and so forth.[1] This growth coalition intensifies land use policy and enhances the revenue of the city by raising property values for members of the growth coalition in what is called 'exchange value'. In doing so, local growth coalitions play a crucial role in electing local politicians and even in scrutinizing administrative details in order to induce mobile investment and create a business-friendly environment.

While maintaining the core argument of the materialistic intensification of local political activities, the contemporary version of growth machine theory has now shifted towards broader, more competitive, and flexible local development politics instead of internal land use patterns (Leitner and Sheppard, 1999). Despite criticisms, the updated theory clearly marks an advancement of the urban politics and growth machine thesis, and related concepts have garnered a good deal of attention from academia for the purpose of comparison.[2] These comparative studies have focused on the conditions necessary for growth machine politics, which can vary significantly from one country to another. In particular, they have paid attention to the extent to which land use powers are delegated to the city level and national governments promote local land use for private interests. For example, British scholars who recognized the lack of a theoretical framework for UK urban politics attempted to address the ways in which urban growth coalitions form in different contexts. The rising level of interest in the growth machine thesis has therefore led to the greater availability of international comparisons in urban studies.

Urban scholars generally find urban regime theory to be a valuable theoretical advance. The concept of the urban regime emerged as a dominant political theory since 1989 when Clarence Stone published his seminal work *Regime Politics*. The theory is slightly different from growth machine theory in that it explains urban politics by emphasizing the interdependence of governmental and non-governmental forces in meeting economic and social challenges. According to urban regime theory, under an increasingly competitive environment, financial difficulties and the growth of non-governmental organisations, governmental bodies are inherently ill-equipped to bring about policy change and effectively govern localities (Stone, 2008). Building governing capacity goes beyond the simple electoral process and politics and requires a governing coalition capable of mobilizing both local and non-local resources. Cities are thus able to achieve important public goals by assembling coalitions of political, business and community elites. The urban regime is an 'informal' yet stable group with access to institutional resources, and has a significant impact on urban policy and management (Stone, 2008). From the urban regime perspective, therefore, urban politics in contemporary cities is largely a consequence of the division of labour between the state (politics) and the market (non-governmental bodies).

Although urban regimes are a type of public–private partnership, a central principle underpinning the theory is that 'business elites' occupy crucial and influential positions within a governing coalition. Stone (1989: 195–6) also admits that business elites within a coalition confer a greater resource advantage than the investor class in the Atlanta case and that various mixes of business interests produce differently biased input regarding urban policymaking because of the scope of resources they command in localities.[3] In this regard, urban regime analysis also largely reaffirms the claim made by growth machine theorists on the role of local economic interests. For this reason, the majority of urban political scientists in the US have portrayed local business-led pro-growth politics in cities as the dominant political paradigm.

From a comparative perspective, these two theories provide a set of important conditions for the analysis of Asian cities – in particular, East Asian – city politics and political economy. For example, South Korea is one of the most rapidly urbanizing countries in the world and its big business conglomerates (*chaebol*) are located in major metropolitan cities. For this reason, above all things, one might expect that the rapid industrialization and development of cities in the region would produce a group of business and local economic elites who could profit from city development. In addition, business interests would be very influential in city policymaking and that government would play a somewhat secondary role and be subservient to business power. In this case, it is easy to imagine that growth machine and urban regime theory might provide a useful basis for understanding urban issues emerging in Asian cities and elsewhere.

However, several comparative analyses of growth politics and urban regimes following the above theoretical models have highlighted important limitations in applying the two theories. For example, European countries such as the UK with more or less centralized systems and strong national political parties show the extent to which limited encouragement for local economic development has often resulted in stricter land-use regulations and anti-growth policies (Cox and Mair, 1989; Davies, 2003). Pierre (2005: 450) also claims that the term 'urban regime' has a non-contextual characteristic that limits its comparability.[4] Regulation theorists have also contended that cities function in a broader economic, social and political context by emphasizing the distinctive role of the state (Stoker, 1998). In this regard, Asian and European cities share common ground in their urban political economies. Yet, as transitional societies moving from the developing to the developed world, Asian cities must be analysed in terms of a broader and more contextual research framework. In particular, global city theorists often argue, it is necessary to scrutinize how the state (or state actors) effects relations among politics, economy and society in the age of industrialization, urbanization and globalization (Hill and Kim, 2000).

Urban growth and city politics in Asia

Before beginning a discussion of the state-centred nature of Asian city politics, it is useful to first provide a brief historical and comparative overview of economic development in the Asian region, focusing on Japan and the East Asian Tigers. While there is little doubt that Asia (or East Asia) has become a prosperous region in a very short period, there has been a robust debate over the reasons for its success. For example, Japan's gross domestic product (GDP per capita) in the early 1950s was less than USD$300, but it later outstripped all the other members of the group of advanced economies.[5] As Vogel (1991) puts it, 'Japan reached number one in almost all areas'. The economic performance of the East Asian Tigers (South Korea, Singapore, Taiwan and Hong Kong) has been almost as impressive.

In explaining this extraordinary success, market theorists adopting the neoclassical view have regarded private investment, a high domestic savings rate, education level and human capital, macroeconomic policies and efficient government administration as important (World Bank, 1993). They have not disregarded the function of the state in spurring tremendous economic growth, but from this perspective, the state and state actors play a 'market conforming' role by incorporating market forces into the governmental policymaking process. Having a market-conforming state in the process of economic development ensures adequate investment in people; provides a competitive climate for private enterprises; keeps the economy open to international trade; and maintains a stable economy. Beyond these roles, it is believed that governments are more likely to do more harm than good unless the intervention is market friendly.

However, the role of government in Asia's economic success involved more than just market-conforming actions. The other perspective known as developmental state theory challenges the aforesaid neoclassical perspective and argues that the government took the initiative in promoting the national economy by deciding 'when, where, and how much to invest in which industries' (Woo-Cumings, 1999). Western Europe and North American countries experienced an era of industrial and technological innovation during most of the eighteenth and nineteenth centuries and their economies depended upon continuous technological development. In comparison, Asian countries were late to the party. They did not have industrial and knowledge bases, skilled workers, or resources. At the same time, they were keen to catch up with advanced economies. Under these circumstances, their choice was between borrowing money and employing Western technologies

(Amsden, 1989). In particular, Japan and the East Asian Tigers recruited the most talented individuals for their core pilot organizations such as the Japanese Ministry of International Trade and Industry (MITI) and the Korean Economic Planning Board (EPB) and orchestrated industry and the national economy through various 'discretionary-basis' public policies. State intervention in the markets of state-led economies was the key to Asian success in subsidizing, monitoring and guiding the private sector. Therefore, with occasional exceptions, strong state intervention through various mechanisms including nationalized banking, pilot agencies, industrial policies and nationalistic concerns were a common feature of economic development across Asia.

This different trajectory of economic advancement in Asia carries with it an important implication for city politics and political economy. First of all, because the developmental state was designed to mobilize, in a top-down manner, the entire society in pursing the goal of achieving rapid economic growth, the state – in particular, central government – could control business activities, financial resources, labour movements and ordinary citizens (Hill and Kim, 2000). For example, Singapore applied social engineering to direct the economy and society and to influence the nation's position in the global economy. Singapore's policy initiatives in almost all fields have shaped people's goals and the way they achieve them. The city-state has provided social goods in various policy fields including housing, retirement, education, health, childbearing, the environment and so forth, and none of these goods were *neutral* (Salaff, 2004). Important groundwork led by the Singapore state has gone into laying the foundation for national development. As a result of the centralized state structure, there has been little room for bottom-up policy initiatives.

Another important feature of a strong state is that sub-national governments and cities have become subservient to the central political world. Local decision-making and administration are tightly controlled by the central government. Authoritarian states like China and Vietnam do not hold popular mayoral elections, and local Malaysian executives are still appointed by the central government.[6] Even in countries that run local elections, the central government prescribes the powers and functions of local government. Decentralization has been a recurring topic in many Asian countries in recent decades yet even when local self-governance is promoted as a national goal, the central government still intervenes to introduce and sustain such reform.[7] For example, until the late 1990s, Japanese central ministries were superior to local government and formulated policies that required local governments to implement vertically delegated functions (*kikan inin jimu*). In addition, the central government compulsorily issued a standardized, detailed guide about the number of, and qualifications for, a considerable number of local public offices and their directors and employees (*hitchikisei*).

While the central government plays a critical role in the local decision-making process, the local political structure emulates the central political structure in many cases – the executive-centred system. For example, local executives in South Korea enjoy a relatively extensive range of formal powers over local affairs, as does the president in national politics. The central government's relationship with local councils, which are supposed to monitor or support local mayoral action, is based on a give-and-take or patron-client model (Park, 2000: 48–9). While Japan has been going through a decentralization process since the mid-1990s, local councils have been placed in a relatively weak position in local affairs (Muramatsu, 1997). As seen in the progressive local movements of the 1960–70s, mayoral policy initiatives are important in the local context in Japan.

Second, the formation of cities and urban development has also been largely dependent upon central government policies. Because Japan and the East Asian Tigers suffered from a lack of natural resources and of knowledge and technology, efficient distribution and investment in limited resources were the key to rapid development. For instance, post-war Japan developed economically through an 'export-oriented strategy' that resulted in the concentration of people and industries in specific industrial areas such as Tokyo, Nagoya, Osaka and Northern Kyushu. This concentration was actually considered to be the fountainhead of the Japanese economic miracle, but after the 1960s, the government's focus shifted from selected and uneven development to a comprehensive developmental strategy that targeted relief of over-concentration and balanced development through a series of Comprehensive National Development Plans (1962–1998 OECD, 2005).[8] While the problem of over-concentration has been ameliorated, Tokyo's 'unipolar concentration' still causes many social problems in Japan as indicated in Figure 6.1. Nevertheless, the formation of industrial cities and regions has been driven by the central government.

In South Korea, the National Economic Development Plan, formulated in 1962, and the Comprehensive National Development Plan, issued in 1963, have brought together packages of government plans aimed at rapid economic

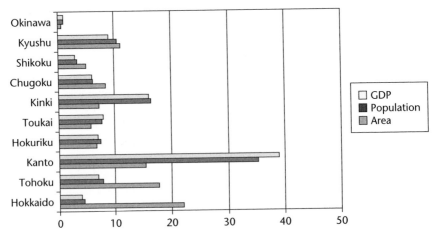

FIGURE 6.1 The pattern of regional disparity in Japan: population and GDP (%)
Source: OECD Territorial Database
Note: Figures are for 2000

PLATE 6.1 Seoul's development
Source: Yooil Bae

development. To modernize the economy and industry, the government strategically encouraged economic development by providing financial and administrative incentives to highly competitive and export-oriented companies, and adopted a 'growth pole policy' that promoted the development of selected cities and regions such as Seoul and other mega-cities in the coastal regions (Plate 6.1). Because the regime's top priority was to distribute limited resources to 'competitive industries' in 'selected regions,' these uneven development policies became the best way of achieving the national goal of modernization in a short period. Table 6.1 shows that a few mega city-regions that grew faster and to a greater extent such as the Seoul Metropolitan Region (SMA) (Chung and Kirby, 2002). Almost half of the national population (47 per cent) is concentrated in the SMA, which constitutes only 11 per cent of the national territory.[9] With occasional exceptions, this pattern has been replicated in many other Asian countries.

For the small city-state of Singapore, attracting people from foreign countries has been a central component of the economic, social and cultural shift that has left its imprint on urban development. The urban system of Singapore has been dominated by internationalism and the big hub city paradigm. Although Singapore's urban development is currently overseen by agencies such as the Urban Redevelopment Authority (URA) and the Ministry for National Development, urban policy

TABLE 6.1 Indicators of regional disparity in South Korea (2000)

Indicators		National	SMA (%)
Population and housing	Area (Km^2)	99,800	11,754 (11.8)
	Population (unit: thousand)	48,289	22,525 (46.6)
	Population density (people/Km^2)	484	1,971
	Housing rates (%)	96.2	86.1
Regional economy	Gross Regional Product (billion won)	465,183	223,081 (48.0)
	Number of manufacturers	98,110	55,874 (57.0)
	Number in the service industry	794,095	360,102 (45.3)
	Savings deposits (billion won)	404,661	275,394 (68.1)
	Amount of loans (billion won)	310,804	202,797 (65.2)
Others	Number of universities	162	66 (40.7)
	Public agencies	276	234 (84.8)
	Medical facilities	42,082	19,471 (46.3)
	Number of automobiles	12,914	5,983 (46.3)

Source: Ministry of Construction and Transportation (2002), *2002 Annual Report on National Land Use*.
Note: Population and housing indicators and number of universities and automobiles are 2001 data.

must be understood within the broader perspective of the global city vision and the goal of competitiveness (Newman and Thornley, 2005). Using its land owner-ship powers, the state decides on the types of development allowed, along with their location, scale and even their timing and space requirements. In this way, the broader economic agenda and plan can easily be reflected in the physical and structural development of the city.

Finally, the power of business and civil society in Asian countries has a rather ambiguous implication for city politics in the region. Developmental state theorists generally agree that the resulting economic miracle was built on the foundation of government–politician–business networks. Richardson (1998: 21) described Japanese post-war development as 'an alliance between Liberal Democratic Party (LDP) government and big businesses'. This indicates that there was mutual agree-ment between the two parties. Big businesses wanted stable and business-friendly government, while the government targeted rapid economic growth. In doing so, members of the business world (*zaikai*) frequently had informal meetings with governmental officials and became important political players. Because of the existence of big business groups such as *keiretsu* and *chaebol*, and despite variations among countries, one might expect business to have exercised a substantial degree of power at the local level as well. Regardless of their power to influence national politics, however, business interests, with occasional exceptions, were less mobilized, found it hard to establish common interests, and operated in a fragmented fash-ion.[10] This is partly due to their tendency to depend on their respective industrial associations at a higher level or on top-level industrial organizations. In addition, local economic associations could not expect to gain much from local government given its limited capacity and authority (Park, 2000: 53).

The strong state thesis argues that civil society and the general public assume less important positions in politics, yet the politics of civil society in Asia cannot be defined in an easy manner. Robert Pekkanen (2006) described Japanese civil society as 'members without advocates' in depicting the state-sponsored nature of civil society politics. According to his view, the growth of civil society in Japan was highly dependent on the central government's institutional arrangements and support, and during the post-war period, only a certain type of civic organization – service-oriented, non-professional, small and localized groups – survived. Under their authoritarian leaders, South Koreans and Taiwanese also experienced weak civil society during the industrialization process. However, increasing distrust in government, economic difficulties and democratization in this region provided the momentum for civil society to develop, although Western-style 'grassroots' local politics are rarely found.

As Linda Weiss (2000) argues, the degree of state intervention and top-down influence differ from country to country and there is no single Asian political-economic model. There is even variance among the East Asian Tigers. Yet the key features of a strong state including state economic policies, local government subordination and relatively weak influence from the bottom-up level were prevalent during the industrialization process and their legacy continues today. However, Asian countries nonetheless face several crucial challenges – democratization, decentralization and globalization – that might have a huge impact on politics and the economy at both the national and sub-national levels, and it is to these issues that I now turn.

Changing context: democratization, decentralization and globalization

The 'stateness' in the developmental state theory was developed in the age of industrialization. East Asian countries are important because they too are undergoing transformation stemming from rapid industrialization, democratization and globalization. From a comparative perspective, Western Europe and North America experienced this transformation in tandem with urbanization before the twentieth century began, whereas 'third world' countries are still firing on all cylinders to catch up with advanced countries. As a rare case, the recent experience of East Asian countries offers implications for the meaning and influence of this transformation process.

One of the most notable transformations in this region was the 'escape from authoritarian rule' in 1980s. Democratization has brought about changes such as the practice of democratic and popular elections and a significant level of political freedom. Although Diamond (2008) noted that there are still few signs of democratic stability and consolidation in many countries, democratization has resulted in far-reaching political consequences and influences at both the national and local levels. Moreover, in the wake of the financial crisis of the late 1990s, central governments have taken on a broader range of tasks in establishing a transparent and efficient governing system. Democratization has thus made possible a certain level of popular control over state power through regular elections and political participation.

The rise of civil society and growth of the private sector is an indicator of devolution of authority from the top, down. The grand democratization in South Korea mobilized people and social organizations which had previously been suppressed by the military regime, and allowed them to become the basis of a new civic movement in the contemporary period (Lee and Arrington, 2008). These civic organizations have expanded their interests and activities to more diverse policy areas such as political and economic reform, the environment and transparent government since the late 1990s. Japan was the exception to this democratization trend, but from the mid-1990s, civil society and voluntary organizations have started to emerge as important political players.[11] Accordingly the politics of civil society have thus penetrated the tightly controlled government policymaking process and the power of the state has started to decline in many respects. Although many East Asian countries have faced political-economic crises, a return to authoritarian rule in this region is not imaginable (Diamond, 2008).

At the local level, a third wave of democracy has flourished in the form of decentralization over the last decade. Countries of varying sizes, economic levels and political systems are moving authority from the centre to localities, and sub-national bodies of government in East Asia now occupy a crucial position in the region's development (World Bank, 2008). Although Japan and South Korea are still characterized by a traditionally strong state and a model in which state power is superior to that of the market with central government maintaining control over local government, neither country is an exception to this wave of decentralization. In Japan, as Table 6.2 indicates, the aforementioned system of central regulation seen in governance arrangements such as agency-delegated functions was abolished in 1999, and the Trinity Reform Package aimed at restructuring local taxation, local allocation tax grants and national government disbursements was passed under the Koizumi administration (2003). South Korea has also undergone a series of decentralization reforms since the early 1990s. The governments of Kim Dae Jung and Roh Moo-Hyun enacted several laws including the Special Law for Promotion of Decentralization (2004). Decentralization is still a focus of reform in both countries but there are nonetheless flaws in the legislation that have been implemented. Yet there can be no doubt that attempts to date represent more than marginal adjustments. It seems clear that the political meaning of these reforms signals a move away from former statist tradition.

As a result of local democratization, the local political arena has become increasingly vibrant in the last decade. For example, in South Korea, mayors and governors have acquired a certain degree of authority over local affairs and have come to play a critical role in local economic development by deciding on the allocation of budgets within their jurisdiction. As regular elections have become institutionalized, local political leaders have gradually been exposed to electoral politics. At the same time, the growth in citizens' voices has made local politics

TABLE 6.2 Flashpoints of decentralization reforms in Japan and South Korea

	Japan	*South Korea*
Major legislation	Decentralization Promotion Law (1995) The Omnibus Law of Decentralization (OLD, 1999) Trinity Reform Package (TRP, 2003)	Law for the Promotion of Transfer of Central Authorities (Jan, 1999) Decentralization Roadmap (2003) Special Law on Decentralization Promotion (SLDP, 2003)
Target of reforms	1. Classification of national and local affairs (clarifies the rules on central government intervention in local affairs). 2. Abolition of agency-delegated functions and creation of new functions for local governments. 3. Establishment of resolution council for conflicts b/w central and local governments. 4. Transfer of central affairs. 5. Guarantee of local government's autonomous organization and personnel authority. 6. Rationalization of national grants and local taxation (increased discretion).	1. Delegation of central affairs to local governments. 2. Delegation of public security (police) powers and creation of local police. 3. Delegation of public education powers to local governments (educational autonomy). 4. Abolition of special administrative agencies (SAA – local offices of central ministries). 5. Rationalization of national land local tax system (raising the rate of local allocation tax). 6. Enhancement of authority of local councils (strengthening the authority of local legislation).

possible since the mid-1990s. While it is hard to observe a 'rentier' class, small property owners who benefited from the rapid rise in property values during the industrialization phase have become increasingly concerned with local policy-making on land zoning, planning, development and school projects. In addition, the urban middle-class who championed the economic miracle gradually opened their eyes to local politics and turned their attention to post-industrial values such as environmentalism (Newman and Thornley, 2005).

The history of strong regulation of land – in particular, farmland – in South Korea since 1950 has nonetheless hampered the formation of big rentier groups who would benefit from the development of land as commonly observed in U.S. cities (Park, 2000: 54–5). The central government's tight control over local decision making and financial resources has also constrained the politics of growth and development at the local level. However, democratization and decentralization has created an opportunity for potential beneficiaries of urban growth policy. In the age of decentralization, local leaders empowered with urban planning authority and budget allocation powers are beginning to formulate urban development policies for their re-election by providing incentives to developers. Citizens eager to raise the value of their property are also gradually being tempted to pressure local politicians. In this context, Bae and Sellers (2007) point to the possibility of the formation of 'local executive-led' growth politics in South Korean cities.

Finally, the influence of economic openness and the globalizing economy also has an important implication for local politics. Faced with the forces of globalization and economic crises, national politicians and policymakers have begun to devolve responsibilities for costly local spending to officers at the sub-national level when financial conditions are tight. In this case, decentralization is a strategic choice for national political elites who seek to consolidate or secure their positions in an economic crisis because this strategy may minimize the blame attached to them for their policy shortcomings (Kahler and Lake, 2004). In many cases, economic shock now undermines national leaders who have dominated the decision-making process and thus puts pressure on the central state from below.

For example, the financial crisis of the late 1990s in many East Asian countries has resulted in neoclassical reforms having a particularly large impact in the Asian region. The crisis and recession during this period paralysed the state, leading to a lower level of trust in government and prompting people to seek an alternative form of governance. Neoclassical theorists have thus sought to discredit the idea that central government should play a crucial role in rehabilitating the economy and transforming a centralized political system. Localities and businesses that have suffered from the uneven distribution of economic resources and overconcentration have also increasingly begun to voice their views on political, economic and social affairs. For example, the support for Japanese decentralization reforms in the early 1990s was initiated by a local economic organization (the *Kansai* Business Association). The central government has had no choice other than to adopt a more market-oriented approach in the face of the 'lost decade' and the economic recession, and empowered and relatively independent local leaders like Tokyo Governor Ishihara have sought greater control over local affairs.

In sum, the trends of democratization, decentralization and globalization have brought about important contextual changes in the urban political economy of the East Asian region. These transformations have created room for political actors including mayors, businesses and citizens at the local level, whereas the strong state-led model adopted in early periods has begun to break apart. Despite the limited nature of local democratization, empowered local executives can now 'do something' for local economic development, and businesses and citizens also are likely to voice their interests through electoral politics. These institutional and contextual changes have therefore made Asian city politics more complicated although the role of the state has not yet withered.

Conclusion: governing without the state?

Given the context of central-local command links, East Asian countries and cities have developed by means of different institutions than those prevailing in Western Europe and North America. The state the East Asian region attempted to develop their economies by borrowing knowledge and resources from the advanced world and in doing so, the state became a dominant player in allocating limited resources and suggesting policy directions. Obviously, economic growth and urban development in these countries were led not by market forces, but by state intervention. For national economic growth and developmental purposes, the state decided on important policies and interventions in the market without having to give much consideration to the preferences of localities and citizens.

Moreover, centrally dominated and planned economic development processes in East Asian cities have produced a highly concentrated population and national infrastructure in specific city-regions. This state-led pattern of development in the post-war period resulted in conditions which ensured the limitations in analysing urban politics in East Asian cities from the perspective of US-based urban political theories such as growth machine and urban regime analysis. As such, while the essence of cities in Asia was about 'growth', this goal was pursued in a different manner. For example, in the process of industrialization, because their resources and technologies were limited, Asian states selected a few industrial sites to be developed. The selected cities and regions became a basis of rapid national economic growth and received significant state support, whereas the remaining regions and cities were excluded from sharing the fruits of the economic miracle. Therefore, unlike cities in the US, seeking supranational support was the top priority for local governments in East Asian countries.

The distribution of power between central and local governments is also important to consider as this relationship was not based on equality. The legal status of local government was dependent on central politics because local governments were not constitutionally guaranteed. Mayoral actions and policymaking were institutionally monitored and supervised by central government. In the era of industrialization, local politics focusing on economic development were restricted. Local businesses and citizens could not therefore expect much from local decision-making and not all politics were local. Quite unlike cities in the US, therefore, political-economic influence from higher level governments remained a critical element of East Asian urban politics. Even in the age of democratization, decentralization, and globalization, which has placed non-state actors in more important positions regarding local economic development, the legacy of the strong state which prevailed in the era of industrialization has been to leave local politicians and state actors with a greater role in urban politics.

From a comparative perspective then, a more detailed and complex analytical framework is needed to consider urban growth and politics in East Asian cities as rare cases of localities which are experiencing industrialization, democratization and globalization at the same time. As I have argued throughout this chapter, comparative analysis of East Asian urban politics from a multilevel perspective linking the global to the national and local is required in order to establish the important elements and contexts shared by countries in this region and to uncover hidden causal mechanisms. Such comparative urban analysis is vital if understanding of urban growth politics in 'developing countries', and more specifically in terms of the rapidly changing environment of East Asian cities is to make a significant contribution to urban theoretical debates.

Notes

1 The growth coalition includes the rentier class and people whose interests are directly connected to development (e.g. developers, financiers, construction companies, real estate firms, etc.), local media and utility companies that benefit indirectly from an increase in the city population, and universities, cultural institutions, sports clubs and labor unions (Molotch, 1976).
2 The criticisms commonly leveled mainly concern the narrow conceptualization of the 'rentier' class in urban politics and the failure to consider much wider structural constraints (Wood, 1996: 1283).
3 In Atlanta's case, business dominance was constantly checked by coalition partners who were members of the black middle class and occupied positions of political leadership. However, the coalition itself was not based on equality.
4 Pierre (2005) also argued that 'urban governance' is a better term for conceptualizing core institutions, context and actors in the urban political arena for the purpose of comparison.
5 The World Bank report 'The East Asian Miracle' (1993) states that 'between 1960–1985, real income per capita increased more than 4 times in Japan and the four tigers and more than doubled in the Southeast Asian NIEs. If growth were randomly distributed, there is roughly one chance in 10,000 that success would have been so regionally concentrated.'
6 In China and Vietnam, local council members are popularly elected, but candidates must go through a pre-screening process (World Bank, 2008: 80).
7 For example, the Thailand Constitution of 1997 requires the promotion of decentralization reform. The Decentralization Plan and Procedures Act was enacted in 1999 and was followed by eight additional enabling laws (World Bank, 2008). Theses decentralizations can be found in the cases of the Philippines (1987) and South Korea (1987).
8 In particular, two laws – the New Industrial City Construction Promotion Act (1962) and the Act for the Promotion of Development of Special Areas for Industrial Development (1964) – both of which were elements of national industrial policies, contributed to the formation of new industrial areas in coastal regions of Japan (OECD, 2005).
9 Korean National Statistics Office (NSO; http://www.nso.go.kr). In 2000, the Tokyo metropolitan area accounted for 32.6 per cent of the national population and Paris housed 18.7 per cent of the French population. Metropolitan areas include the core capital city and surrounding cities and suburbia.
10 Curtis (1999) argued that historically fragmented business interests are not unrelated to the difficulty of forming one big economic organization in Japan.
11 Few Japanese were surprised to see the endless media exposure of the structured corruption at the heart of the nation's most prestigious and professional elite society. A series of bureaucratic mishaps in national and local crises such as the Great Hanshin Earthquake of 1995, the Aum Shinrikyoō sect releasing sarin gas at the Kasumigaseki Subway Station in 1995, and HIV-contaminated blood in 1996 brought Japanese bureaucrats into discredit (Pempel, 1999).

7

PUBLIC PARKS IN THE AMERICAS

New York City and Buenos Aires

Nora Libertun de Duren

Introduction

Focusing on stories behind the development of urban parks helps us to understand how state institutions contribute to shaping the city. The programming of urban parks reveals an underlying conception of who should inhabit public spaces. By comparing the development of iconic public parks in New York City and in Buenos Aires, this chapter explores the normative ideals of urban government towards people and parks. In the nineteenth century, both cities created a rectangular iconic park of 4,000 metres by 800 metres, New York City's Central Park and the San Isidro Jockey Club in Buenos Aires. At the time, both cities were important industrial and commercial hubs experiencing explosive population growth thanks to an intense and rapid flow of immigrants. In both cities local governments were actively managing this growth; providing tap water by the 1890s, subways by the 1900s, and regulating building footprints and maximum heights by the 1920s. However, there were significant differences in the development of parks in each city. Urban parks in Buenos Aires and in New York City importantly highlight national differences between the United States and Argentina. The design of New York City's Central Park 'comes from the people, and to them, in all phases of society it must necessarily be devoted' (*NYT*, 1857). Conversely, San Isidro Jockey Club Park was to be 'for the larger land owners, of aristocratic pretensions and sumptuous fitting' (Clemenceau, 1911).

What explains this divergence in the planning of urban parks in two democratic societies? In the United States, urban growth followed industrialization (Mumford, 1984) and accordingly, one of the main goals for urban public parks was to provide solace from the struggles of city life for the growing mass of industrial workers (Gordon, 1984). In Argentina, industrialization was also concentrated in urban centres with Buenos Aires accounting for the largest share (Dorfman, 1983). However, the economic structure of the country was still dependent on the exploitation of agricultural and agrarian resources, which in its turn impinged on the city's ability to capitalize on its leading role in the national industrialization (Diaz Alejandro, 1970). This chapter thus explores how differences between an industrial city and an agrarian national economy impacted on the urban public realm.

Urban parks in New York City and Buenos Aires

Given the rapid influx of foreign residents that characterized metropolitan growth in the Americas, divergences between urban and national culture were also a source of tension (Tilly, 1988). In the two countries, both national and urban governments were concerned about the influence of urban culture on the formation of a modern national identity. Even for us today who have grown accustomed to big and porous metropolises, the scale of these migration flows was daunting. The United States, which by the 1920s surpassed 106 million residents, received more than 24 million immigrants between the 1900s and 1920s. About 20 per cent of the newcomers settled in New York City (US Census Bureau, 1920). In Argentina the number of immigrants is smaller but its impact on the urban society more dramatic. By 1914, about a third of the Argentina's eight million residents were born abroad, but more than 50 per cent of all newcomers to the country settled in Buenos Aires (INDEC, 2001). Thus, the concentration of immigrants into the capital contributed to the enlarging of the cultural gap between Buenos Aires and the rest of the country.

The rapid inflow of migrants particularly raised concerns over the sanity and physical health of urban residents (Milgram, 1970) and the notion that nature could heal the evils of modernity (Mumford, 1961). In fact, urban parks become a way to naturalize the residents of the city, both by making them 'healthier' beings and making them 'less foreign' beings:

> Public opinion has within the last few years been awakening to a sense of importance of open spaces for air and exercise as a necessary sanitary provision of inhabitants of all large towns, and the extension of rational enjoyment is now regarded as a great preventive of crime and vice. Yet it was not until recently that any official step was taken towards providing for the City of New York City that which every city of Europe regards as a public necessity.
>
> (NYT, 1857)

Contrasting with this pro-naturalistic stance, in Buenos Aires the concern around immigration sparked a renewed interest in the promotion of 'high culture'. The Buenos Aires government targeted its parks as a space to create a city where 'the pride of wealth and luxury flaunts itself as bravely as in Paris or London' (Needell, 1995: 146) and where 'elites constructed public spaces directed to themselves and the foreign tourists, diplomats, and businessmen' (Needell, 1995: 538). In 1882, a group of notable Argentines that included Carlos Pellegrini, the son of a French engineer who was president of Argentina between 1890 and 1892, funded the Argentine Jockey Club in the heart of the City of Buenos Aires with the aim of 'refining the education of the people and the tastes of the elites' (Gayol and Madero, 2007). In 1926, the institution bought 316 hectares in the northern end of the City, in San Isidro and in a few decades the site included a first class hippodrome, two golf courses, an English Club House, seven polo fields, tennis courts, football fields, a kindergarten and a library. However, in 1946, thanks to the support of the trade unions, Juan Domingo Peron became the president of Argentina. In 1953, a paralegal, arsonist, peronist group 'marched to the Jockey Club, world-famed citadel of Argentina's top society, wealth and culture. They burst into thick-carpeted clubrooms hung with Goyas, Corots and Monets. Out windows and into bonfires went books and furniture' (*Time* magazine, 1953). The same year, through a national law, the national government dissolved the institution and the San Isidro Park was nationalized. In 1978, a rightist, non-democratic, military government nullified the law and reinstituted the Jockey Club and its privileges over the site (Rock, 1985).

Urban economy and parks programmes

In 1958 Max Weber argued that urban settlements could be classified into consumer and producer cities, and that the economic activity of the local elites – either commerce or agriculture – defined urban landscapes. Leaving aside the question of how changes in technology modify this categorization, his insight implies that space in the city gets allocated to preserve and reproduce the power of the leading elites. This notion is useful for understanding the different dynamics at play in New York City and the City of Buenos Aires at the time of the creation of major public parks.

In New York City, the presence of industrial labour was an incentive for public works in general and urban parks in particular. By the turn of the century it was clear that rural wealth did not feed the growth of the city, which instead depended on the prosperity of its industrial establishments. Therefore, enabling population growth as well as providing the wellbeing of the industrial workers was important for the success of the city as a whole (Mollenkopf and Castells, 1991). It is within this larger context that the design of a 'democratic park' made sense: 'our Park, like the Government, comes from the people, and to them, in all the phases of Society, it must necessarily be devoted' and this notion promoted the idea of the development of 'progressive programs of increased public recreational activities' (Rosenzweig and Blackmar, 1992: 424). Having an inclusive project for the park was important both as a way to improve the liveability of the city, as well as a means for making New York City competitive vis-à-vis other industrial centres like Philadelphia or Boston (Warner, 1974). Unlike the case of Argentina, where all roads converged into the City of Buenos Aires (Dorfman, 1983), the many growth nodes and the fairly developed transport network of the United

States implied that residents were highly mobile (Tiebout, 1956). The threat of losing residents to another city was one of the important forces driving urban planning decisions in the national context.

The City of Buenos Aires was also a major hub of industrial growth. By the 1930s it accounted for the largest share of industrial production, industrial workers and number of establishments in all Argentina (INDEC, 2001). However, in terms of international trade and total gross domestic product, agricultural production was still the most relevant national economic activity (Diaz Alejandro, 1970) Furthermore, most of the urban industries added value to agricultural products, as with the production of oil, leather goods and frozen meats (Gallo, 1970); hence increasing the importance of the agrarian activities for the national project. It was only after Peron's presidency that unionized industrial workers become a relevant social force in national politics (Halperin Donghi, 1994; Di Tella and Dornbusch, 1986). Before then, the challenge of the government was to integrate the rural provinces and the metropolis into a single national project (Germani, 1980).

In Argentina, the majority of industrial entrepreneurs were from European origins, primarily Italians who located their industries and families in Buenos Aires (Klein, 1983). The wealthier Argentines rarely participated in urban industries; more often than not they continued the agricultural exploitation of their rural 'estancias'. They made their living out of the immense fields of the Pampas and cattle, and viewed the city as a site for consumption (Barsky and Gelman, 2001). Accordingly, and unlike the case of New York City, the industrial workers who concentrated in Buenos Aires were not the imagined patrons of the city public parks. This lack of direct concern for the industrial worker should not, however, be taken as a disregard for the residents of the city. At the time, Buenos Aires was building a system of sewerage and water pipes as well as creating public hospitals and schools. Rather, it reveals the concern of the national government to consolidate national unity, and its consequent project of engaging the rural elites with the city (Hodge, 1996). By the 1900s, Argentina was finally coming out of a long power struggle between Buenos Aires 'unitarios' and the provinces 'federales', with the former party associating the urban with Europe as a symbol of high culture, the French encyclopaedistas, the British free thinkers, international commerce and a republic based on the universal suffrage; while the latter aligned themselves with the indigenous, the rural, the Spanish church, the rural wealthy and paternalistic approach to government (Goodrich Sorensen, 1996).

As Argentina eased these deep differences, it needed to negotiate the fact that the city could not progress without the support of the landowner's wealth (Davis, 2004). Accordingly, the urban and national governments believed that it was imperative to educate the landed elite by exposing them to urban culture. The leading thinkers of the country (such as Domingo Faustino Sarmiento and Juan Carlos Pellegrini) believed that the urban culture was an antidote to the 'backward mentality' of the landlords (Ingenieros, 1963). In their goal to the develop Argentina, they looked at the progress of the United States and Europe and concluded:

What is left for this America for following the prosperous and free destiny of the other (America)? To level up; and it is doing so by mixing with other European races, by making up the indigenous blood with modern ideas, finishing with the middle ages (. . .) The City of Buenos Aires believes to be a continuation of Europe, and she frankly confesses to be North American and French in her spirit and tendencies (. . .) In The City of Buenos Aires all is new, all is revolution and movement. It is a civilized city fed by Bentham, Rousseau, Montesquieu, and all the French literature.

(Sarmiento, 2010)

Thus, the design of the open spaces of Buenos Aires was a conscious effort to mimic European capitals and recreate its society. Accordingly, the government sponsored public parks programmes for the most affluent residents, such as exclusive horse races and charity meetings. Besides, and unlike in the United States, the uneven geographical development of the country and its railway infrastructure ensured little intercity competition (Dorfman, 1983). Hence, and in contrast with New York City, the urban government was not concerned about losing urban labourers but about the indifference of the affluent landowners. In this way, the elite who controlled agriculture, the main export product of the nation and not intercity competition, were the key agents behind the urbanization of the city of Buenos Aires.

Nature, urbanization, civilization and backwardness

The differences in economic activities of the national elites thus allow the differences between public parks in New York City and Buenos Aires to be understood. The dominance of the agrarian over the industrial elite in Argentina justified the absence of a national agenda focused on the wellbeing of urban labour. Moreover, the differences in the urban landscape are also a reflection of the divergent theories about nature that informed these societies. At the time, western European design was a reference point for both North and South America. However, while in Buenos Aires, European design was understood to be of the highest standard, in the United States it became associated with a classist society, one that worked against democratic principles. The semantics of European design thus encouraged the search for a 'genuinely American' landscape, even amidst a metropolis. In the words of *The New York Times*, Central Park was to be 'a reflection upon our national state, which find so much food for study in the ever-changing scenery for which this country is so remarkable. Besides, those Parks, in many instances, are but appendages of grandeur to rank' (NYT, 1856). Conversely, in the City of Buenos Aires, there was an intentional reproduction of European architecture in the city as a mean for imposing 'high culture'. Local traditions were marginalized and the colonial style consciously rejected. The urban government did not want to create a park that helped to perpetuate a rural imaginary but rather wanted to educate local residents into European cultural patterns (Hodge, 1996).

In addition to these divergent views about the value of reproducing European culture in America were also different conceptions and appreciations of urbanization. In North America, evident in the design of New York City's Central Park, is an ideology that expresses a desire to return to nature (Howard, 1996). Being in contact with the natural realm was considered to help achieve a state of balance and protect city dwellers from the evils of urban life, and this justified the building of public parks for the populace (Taylor, 1999). In this way the paradox of progression, via human-made techniques to recreate nature, became one of the defining features of American culture (Mumford, 1961). Likewise, the view of the city as a source of distress, of social malaise, was aligned with the core of the American intellectual tradition. From Thoreau to Emerson, to the utopian cities of Frank Lloyd Wright, cities were a necessary evil likely to corrode individual character and promote disreputable practices among urban masses (Wright, 1996). In the words of Waldo Emerson:

> Whilst we want cities as the centres where the best things are found, cities degrade us by magnifying trifles. The countryman finds the town a chop-house, a barber's shop. He has lost the lines of grandeur of the horizon, hills and plains, and with them, sobriety and elevation. He has come among a supple, glib-tongued tribe, who live for show, servile to public opinion. Life is dragged down to a fracas of pitiful cares and disasters. You say the gods ought to respect a life whose objects are their own; but in cities they have betrayed you to a cloud of insignificant annoyances ... Nature as an answer to the challenges that modern life imposed on men, and cities as the necessary evil.
>
> *(Emerson, 2006)*

We can compare such reflections with the words of Sarmiento, in which 'the wild' brutalized men and cities become the sources of all progress:

> The man of the city wears European dress, lives a civilized life as we know it everywhere; in the city, there are laws, ideas of progress, means of instruction, some municipal organization, a regular government, etc. Leaving the city everything changes in aspect. The man of the country wears other dress, which I will call American, since it is common to all peoples; his way of life is different, his needs specific and limited. They are like two distinct societies, two peoples strange to one another. And more still: the man of the country, far from aspiring to resemble the man of the city, rejects with scorns his luxuries and his polite manners.
>
> *(Sarmiento, 2010)*

The expansion of urbanization was the goal of Argentine leaders who envisioned city life as a means for improving national culture. Furthermore, the city was essential for the organization of civic life, and a geometrized plaza not a naturalistic park was to be at its heart. The Spanish city was organized around a plaza, with the church, the school, government and a central bank at each of the four sides of the central square (Schnore, 1965). The plaza thus became essential to the grid of South American cities, and become the axis around which civic life revolved. But this plaza, this place, was mostly a flat, paved square with no greenery; it was the interaction with other citizens and not with 'nature' which characterized these spaces (Kostof, 1992). Moreover, 'nature' was defined in terms of emptiness and an absence of civilization. The influence of the Catholic Church also ensured a fundamental distrust on 'the sensory' as a way to reach spirituality (Benevolo, 1980).

In Argentina in the nineteenth century, where untamed nature abounded, the urban versus rural confrontation thus resonated through a political confrontation between the landed wealth of the provinces and the commercial prosperity of the city. Unlike the United States, in Argentina the city was not seen as a source of social anomalies but as the best product of civilization (Ingenieros, 1963). The intellectual elite therefore had a profound disregard towards the culture of those disengaged from city life. The rural was to be tamed or abandoned and city culture would guide the evolution of the new nation. Indeed, prominent thinkers such as Domingo F. Sarmiento believed that a relationship with nature was what drove humans back to an uncivilized state, and that it was only in the city, where a multitude congregate, that the ideals of a new republic could come to fruition:

> This study that we are not yet ready to perform because of our lack of instruction in philosophy and history, done by competent viewers, would have revealed to the amazed eyes of Europe a new political world, a naive, honest, and primitive confrontation between the latest conquest of the human spirit and the rudiments of the wild life, between the populous cities and the dark forests.
>
> *(Sarmiento, 2010)*

The economic downturn that followed the deindustrialization of New York City led to the decline of Central Park in the 1970s. As the city almost went bankrupt, the resources to keep the park as a 'respite for the working classes' faded. In a similar vein, a change of the economic model (albeit towards further industrialization) triggered a crisis in the parks of Buenos Aires. The expansion of the industrial economy in Argentina and the empowerment of trade unions promoted the downfall of San Isidro Park. In a symbolic gesture against 'the landed elites', the president Juan Domingo Peron expropriated the park. By the 1980s, however, both parks were being reinvigorated. New York City Central Park recreated its management and funding structure through The Central Park Conservancy, a public and private partnership. The creation of this organization allowed the maintenance of the public park by raising funds from private sources. In Buenos Aires, the Jockey Club reopened its doors, but it also added a significant number of events and attractions for the general public. While belonging to the Jockey Club remains expensive, entrance to the horse races and to the park itself is now free for everyone.

Conclusion

The comparison of New York City's Central Park and San Isidro's Park in Buenos Aires exemplifies how national economic structures influence the programming of urban parks. Governments relied on the programming of urban parks to cater to those groups whose demands were more likely to destabilize their economic projects. In the case of New York City, providing access to nature for the crowded masses of urban workers was one way of preventing the emergence of social unrest. The fear of urban riots was very real in a city that boasted the highest population density in the Americas. Nature was to be an antidote to the evils that urbanization might trigger. Conversely, in Argentina, urban parks were designed to keep the rural elite engaged with the project of the Nation, disregarding 'the native' in favor of the European, and rejecting the celebration of Nature as a path to creating a national culture, for closeness to nature was associated with ignorance and backwardness. Rather, the concern of the government was not to bring nature to the city but to bring the city to the peasants.

Once the economic models under which these parks were created stopped working, these parks collapsed, too. The economic downturn that followed the deindustrialization of New York City was behind the decline of Central Park in the 1970s. As the city went almost bankrupt, the resources to keep the park as a 'respite for the working classes' disappeared. A change of the economic model (albeit towards further industrialization) also triggered a crisis in Buenos Aires parks. The expansion of the industrial economy in Argentina and the empowerment of unions promoted the downfall of San Isidro Park. In a symbolic gesture against 'the landed elites', the president Juan Domingo Peron expropriated the premises of the park. In both cities public space became hostage to the government's needs. Moreover, it was only after the organization of these parks was realigned with new governing forces that parks become prosperous again. In the 1980s New York City Central Park recreated its management and funding structure through The Central Park Conservancy, a public and private entity. The creation of this ad-hoc organization has allowed the maintenance of the public character of central Park while also raising funds from private sources. In Buenos Aires, the Jockey Club reopened its doors, but it also added a significant number of public events and attractions to the general public. While belonging to the Jockey Club is quite expensive, entrance to the horse races and to the premises is free for everyone.

Late changes in local economic structure have also impacted on urban park planning today. In the case of New York City 'The High Line', its latest most notable park, reflects how the city is one of the central nodes of the global economy (Fainstein, 1992; Marcuse, 1993; Friedman, 1986; Sassen, 1991; Castells,

1989; Hall, 1998). It turns a piece of obsolete industrial infrastructure into a linear park at the same time as it becomes a catalyst for the design of new 'star-architecture' and international hotels in its surroundings. Similar to the case of Central Park, a public-private partnership is responsible for the preservation and maintenance of the public park, which it is intended as a site for the reintroduction of nature into the city. In Argentina, private developments have taken the lead in the development of parks. Increasingly, these spaces facilitate the increase in real estate value but do little to ease social differences among those living in the city. The design of open spaces continues to take a backseat to the development of major real estate developments.

In the light of these dynamics between urban parks and national economic structures, a few conclusions emerge. Firstly, cultural attitudes towards the value of urbanization seem to be more perennial than economic structures. Secondly, adequately programmed and maintained urban parks are valuable goods for all classes. This goes beyond the secondary benefits that all enjoy when urban violence is appeased. The case of Buenos Aires shows that even most affluent groups value having access to public parks. Thirdly, urban parks that cater to only one sector of the population are likely to suffer when there are changes in national power structures. That is, when the economy that sustains them fails they are likely to fail too. And if there is a symbolic association between certain group and the park, the park's fate is tied to that group. In the long term, a park that serves many social groups is more resilient to changes in government and power. Fourthly, the delinking of the wellbeing of the urban population from that of governing elites is likely to lead to disinvestment in public spaces. For if the connection between the living conditions in the city and the production of wealth for those who rule the city is not evident, it is difficult to elicit support for improving the material conditions of urban life.

8

AN ILLNESS CALLED MANAGUA

'Extraordinary' urbanization and 'mal-development' in Nicaragua

Dennis Rodgers

Introduction

The relationship between cities and development processes has long been recognized as historically important. Charles Tilly, for example, observes that less densely urbanized regions gave rise to different kinds of states than areas with large numbers of cities because metropolitan centres played a key role in the transformation of states from feudal 'war machines' to 'multipurpose organizations' of a more developmental nature. Drawing on the European historical experience in particular, Tilly argues that strategic collaboration between urban-based capital and rural-based coercion specialists (i.e. warlords) – the former financing the latter's warring, but also paying for their own security – symbiotically welded these two politically important groups together into a unit that over time coalesced into the modern developmental nation-state due to the fact that this 'bargaining, in its turn, created numerous new claims on the state: pensions, payments to the poor, public education, city planning, and much more' (Tilly, 1994: 9). Largely as a result of such analyses, urbanization has perhaps not surprisingly frequently come to be considered almost synonymous with progress, to the extent that the proportion of a country's population that is urban and its urbanization rates are often presented as key indicators of its level of development (Beall and Fox, 2009).

As Jennifer Robinson (2006) has pointed out – and as Tilly's ideas illustrate – the nature of the connection between urbanization and development has been generally theorized in a way that is fundamentally grounded in the historical experience of the West, and it is legitimate to ask whether contemporary hegemonic epistemologies derived from this particular historical experience are universally applicable. Certainly, the experience of Managua, the capital of Nicaragua, challenges the idea that urbanization is synonymous with development. For example, despite being the most urbanized country in Central America, Nicaragua also remains the most underdeveloped in the region. The reasons for this are both fundamentally reflected by and related to Managua's particular urban development over the past two decades, which has directly contributed to the consolidation of a neo-oligarchic configuration of power that has implemented a low-growth economic model reminiscent of nineteenth-century hacienda-style feudalism, a situation that contrasts starkly with Tilly's Eurocentric narrative of city-driven progress.

Managua's distinct trajectory raises the question of the extent to which it constitutes an 'ordinary' or 'extraordinary' instance of urbanization (see Robinson, 2006); or in other words, whether it is a unique case study or a city representative of more general urban dynamics. This chapter seeks to address this basic issue by exploring Managua's changing morphology, its determinants and the key actors involved in order to illustrate how the city's urbanization process has underpinned the emergence of a particular societal model in contemporary Nicaragua. In doing so it points to the critical need to engage with urban experiences beyond those of cities in the 'West' if we are to properly get to grips with the possibilities and pitfalls of urbanization, suggesting that urbanization is a fundamentally embedded process, in many ways determined by contextually specific factors. The chapter begins by exploring Managua's urban development in a historical perspective, before then moving on to a characterization of the city's contemporary transformation, and finally relating this to the broader political economy of development in Nicaragua.

Managua's urban development in historical perspective

Sprawling along the southern shores of Lake Xolotlán, Managua presents a strange landscape: a humid basin filled with foliage and vacant lots, with sporadic developments thrown up apparently at random, punctuated by low hills and lagoons formed in the craters of extinct volcanoes. Designated in 1852 as a compromise capital of Nicaragua over the feuding cities of León and Granada, respective headquarters of the Liberal and Conservative factions of the ruling oligarchy, the city remained a relatively marginal settlement until the middle of the twentieth century. Managua exploded to prominence with the mechanization of Nicaraguan agriculture from the 1940s onwards, which led to significant rural–urban migration. The city's population growth rate averaged 6.2 per cent a year between 1950 and 1971, compared to 3.1 per cent in other urban centres, and 1.6 per cent in rural areas, and over the course of the next two decades the metropolis grew to become the largest in Central America, with over a quarter of a million inhabitants in 1960 and almost 450,000 by 1970 (Higgins, 1990: 380).

The overwhelming majority of this growth occurred in the ever-expanding informal settlements on the city's periphery that Reinaldo Téfel Vélez (1974) famously described as 'the Hell of the Poor', implicitly contrasting them with the bustling metropolitan centre of bars, dance-halls and cinemas which during the 1960s became a magnet for wealthy tourists from all over North and South

America, a playground known as 'Salsa City', where 'every day is made for play and fun, 'cos every day is fiesta', as Guy Lombardo sang in his classic song 'Managua, Nicaragua'. The party in 'Salsa Managua' was rudely interrupted on 23 December 1972, however, when the city suffered a devastating earthquake that killed 20,000 people, destroyed 75 per cent of the city's housing and 90 per cent of its commercial buildings – including most of the city centre – and left 300,000 homeless (Black, 1981: 57). Although international aid poured into Nicaragua to help rebuild its shattered capital, very little systematic reconstruction took place as Anastasio Somoza, the dictatorial President of Nicaragua, appointed himself head of the reconstruction committee, attributing over 80 per cent of reconstruction contracts to his own companies, but completing less than 20 per cent of them, with most rebuilding efforts directed to land that Somoza and his family and cronies owned on the periphery of the city. As a result:

> the destroyed central part of Managua was not rebuilt and ... was virtually abandoned. Only a few buildings survived the earthquake, and the central core took on a post-apocalyptic look. . . . The rebuilding effort that did take place ... created new residential areas east-south-east of the city centre. . . . This gives the city the appearance of a deformed octopus. The tentacles of the octopus reach out along major transport arteries away from the old centre, but the octopus's body is riddled with gaping holes.
>
> *(Wall, 1996: 48–9)*

The lack of post-earthquake reconstruction and obvious venality of the Somoza regime played a major role in intensifying resistance against it. The ranks of the Sandinista National Liberation Front (FSLN – *Frente Sandinista de Liberación Nacional*), which had been (unsuccessfully) waging guerrilla warfare against the dictatorship since the early 1960s, swelled and led to a full-blown insurrection that overthrew the dictatorship on 19 July 1979. The revolutionary junta that subsequently took power rapidly drew up plans to reconstruct and transform Managua, proposing a series of new strategies to rebuild destroyed infrastructure, regulate land use and upgrade informal settlements. These were never implemented, however, as the trade embargo imposed by the US from 1981 onwards, and the civil war against the US-sponsored Contra, meant that the revolutionary state's meagre resources were re-channelled towards more pressing needs such as food production and national defence (Massey, 1986).

The urban development of Sandinista Managua was further complicated by the influx of displaced persons from the countryside due to the war, which overburdened already deficient urban services and infrastructure, and led to an explosion of marginal squatter settlements. While the annual rate of rural–urban migration to the city averaged some 28 per thousand during the decade prior to the revolution, by 1983 it was estimated to have risen to 46 per thousand – equivalent to almost 30,000 new inhabitants per year (Chavez, 1987: 234). The city's predicament was not improved by the return of large numbers of Contra rebels and their families from Honduras and Costa Rica following the Sandinista's electoral loss in 1990 and the end of the civil war. This led to widespread social conflict and processes of spatial polarization in the city, most evident in the emergence of new settlements with explicit political associations, such as *barrio* '3–80' in south-central Managua, named after the *nom de guerre* of Enrique Bermúdez, commander of the Contra Northern Front during the civil war, or the 'Omar Torrijos' squatter settlement in the north-east of the city, with obviously different political affiliations.

Regime change also led to the return of many wealthy Nicaraguans who had left Managua for the US in 1979, and these had a significant impact on the city's morphology through their frenetic conspicuous consumption. As David Whisnant (1995: 448) observed:

> determined efforts by the 'Miami boys' (as they are called) . . . to recreate their cherished Miami social and cultural 'scene' . . . transformed the Managua night: neon-lit bars and exclusive clubs, designer clothing, Nicaragua's first surf shop, one-hour photo processing, expensive cars cruising the scene, and pervasive preening, posturing, and dalliance.

Global franchises such as Pizza Hut and McDonald's rapidly set up shop in post-revolutionary Managua in order to cater to the new Americanized elite, as did numerous luxury hotels, and four US-style malls have been built in the city since 1990: the Plaza Inter, the Centro Comercial Metrocentro, the Galerias Santo Domingo, and the Multicentro Las Americas. The last of these malls was built in 2006 at a cost US$25 million by a private real estate development company called Desarrollos Sooner, also one of the biggest players in the booming condominium and gated community construction market that emerged in post-revolutionary Managua from the end of the 1990s onwards (*Confidencial*, 2006). One of their flagship developments, for example, is the Portal San Cayetano gated community, 'located in Managua's most exclusive living area of Las Colinas', which offers two-storey, 230 square-metre houses providing 'an enchanting union between the elegance of the façade and the comfort of the interiors in a global refined neo-colonial mood' for US$160,000 each (Sooner Real Estate, n.d.). Overall, some 60 different private residential developments, big and small, have sprung up in and around Managua over the past decade.

These new urban developments clearly cater for a limited number of Nicaraguans considering that 90 per cent of the economically active population of the country earns less than US$160 a month (Gobierno de Nicaragua, 2005: 27). Despite the fact that 58 per cent of urban households in Nicaragua were estimated to be over-crowded in 2002, and some 40 per cent 'constructed with scrap or other imper-manent materials' (Inter-American Development Bank, 2002: 1), state-sponsored housing projects have been few and far between since 1990. Although a US$22.5 million Inter-American Development Bank low-income housing programme was approved in September 2002 (Inter-American Development Bank, 2002: 1), this was not implemented, partly because potential beneficiaries of the programme are required to possess a full legal title to their property, which is frequently not the case in contemporary Nicaragua (Sánchez Campbell, 2004), but also because the post-revolutionary state has neither the capacity nor the will to implement large-scale public urban development programmes.

This is particularly obvious from a planning perspective, and more specifically in relation to compliance to the land use provisions stipulated in the Managua Master Plan (PRM – *Plan Regulador de Managua*). This document – originally drawn up in 1982 – categorizes the city into different types of zones (residential, commercial, industrial, parkland, mixed etc.), and establishes rules and restrictions concerning construction, population density, right of way, and so on, determined on the basis of criteria relating to environmental risk factors, as well as notions of 'rational' city organization, particularly with regards to transport. The Managua Master Plan was never properly implemented due to lack of funds, but it underwent significant modification in 1998 as a result of a series of Partial Urban Organization Plans (PPOU – *Planes Parciales de Ordenamiento Urbano*), which reclassified land in certain urban areas. In theory, this was to update obsolete categorizations in order to prepare for the elaboration of a new General Municipal Development Plan (PGDM – *Plan General de Desarrollo Municipal*); but as Gerald Penske, the Director the managua municipality's Department of Urban Development, admitted frankly in an interview in July 2007, in practice the Partial Plans 'simply legalized what had happened on the ground' and reflected 'the limited urban control exercised by the municipality over the city's territory'.

Penske illustrated this relative to the evolution of the affluent neighbourhood of Los Robles. The 1982 Master Plan had originally designated this area as exclusively residential, but in the late 1990s and early 2000s it became very much mixed in terms of land use, as expensive art galleries, exclusive bars, clubs, casinos and fine restaurants all opened there, creating an nightlife area known as the 'Zona Rosa'. Although in theory the municipality had the right to challenge and fine any new developments that contravened existing land use regulations, Penske contended that it lacked the resources to monitor these. 'Thirty years ago,' he told me, 'the Vice-Ministry for Urban Planning had 40 inspectors and 40 Volkswagens that went all over the city to make sure that regulations were respected. Now I have just three inspectors who only have motorbikes to cover a city that's three times the size it was in the 1970s'. As a result, the city had little choice other than to accept new developments as *fait accompli*.

At the same time, it was also obvious that infringements to planning rules and regulations were much less tolerated when they occurred in poorer neighbourhoods. On 1 February 2006, for example, a group of 43 families that had occupied vacant land between the Huembes market and the Isaías Gómez neighbourhood was violently expelled by riot police at the behest of the Managua Municipality, on the rather contradictory grounds that 'this land was not adequate for the construction of housing because it is near a source of potential contamination, being near an open sewage channel', but also that 'this land is classified as part of

Managua's parkland' (Cruz Sánchez, 2006). Penske argued rather disingenuously that this different treatment was due to the variable economic power of businesses in Los Robles and squatters, contending that if the Municipality attempted to impose fines on businesses in Los Robles that did not respect land use regulations, these would be able to absorb them as part of their operating costs and so nothing would happen, but that illegal squatters could be dealt with effectively because they did not have the means to buy their way out of respecting the rules.

The influence of market forces on Managua's urban development is by no means surprising, but it only tells part of the story, however, as the blatant bias that can be observed in the evolution of Managua's road network over the past decade highlights well. Certainly, one of the most striking transformations to have occurred in the city is the overhauling of its legendarily abysmal road infrastructure. As late as 1997, potholes were a chronic driving hazard, traffic was chaotic, carjackings frequent, and there was no discernible logic to the city's byzantine road infrastructure. By 2000, the municipality had carried out a large-scale programme to fill in the potholes, resurfaced and widened the major arteries of the metropolis, built a suburban bypass road in the south-west of the city, and replaced traffic lights with roundabouts. These works were ostensibly to speed up traffic and reduce congestion, but when considered on a map, however, a definite pattern emerges whereby the new roads predominantly connect locations associated with the lives of the urban elite, for example linking the (newly re-modelled) international airport to the Plaza Inter mall to the Metrocentro mall to the Zona Rosa to Las Colinas and so on.

This particular pattern of infrastructural development is the keystone to a broader process of socio-spatial 'disembedding', whereby a whole layer of Managua's urban fabric has been purposefully 'ripped out' from the general patchwork quilt of the metropolis. The living and working spaces of the wealthy – protected by high walls and private security – have been joined together into a 'fortified network' by means of the new roads, which the elite can cruise at breakneck speeds in their expensive four-by-four cars, no longer impeded by potholes, crime, or traffic lights (Rodgers, 2004). When I confronted Guillermo Icaza, Deputy Director of Municipal Planning, about this during an interview in July 2007, he contended that:

> all we do is implement the pre-existing road plan which was determined in 1982. . . . There are different types of roads, for example you can have primary roads, which are big, metropolis-cross-cutting roads, and then you have more secondary roads, which might connect different parts of the city to each other, and then you have tertiary roads, which are small and only concern a particular locality. . . . The problem is that we're still a long way from completing the Plan, and we're still focused principally on primary roads, because that's what it calls for us to finish first.

Icaza was however unable to explain why certain primary roads in the 1982 Master Plan had been built and improved while others had not, and why the latter seemed to be particularly concentrated in poorer areas of the city. Perhaps even more damning, many of the new roads built or improved since 1997 were not actually planned for in the 1982 Master Plan, including for example the suburban bypass in the south-west of the city that connects the exclusive Las Colinas neighbourhood with the road to the south (*Carretera Sur*) in Western Managua, which has numerous exclusive private real estate developments. Perhaps the most notorious example of an unplanned road, however, is the 18-kilometre stretch of country road going from the road to the south to the rural settlement of El Crucero that was paved in 1999 at a cost of US$700,000, clearly in order to provide the then President Arnoldo Alemán with easier access to his hacienda in El Crucero (Bodán, 1999). A similarly parochial logic can be surmised regarding the provision in the 2007 municipal budget for two roadwork projects to repair 1.74 kilometres of roads in the exclusive Las Colinas neighbourhood, at a combined cost of US$3.3 million, equivalent to almost two thirds of the US$5.3 million that the municipality had budgeted to spend on maintaining the city's 1,157-kilometre road network (Alcaldía de Managua, 2006).

Neoliberal urbanization?

Florence Babb (1999: 27) has characterized Managua's post-revolutionary develop-ment as representing 'the making of a neoliberal city', linking this process to broader transformations. She highlights how the successive Conservative and Liberal governments of Chamorro, Alemán and Bolaños actively encouraged the desires of market forces suppressed during the Sandinista period to remodel the post-revolutionary metropolis according to the consumption needs of the wealthy. Arnoldo Alemán, for example, explicitly sought to 'beautify' the metro-polis in order to make it more attractive to private investors through large-scale construction projects of various sorts (see Babb, 1999). The Chamorro, Alemán and Bolaños governments furthermore also all blatantly facilitated private real estate development initiatives by fostering highly suspect tax breaks such as the notorious Law 306 on incentives for the tourist industry (*Ley de Incentivos para la Industria Turística*), passed by the Nicaraguan Parliament in June 1999, and offering a large palette of tax benefits to tourism-related businesses that invest a minimum of US$50,000 in their activities, including, for instance, an 80 to 90 per cent exemption on all income tax for up to ten years, as well as a 100 per cent exemp-tion on property tax, VAT and import duties for the same period (Gobierno de Nicaragua, 1999).

When considered in light of such schemes, it is tempting to lay the blame for the iniquitous development of Managua over the past two decades on an unholy alliance between market capitalism and the successive right-wing post-revolutionary governments, together incarnating an elite-led backlash against the Sandinista revolution's (frustrated) attempts to foster more egalitarian forms of social organ-ization in Nicaragua. In actual fact, however, the laws establishing tax exemptions such as the ones described above have all been voted for not only by Liberal and Conservative MPs, but also Sandinista MPs. Indeed, one of the most pro-active figures in debates in favour of Law 306 was Tómas Borge, the last living founding member of the FSLN who was at the time President of the Parliamentary Tourism Commission, and also – coincidentally – owner of a hotel that stood to benefit from the law (Babb, 2004: 551).

The approval of this piece of legislation arguably give us 'some insight into the huge interests at stake, which seem to involve all the power groups', and starkly highlights how associating the 'neoliberal' urbanization of Managua solely with Liberal and Conservative politicians and entrepreneurs is overly simplistic (Báez Córtes, 2006). Indeed, many of the most iniquitous urban developments in Managua over the past two decades have occurred not under Arnoldo Alemán's mayorship (1990–5), or that of his hand-picked successor, Roberto Cedeño (1996–2000), but under those of the Sandinistas Herty Lewites (2000–4) and Dionisio Marenco (2005–9). This includes the initiative perhaps most blatantly associable with 'neoliberalism': the plan to build a rapid busway system in northern Managua in order to ferry workers from poor neighbourhoods to the city's Free Trade Zones (FTZs). This project calls for the creation of a bi-directional busway through the middle of the busy northern road (*Carretera Norte*) that runs alongside Lake Managua, at an estimated cost of over US$55 million. Once completed, buses will carry up to 200,000 passengers a day from the populous working class Ciudad Sandino satellite city to the west of Managua to the labour-hungry Free Trade Zones (FTZs) to the east, via several other poor lakeside neighbourhoods (Banco Interamericano de Desarrollo, 2005).

Admittedly, constructing a busway to feed labour to the FTZs makes some sense considering that these are the single fastest growing employer in the country. The first FTZ opened in Managua in 1992, and a further 34 have since been established, mainly in and around Managua. The number of companies operating in the FTZs has similarly grown, from 5 in 1992 to 99 in 2006, while the number of workers employed has expanded from 1,000 in 1992 to 80,000 in 2006.[1] Export production through the FTZs has multiplied from US$3 million in 1992 to US$900 million in 2006, equivalent to 46 per cent of total Nicaraguan exports, and a staggering 87 per cent of manufactured exports (Banco Central de Nicaragua, 2007: 25). By far the largest activity is apparel assembly, with the companies involved mainly South Korean (25 per cent), US (24 per cent), and Taiwanese (21 per cent) (Comisión Nacional de Zonas Francas, 2010). More directly significant, however, is the fact that some two-thirds of the infrastructure of the private FTZs belongs to Nicaraguan entrepreneurs, who rent space within their compounds to foreign companies, while also receiving large tax exemptions (Laguiazf.org, n.d.). The owners of these compounds unsurprisingly include big Nicaraguan economic conglomerates, such as the Pellas Group, but it is widely known that several major Sandinista politicians, including Dionisio Marenco and Daniel Ortega, are owners of FTZ infrastructure,[2] which perhaps explains their enthusiasm for the busway scheme.

The political economy of post-revolutionary Nicaragua

In order to understand this apparently paradoxical Sandinista 'neoliberalism', it is necessary to go back to the roots of the 1979 revolution. The conventional story is that it was a mass popular uprising led by the vanguard FSLN that over-threw a terminally corrupt dictatorship after 45 years of oppressive rule. The truth is that for the first 38 years of the dictatorship, there was very little effective opposition to the Somoza regime. Although the FSLN was founded in 1961, it had little success as a guerrilla force and in fact ceased all armed actions for several years in the wake of the disastrous 1967 Pancasán offensive during which 13 senior members of the organization were killed.

Part of the reason for this ineffectiveness was that the dictatorship was largely based on a tacit pact between the Somoza family and the traditional Nicaraguan oligarchy, composed of Liberals and Conservatives. Each group was associated with particular economic interests which as a general rule did not overlap. When they did, the three groups tended to cooperate rather than compete with each other, sharing markets and ensuring that all profited (Mayorga, 2007: 37–9). The three groups cooperated happily during the growth years of the 1950s and 1960s, but the worldwide crisis of the early 1970s caused serious strains, as did Somoza's securing of the lion's share of lucrative reconstruction contracts in post-earthquake Managua for his companies. Conservative sections of the traditional elite with interests in the construction industry accused the regime of 'disloyal competition' (Núñez, 2006: 130).

In 1974, Pedro Joaquín Chamorro Cardenal, scion of one of Nicaragua's leading Conservative families and editor of the newspaper *La Prensa*, founded the Democratic Electoral Union (UDEL – *Unión Democrática Electoral*), in order to provide a focus to the burgeoning elite opposition to the Somoza regime. The UDEL was not very successful in mobilizing cross-party support, partly because of the enduring suspicion between Liberals and Conservatives. It did, however, provide the impetus for many young Conservatives to join the ranks of the FSLN during the early 1970s, and opened up channels of communication between the Sandinistas and the Conservative elements of the traditional oligarchy which became very important as the former began to gather strength and plan renewed armed struggle. As Orlando Núñez (2006: 132) put it:

> it is difficult to imagine the revolutionary insurrection without taking into account the participation of the Conservative capitalist class, both eco-nomically and politically: the money they provided, the use of their remote haciendas for training, the safe houses in the cities, their children joining the ranks of *Sandinismo*, [and] the public legitimisation that they provided to the struggle against Somoza.

Perhaps the best exemplification of this alliance that emerged between the Con-servative elite and the FSLN is the so-called 'Group of Twelve', which brought together prominent industrialists, businessmen, priests and academics who publicly backed the FSLN's struggle against Somoza in 1977, some of whom later became members of the revolutionary government during the 1980s, including Sergio Ramírez Mercado, who was Vice-President of Nicaragua between 1984 and 1990, and Miguel D'Escoto Brockmann, who was Sandinista Minister of Foreign Affairs. Other figures emanating from the Conservative elite who participated in the revolutionary regime include Luis Carrión Cruz, one of the nine *comandantes* of the FSLN's National Directorate and Vice-Minister of the Interior during the 1980s, and son of Luis Carrión Montoya, a descendent of one of the oldest families in Nicaragua and one of the most prominent bankers in Nicaragua during the 1960s and 1970s. Similarly, Ernesto Cardenal Martínez, Minister of Culture under the Sandinistas, and his brother Francisco, who was Minister of Education, come from one of the country's most prominent Conservative families, as does Reinaldo Antonio Téfel Vélez, director of the Nicaraguan Institute of Social Security during the 1980s, among many others (Vilas, 1992: 325–8).

Few businesses belonging to major Conservative families were affected by the Sandinista regime's nationalizations during the 1980s, which mainly focused on assets that had either been in the hands of Somoza and his cronies, or else prominent Liberal families, many of whom had fled Nicaragua following the triumph of the revolution (Everingham, 2001: 72). Although the mixed nature of the revolutionary economy meant that Conservative businesses were able to continue operating unimpeded, their margin of operation was constrained by the Sandinista government's attempt to regulate certain economic activities, competition from newly formed state enterprises, the new regime's demands that labour rights be respected, and, from the mid-1980s onwards, the militarization of the economy and hyperinflation. Conservative businesses consequently began to develop their activities beyond Nicaragua, including in particular in other Central American countries (Mayorga, 2007: 42–50). This was particularly the case of the Pellas banking group, for example, which extended its financial activities to the Caribbean and the US during this period, setting up a bank in the Cayman Islands and buying another in the US which became the nucleus for their now extremely powerful Banco de América Central (BAC) international financial conglomerate, the fifth largest financial group in the region with US$3.1 billion in assets, and also the most profitable according to Standard and Poor's 2005 Central America ratings (BAC, 2006).

This transnationalization of their economic interests increasingly put the Conservative elite at odds with the Sandinista regime's autarchic economic policies, and led to its underwriting the National Union of Opposition (UNO – *Unión Nacional Opositora*), an eclectic coalition of Liberals, Conservatives and a hodge-podge of extreme left- and right-wing mini-parties that defeated the FSLN in the 1990 elections and brought the revolutionary experiment to a close. The Sandinista party had not expected to lose the elections and faced a critical dilemma. As Sergio Ramírez (1999: 55) described:

the fact is that *Sandinismo* could not go into opposition without material resources to draw upon, as this would have signified its annihilation. The FSLN needed assets, rents, and these could only be taken from the State, quickly, before the end of the three month transition period [before formally handing power over the UNO]. As a result there was a hurried and chaotic transfer of buildings, businesses, farms, and stocks to third persons who were to keep them in custody until they could be transferred to the party. In the end, however, the FSLN received almost nothing, and many individual fortunes were constituted through this process instead.

Certainly, much has been written about the fact that Daniel Ortega drives a Mercedes Benz or that his brother Humberto lives in a huge mansion on the road to Masaya that has its own private baseball field, for example (Rojo, 1999).

This unedifying episode in the FSLN's history is known in Nicaragua as the '*piñata*'.[3] Much more significant than any example of individual personal enrichment, however, the *piñata* effectively created the nucleus of a Sandinista economic group. Media reports have identified over 50 businesses associated with the FSLN, including financial service providers such as Fininsa or Interfin, the Victoria de Julio and Agroinsa sugar refineries, INPASA printers, media outlets such as the Canal 4 television station or the 'Ya!' radio stations, as well as Agri-Corp, the biggest distributor of rice and flour in Nicaragua with a US$100 million turnover (Loáisiga Mayorga, 2003; Mairena Martínez, 1999; Everingham, 2001; Mayorga 2007). During the 1990s there emerged an organized 'Sandinista entrepreneurs bloc' ('*bloque de empresarios sandinistas*'), led by the former FSLN *commandante* and member of the National Directorate Bayardo Arce Castaño, who is a major stakeholder in Agri-Corp – run by his brother-in-law, Amílcar Ibarra Rojas – and is also associated with the real estate development company Inversiones Compostela, run by his wife, Amelia Ibarra de Arce, which has over US$4 million worth of investments in Managua (Loáisiga Mayorga, 2005). Other members of this group include Dionisio Marenco and Herty Lewites, for example (Marenco Tercero, 2005). The group has been the financial lifeline of the FSLN, particularly during election time. In 2000, for instance, it raised almost US$2 million to finance Lewites' campaign to be elected mayor of Managua (Bolsa de Noticias, 2011).

Many of the enterprises associated with this new Sandinista bourgeoisie have their origins in Liberal assets expropriated in the 1980s, and not surprisingly the return of former proprietors after regime change in 1990 led to legal conflicts over their ownership. Although some Liberal families in exile successfully started businesses anew whilst abroad including, for example, the Morales Carazo family whose Jaremar group is now a major player in the Central America agro-industry (Mayorga, 2007: 50), many returned intent on reclaiming the expropriated sources of their past wealth and status, and joined forces with dispossessed Liberals families that had remained in the country to claim what they perceived as their birthright.

The ensuing conflicts ensured the election of the stridently anti-Sandinista Arnoldo Alemán as President of Nicaragua in 1996.

Alemán's accession to power initially led to an intensification of conflicts over expropriated property as he sought to push the claims of the Liberals that had brought him to power, but the FSLN retained enough mobilisatory power to block any attempts to overturn what they considered to be a reasonable post-revolutionary economic settlement. The need to regain expropriated properties became steadily less urgent during the course of the Alemán presidency, however, as it became hugely associated with sleaze and corruption (Nitlapán-Envío team, 1999). Controversial privatization processes whereby numerous state enterprises were sold off to friends and relatives of Liberal government officials at rock-bottom prices effectively contributed to replacing their expropriated patrimony, as did siphoning off a significant proportion of the US$1 billion of international aid that was channelled to Nicaragua following its battering by Hurricane Mitch in 1998 (Rocha et al., 1999). Arnoldo Alemán, for example, is estimated to have personally embezzled up to US$100 million during his presidency (Transparency International, 2004: 13).[4]

The widespread corruption meant that many expropriated Liberals were able to reconstitute their patrimony and no longer needed to press Sandinista entrepreneurs for the return of expropriated properties. This changed state of affairs allowed for a *rapprochement* between Alemán's Constitutionalist Liberal Party (PLC – *Partido Liberal Constitucionalista*) and the FSLN led by Daniel Ortega, which become a 'pact of co-governance' in 1999, whereby the two parties formally divided up the Nicaraguan state apparatus between them and agreed to support each other's appointments. The pact also reformed electoral law, reducing the threshold for a first-round victory in presidential elections from 45 per cent to 35 per cent, and also granting former Presidents a seat in the National Assembly for life. These two measures clearly aimed to facilitate an Ortega victory in the forthcoming 2001 elections and to obtain parliamentary immunity for Alemán after he left office. However, Nicaragua's increasingly poor business ratings as a result of corruption, as well as emergent tensions between the economic priorities of the nascent Liberal-Sandinista national bourgeoisie and the Conservative–Liberal transnational elite, meant that the latter flexed their political muscle and imposed a Conservative, Enrique Bolaños, as the right's unified presidential candidate.

Believing Bolaños to be a controllable yes-man, Alemán acquiesced, but once duly elected with the financial backing of the five richest economic con-glomerates in the country, he proved to be unexpectedly independent-minded. Bolaños turned on Alemán and sought to have him indicted for corruption, an opportunity that the FSLN were quick to seize on to obtain greater leverage on Alemán within the context of their shaky pact with him. The FSLN helped strip Alemán of his parliamentary immunity and he was condemned to 20 years in jail for fraud and embezzlement in December 2003. The FSLN, however, quickly reverted back to their pact with the PLC – which Alemán continued to control, first from prison and subsequently whilst under house arrest (for health reasons) – as the Bolaños government pushed policies favouring the banking and financial service sectors that constitute the foundation of the Nicaraguan transnational economic groups' wealth (Mayorga, 2007: 69–72), at the expense of agriculture, urban manufacturing and property ownership from which both Sandinista and Liberal economic groups drew their wealth.

As a result, although the November 2006 elections seems to have been a hotly contested four-way contest between Daniel Ortega, the dissident Sandinista Herty Lewites (replaced by Edmundo Jarquín when he died suddenly of a heart attack in July 2006), the PLC candidate José Rizo (anointed by Arnoldo Alemán), and Eduardo Montealegre, Bolaños' Minister of the Economy who ran representing the Conservative Party–Nicaraguan Liberal Alliance (ALN-PC – *Alianza Liberal Nicaragüense-Partido Conservador*), it in actual fact pitted a Liberal-Sandinista national bourgeoisie against transnational capitalist interests. Ortega's victory with 38 per cent of the vote, coupled with the fact that the FSLN and the PLC won enough seats to control Parliament, constituted a significant shift in political economy terms, well reflected in the composition of Ortega's cabinet, which reads like a veritable who's who of both Sandinista and Liberal businessmen.[5] Perhaps not surprisingly, one of Ortega's first acts on assuming office was to change Arnoldo Alemán's detention regime from 'house arrest' to 'country arrest', thereby allowing him to return to public life, and in April 2007, FSLN and PLC members of parliament voted together to pass a law reducing the maximum penalty for corruption to 5 years, applicable retroactively, meaning that Alemán's prison sentence was shortened by 15 years, and he was released in early 2008.

The obvious question concerning this Liberal-Sandinista national bourgeoisie is whether it is in any way progressive, in the pro-nationalist, anti-imperialist Leninist sense of the term, that is to say, whether it can be the lynchpin for a process of productive capital accumulation and national development that is sorely lacking in a post-revolutionary Nicaragua that has been 'mal-developed' along lines that favour transnational rather than national interests, as the country's huge trade deficit dramatically illustrates (Robinson, 1998). The answer is clearly no. Far from seeking reform, 'liberal *Sandinismo*' seems instead to be angling for an economic settlement whereby FSLN businesses and those of its PLC partners can derive a low-level of profit from exclusive monopolies over protected sectors of the domestic market, disconnected from transnational imperatives. This is a strategy more associated with nineteenth-century hacienda-style feudalism than any form of progressive capitalism, and it is clearly both reflected and underpinned by Managua's particular urban development, as the investments in FTZ infrastructure by recent municipal regimes demonstrate very well. To this extent, far from constituting the accession to power of a progressive national bourgeoisie – and even less a return to the utopian politics of Nicaragua's past – Ortega's electoral return arguably ultimately constitutes little more than a wry illustration of Karl Marx's (2004: 3) famous aphorism that 'great historic facts and personages recur twice … once as tragedy, and again as farce'.

Conclusion

In a classic article considering the value of urban research, Charles Tilly (1996) famously argued that the study of urban contexts potentially provided privileged insight into the dynamics of large-scale socio-economic processes. He was referring of course to his own grand theorizing about the historical relationship between cities and development, and in particular the way in which cities played critical roles in promoting developmental states in Europe, but his comment in many ways echoes one made much earlier by the Nicaraguan poet Pablo Antonio Cuadra in an essay entitled 'An Illness called Managua', where he contended that the city was paradigmatically 'the reflection of society, of its grace and its bitterness, of its vice and its beauty, of its history and its community' (quoted in D'León Masís, 2002). Although Cuadra's comment could be interpreted as suggesting that Managua is more a 'symptom' than an 'illness', that is to say an epiphenomenon of an underlying societal political economy, as the above discussion will hopefully have made clear, the metropolis itself has in fact actively contributed to the construction of Nicaragua's particular post-revolutionary political and economic order in a rather pathological manner, thereby promoting a process of 'mal-development' rather than the progress that Tilly inherently associated with urbanization.

As such, Managua's evolution starkly highlights the need to go beyond historically specific experiences such as those described by Tilly if we are to truly understand the possibilities and pitfalls of global urbanization. It is only by recording, comparing and juxtaposing the urban experiences of cities worldwide that we can expect to gain insight into more general urban trends and dynamics. At the same time, however, Managua's distinct trajectory also raises the question of the extent to which it constitutes an 'ordinary' or 'extraordinary' instance of urbanization (Robinson, 2006). In other words, to what extent can it be seen as representative of broader, general urban tendencies? Certainly, there is a definite sense in which the city's particular transformation over the past four decades represents an exceptional process, linked to the quite unique dynamics of Nicaraguan society during this same period. When seen from this perspective, though, the urban development of a city emerges very much as a fundamentally embedded process, and to this extent, all forms of urbanization can plausibly be seen as 'extraordinary' in the sense that they are always deeply contextually specific. This, however, perhaps poses less of a challenge for a renewed urban theory than may initially seem, since – to paraphrase the poet William Blake (1988: 250) – 'general forms have their vitality in particulars', and identifying any underlying universal dynamics to global urbanization processes arguably depends first and foremost on understanding the particular circumstances that underpin them.

Notes

1 See COHRE (2004) and Everingham (2001). A further 240,000 are employed indirectly, meaning that over 15 per cent of the Nicaragua labour force is in employment related to FTZs (BCN, 2007: 9).

2   Interview with Francisco Mayorga, Managua, July 2007.

3   A *piñata* is a decorated *papier-mâché* figure filled with sweets, which is often featured at parties in Latin America. It is struck with a stick by blindfolded persons until it breaks open and its contents spill out, at which point a scramble ensues as people attempt to grab as many treats as possible.

4   Most of those partaking in this corruption, as well as the Sandinista *piñata*, clearly secured assets on a much smaller scale. Francisco Mayorga notes that there are just 350 individual accounts in the Nicaraguan banking system with deposits amounting to more than US$1 million, and suggests that only 12 families groups own assets over US$100 million: the Pellas Chamorro, Chamorro Chamorro, Lacayo Lacayo, Baltodano Cabrera, Ortiz Gurdián, Zamora Llanes, Coen Montealegre, Lacayo Gil, Fernández Holmann, Morales Carazo, González Holmann and Montealegre Lacayo (Mayorga, 2007: iii, 125).

5   For example, Bayardo Arce Castaño was appointed Daniel Ortega's economic advisor with Cabinet rank, while his business partner Samuel Santos López, became the new Minister of Foreign Affairs. The Ministry of Transport and Infrastructure went to Fernando Martínez Espinoza, owner of one of Nicaragua's biggest construction company, while Mario Salinas Pasos, president of the private real estate development company Desarrollos Sooner, was appointed the new director of the Nicaraguan Tourism Institute (INTUR – Instituto Nicaragüense de Turismo), with Cabinet rank (Guevara Jerez, 2007).

9

THE CONCEPT OF PRIVACY AND SPACE IN KURDISH CITIES

Hooshmand Alizadeh

Introduction

Privacy is not a topic that has received much attention from urban theorists. When discussed, privacy has tended to be addressed through concepts such as territoriality and personal space; 'two ways in which people create different types of boundaries to regulate their interaction with others in their environment' (Bell *et al.*, 2000: 253). Drawing on behavioural studies, such definitions have tended to view privacy through both spatial and social dimensions, or as Altman (1975) suggests as a boundary regulation mechanism to indicate desired level of personal and group privacy. However, what generally characterizes research into privacy is a focus on the private and domestic domains rather than with reference to public space. In these terms, writers have been most concerned with defining and explaining the social construction of privacy with reference to culturally specific aspects of the home and domestic life (Margulis, 2003). In the context of a general omission of privacy from urban theoretical debate there are nonetheless a handful of theorists who believe that the concept is important to understanding urban life (Lawson, 2001; Cowan, 2005; Caves, 2005; Parker, 2004; Saegert, 2010).

The lack of direct reference to privacy in urban theory can perhaps be best explained through the nature of built environment, and more specifically the interface between the private and public sphere in Western cultures since the eighteenth century onwards. Urban change at the expense of private sphere where 'the institution of the family loses monopoly control over its privatized domain' has ensured the emergence of a dichotomy between public and private space (Parker, 2004: 140). For this reason, direct reference to the concept of privacy has been generally ignored by urban theory dominated by Western accounts of urban life.

In contrast, in terms of the urban context of Eastern culture, and especially the Middle East, privacy has been a feature of academic concern due to the nature of everyday life based around Islamic law. In the holy *Koran*, as the main source of the Islamic law, the private domain has acquired prime importance to the family unit as a space recognized as 'the base of entire socio-cultural structure and a self-sustaining institution which ensures ideological and cultural stability over the entire spectrum of society' (Saleh, 1998: 541). Such religious and cultural belief is particularly pronounced in respect to visual privacy among Muslims as an essential element of a refined life of goodness and purity, in relation to both indoor and outdoor living space (Saleh, 1997). For this reason, as subsequent sections of this chapter will reveal, there is a significant amount of theoretical work which directly deals with privacy and its effects on the urban form and everyday life in Islamic cities. The importance of privacy has been recognized as a fundamental controlling feature of the physical organization of the Islamic city (Monteguin, 1983; Abu-Lughod, 1983; Madanipour, 1998a; Memarian and Brown, 2003; Stewart, 2001).

However, even with regard to the literature on privacy which focuses on the Islamic city, it is important to note the lack of attention paid to Kurdish cities. Kurdistan, a mountainous region sandwiched between two rival empires – on one hand the Ottoman Empire, which is Turkish and Arabic-speaking as part of the Hanafite (Sunni tradition), and on the other hand the Safavid Empire, which is Persian-speaking and part of the Shiite sect. While the majority of Kurdish Muslims adhere to the Shâfi'ite Sunni rites there has nonetheless been a desire to define Kurdish identity as being distinct from both nearby traditions. Politically, Kurdistan remains of peripheral concern to each of the empires, but has also had an important role to play in term of mediation between them (Bruinessen, 1999). Nonetheless, an important point of history was the division of Kurdistan region among four countries which has ensured that Kurdish cities have tended to be ignored in theorizations of the Islamic cities. Such comments and geo-physical conditions of the Kurdistan landscape notwithstanding, everyday life of Kurdish cities is to a certain degree distinctly different from the surrounding cultures. In these terms, the Kurdish city has remained unnoticed in the urban literature on the Middle East.

Cross-cultural studies of privacy

In this section, I discuss privacy in terms of cross-cultural comparisons in order to highlight how the ways in which study of Kurdish cities can contribute to urban theory through an exploration of the *concept of privacy* in place-making. Privacy is a highly subjective term and has been used conceptually to define both private and public spaces in a manner that is context and culturally specific. The concept has its historical origins in well known philosophical discussions, most notably Aristotle's distinction between the public sphere of political activity or appearance, the *polis/agora*, and the private sphere associated with family and domestic life or the hidden world of the household, the *oikos*. In terms of popular usage, dictionaries provide an insight into the way the word is commonly used. For example, the *Oxford English Dictionary* (2009) defines privacy as 'the state or condition of being alone, undisturbed, or free from public attention, as a matter of choice or right; freedom from interference or intrusion'. For some, privacy extends to 'a right to prevent being contacted or approached by parties without consent' (Kaminsky, 2003, cited in Gordon, 2004). Moreover, in summarizing literature on privacy, Schoeman (1984) distinguishes between privacy as a moral value in individuals' lives and privacy as a means of preserving human relationships. Similarly, for Newel, privacy is defined as 'a voluntary and temporary condition of separation from the public domain' (Newell, 1998). The term is thus correlated with seclusion and contrasted with the worlds of the communal and public (Georgiou, 2006).

In these varied definitions of privacy, a distinction between the two realms of public and private life is paramount with privacy not only being associated with intimacy but with ways of managing interpersonal relationships. Such definitions thus rest on a connection between privacy, behaviour and activities that allow or restrict bodies, places and information. Generally speaking then 'privacy is the process of regulating interaction with others ... a dialectic process of increasing and decreasing interaction' (Witte, 2003: 8). In these terms, as Abu-Gazzeh (1995: 95) summarizes, 'privacy serves three main functions; the limiting of social interaction; the establishment of plans and strategies for managing interaction, and maintenance and development of self-identity'.

However, while such broad definitions are useful, it is important that urban theory accounts for the dynamic and varying nature of privacy across time and space. For example, the societies of Bali and Java in Indonesia would appear to be opposites with regard to notions of privacy. The people of Java are renowned for having open homes, while the people of Bali have more bounded domestic spaces. Javanese houses are generally small, detached and made of bamboo, occupied by single families and have no walls or fences around the property. Moreover, the walls are thin and loosely woven; and commonly houses do not have doors. In relation to Western conceptions it would be easy to conclude that the Javanese do not value privacy. This could not be further from the truth, for Javanese people maintain privacy through politeness and emotional restraint in speech and behaviour.

In developing a comparasive analysis Hall (1966) asserts that Germans are more concerned with physical separation than are Americans; and go to considerable lengths to maintain privacy by means of doors and physical layouts. Hall explains that the English are also private people, but manage their psychological distance from others via verbal and nonverbal means (such as voice characteristics and eye contact), rather than by physical and environmental means. Amongst the Muslims of Malay in Malaysia, it is also argued that there is a 'different perception of privacy in comparison with Islam. Malay society has emphasized community intimacy over personal and family privacy. The low priority given to privacy is reflected in the flexible and open planning of the traditional Malay house' (Hashim and Rahim, 2006: 305). However, the distinctive zoning of spaces in Malay houses is achieved through minor changes in the height of walls or partitions which provides a required level of privacy. Indeed, the home has a guest room (public domain) which is male dominated and a family room (private domain) that is the domain of women and children in the family (Hashim and Rahim, 2006).

Such examples and others from ethnographic and environmental-behaviour writers (see, for example, Rapoport, 1962, 1977, 1990; Altman, 1977; Kent, 1984; Lawrence, 1990; Khazanov and Crookenden, 1994; Oliver, 1997; Saleh, 1997; Al-Kodmany, 1999, 2000; Memarian and Brown, 2003; Hashim and Rahim, 2006) highlight how the 'norms and customs' of different ethnic and cultural groups are reflected in their use of space, through the configuration and design of the home, and even that the distances and angles of orientation that people maintain from one another varies cross-culturally (Abu-Gazzeh, 1995: 94–95). Indeed, Altman posited two ways in which privacy is regulated through behavioural and environmental mechanisms (Altman, 1977). Behavioural mechanisms include verbal and non-verbal behaviour which are influenced by socio-cultural factors, where: 'people in all cultures engage in the regulation of social interaction through behavioural mechanisms by which accessibility is controlled and are probably unique to the particular physical, psychological and social circumstances of a culture' (Hashim and Rahim, 2008: 94). Regulating mechanisms also vary considerably across cultures with the concept of privacy being an important parameter influencing 'the level of socially acceptable space consumption and proximity' (Burgess, 2000: 14).

Visual privacy

Altman and Chemers (1987), Rapoport (1977), and others have also suggested that in environment–behaviour research any variable or sub-variable should be the focus of sustained study (Al-kodmany, 2000). Here, then, I wish to spend some time in discussing visual privacy as a key factor in the organization of space in both the architecture and urban design of Islamic cities. In particular, I show how as a ritualized and societal concept visual privacy has had a great influence on the built form of the Islamic cities and especially 'the delineation between public and private spheres' (Abu-Lughod, 1983; Madanipour, 1998b; Memarian and Brown, 2003; Stewart, 2001: 177). As Al-Kodmany (2000: 285) suggests, Islamic teaching demands that 'residential visual privacy is essential to a refined life of goodness and purity'. The importance of this concept is also recognized in Iranian architecture, practised through the maintenance of social concepts of *Mahram* and *Nâ-Mahram* which contextualize the thresholds of interaction between male and female in the social structure of urban settings (Mazumdar and Mazumdar, 2001).[1] The non-verbal manifestation of spatial concepts can also be observed in the structure of houses, especially those of wealthy people, within which a multi-courtyard structure defines two separate parts of the house; the *Andaruni* (inside), the private sphere, and the second, *Biruni* (outside), the public realm. In line with these concepts, larger houses are equipped with two courtyards in order to increase a sense of privacy through the separation of the quarters of men from those of women.

In these terms, the concept of privacy can be recognized as a fundamental formula for controlling the physical and social organization of Islamic cities. The private domain (dwelling) has priority over public spaces as the home contains most aspects of everyday life in its courtyard structure; where Muslims can 'enjoy the pleasures of life in the open air and in strict seclusion' (Monteguin, 1983: 48). The importance of the private domain can also be recognized in Islamic ideas of family life as 'the base of entire socio-cultural structure and a self-sustaining institution which ensures ideological and cultural stability over the entire spectrum of society' (Saleh, 1998: 541). Thus, preference is given to the inner rather than outer façades, with the outside façades normally being left blank with small openings above the eye level of the passer-by to prevent them from looking in. The only exception is the doorway which is usually decorated to indicate the habitation within. As Hakim (1986) points out even the location and position of windows and doors in Muslim houses is influenced by the need to avoid direct views into the house.

The interaction between public and private domains in homes in the Islamic city is thus characterized by 'successive hierarchical sections which herald increasing degrees of privacy' (Bianca, 2000: 37). This further leads to the organization of space where 'the private realm is separated from the public by a semi-private space' (cul-de-sac pattern), as a transitional passage between the spaces to create a protected area under the control of immediate neighbours 'within which kin-like responsibilities (and freedom) govern' (Madanipour, 1998b: 243; Abu-Lughod 1983: 67). Privacy is maintained to allow communal socialization among women and children (Saleh, 1997: 171). The desire for privacy also extends to the access route to the private realm to the degree that a semi-private space is aligned with the entrance to the house in the form of a semi-vestibule, just before the entrance door, and a vestibule linked to a bent corridor intended to restrict the view from the outside and defining a transitional space between the interior and guest areas.

Through its spatial and social organization the courtyard structure is thus well suited to Islamic culture, with 'a high level of domestic visual privacy' as a main feature of 'inward-facing rooms' (Memarian and Brown, 2003: 188), which are 'shut off from its surroundings by high and solid walls' (Warren and Fethi, 1982: 44). From this perspective, the spatial and social organization of the home has important symbolic value which conveys in microcosmic the order of the universe which responded ideally to the requirements of the Islamic social order (Petherbridge, 1984: 201; Bianca, 2000: 79). To some extent, its application is observed even in regions such as Kabul in Afghanistan and Sanandaj in Iran, where 'heavy snowfall is experienced rendering this model inefficient and cumbersome for its users' (Hakim, 1994: 126).[2]

One of the main features of the privacy principle is, of course, to control social interaction between men and women. Islamic law requires that 'sexual and emotional activities should be centred on the family nucleus to consolidate family life and reduce social stress', and that one avoids possible 'illegal sexual intercourse' (Memarian and Brown, 2003: 188). Public space is thus regarded as unsafe for Muslim women and to some extent to be avoided. In these terms, the form of the city has been structured in such a way that minimizes contact between men and women outside the kin group (Abu-Lughod, 1983). The logical consequence of the determination of visual privacy in emphasizing designs is to 'protect' women from the eyes of strangers and to create clear distinctions and hierarchical spaces through the gendering of the Islamic city.

In summary, it is evident through the examples presented in this section that privacy is a ritualized and societal concept which has had a great influence on the urban form of the Islamic city. Its diffusion throughout Islamic lands and its continuity from the pre-Islamic era has been transmitted through successive generations. In contrast to the thinking of some 'orientalist' scholars, I argue that Islamic settlements are neither fortuitous nor amorphous in their organization, but they are expressions of social intercourse and allegiance to Islamic society (Petherbridge, 1984: 195). However, in the remainder of this paper I interrogate the extent to which there is diversity in how privacy is conceived in heterogeneous ways in Islamic cities. More specifically, the chapter now turns to attitudes towards privacy in Kurdish cities.

Privacy in Kurdish culture

As noted earlier, central to the concept of privacy in Islamic cities is gender segregation out of kin groups, especially in relation to physical approaches to the private domain from public thoroughfares and vice versa; the matter of over-looking and being overlooked; and the presence of women in the public space. In this section I highlight how in Kurdish culture, tension between the two realms of public and private space tends to be mitigated through hospitality, a sense of neighbourliness among inhabitants, and the specific form of urban settlements. Hospitality, as Barth (1953: 109) indicates, 'is an integral aspect of the Kurdish ideal ... conversely, a person with a reputation for miserliness suffers correspond-ing loss of prestige'. Rich (1836: 70) in his first contact with the Kurds in 1820, describes 'the manners and customs of the Kurds [are] unlooked-for honour, and [are] great proof of his friendly and hospitable disposition' and can be traced in the social behaviour and ceremonies of the Kurdish people. For example, the most popular Kurdish dance called *Rash-balak* is usually performed in a circle with the young boys and girls taking each other's hands (see Plate 9.1a and Plate 9.1b) and the dressing of Kurdish women in vivid colours highlights harmony with the natural landscape (Alizadeh, 2007).

In the view of Minorski (quoted in Nikitine, 1987: 223), 'Kurds are the most liberal-minded in relation to the status of the women among all Muslims' and women 'enjoy a respectful position not observed among their Arab, Turk or Persian neighbours' (Kasraian and Arshi, 1990: 26). Nikitine (1987: 224) also stresses that

PLATE 9.1   (a) and (b): Popular Kurdish dance (*Rash-balak*) which is performed by girls and boys holding hands
Source: Hooshmand Alizadeh

Kurdish women usually socialize with men, to the extent that 'they entertain the guests in absence of their husband and they are quite free and confident in their presence and do not cover their face as do the other Muslim women'. Mela Mahmūd Bayazīdī (1963: 10), a learned mullah, in his writings titled *Customs and Manners of the Kurds (1858–9)*, also noted that 'women did not veil, and together with men participated in production work as well as singing, dancing, and other entertainment'. Similarly, Galletti (2001: 209) indicates that, 'western visitors have often described the strong character of the women whose role has always been relevant in Kurdish society as mother, partner, political chief and sometimes fighter and bandit'. For example, in the words of Pietro della Valle (the well-known seventeenth-century Italian traveller quoted in Gallatti, 2001: 210), women inhabit public space freely and unveiled and talk spontaneously with men both local people and foreigners, and entertain guests as the head of the house.

Throughout Kurdish history it is thus possible to find women reaching high positions and becoming political, and in some cases even military heads of tribes.[3] At the end of the nineteenth century, for example, the first female writer on Islamic doctrine and historiographer[4] in the Middle East *Mâh Sharaf-Khânoum*, known as Mastura Kurdistani, was Kurdish. It is also important to mention the name of *Adela Khânum* of *Halabja* who occupied the headship of the *Jâf* tribe of southern Kurdistan (Bruinessen, 2001: 96). In line with the prominence and status of Kurdish women there is a strong sense of neighbourliness with 'hospitality as one of the finest features in the Kurds' character' (Hay, 1921: 49). By highlighting the importance of hospitality and the importance of women to public life it is clear that privacy in the Kurdish city is somewhat different other Islamic cities (as noted by Abu-Lughod, 1983; Hakim, 1986; Madanipour, 1998c; Al-Kodmany, 2000).

The urban form of the Kurdish city

According to Alizadeh (2006) and Izady (1992), the underlying structure of the Kurdish cities is based on the concept of mound cities used in the Zâgros–Taurus mountain ranges prior to the Median dynasty (728–550 BCE). Owing to the geo-political situation of Kurdistan occupying the frontier between two competing empires, west and east, a specific political dimension of place-making is evident in the structure and form of Kurdish cities and the strong sense of neighbourliness and social cohesion among its inhabitants. Although such a unified community structure is recognized by Muslim and non-Muslim scholars as flowing from the lessons of Islam extracted from the *Quran* and *Sunnah*, in the case of the Kurdish city, unity also emerges from socio-political and environmental circumstances that have dominated Kurdish society for more than 2,000 years. In explaining these conditions, it is necessary to look at urban history, especially when the Median Empire, as the first unified Iranian government and the main descendant of the Kurdish people, was founded in 728 BC and lasted until 550 BC. The importance of this empire is such that Kurds align contemporary national identity and calendar with the date of 612 BC.[5]

This important point in history contrasts to the later removing of autonomous government from the Kurdish people after the Median Empire and, consequently, a physical movement to the mountain areas of the *Zâgros* in order to find refuge:

> Kurdistan, the mountainous region where most of the Kurds lived, has long been a buffer zone between the Turkish-, Arabic- and Persian-speaking

PLATE 9.2 Typical urban settlements in the Zâgros mountains: (a) Mardin in Turkey
Source: Hooshmand Alizadeh

regions of the Muslim world. Politically, Kurdistan constituted a periphery
to each of these cultural political regions.

(Bruinessen, 1999)

With the harsh conditions of the mountains and exclusion from governments in
the surrounding lands Kurdish society become isolated within the Zâgros mountains. In this context, Kurdish cities can be characterized by features of self-reliance,
being outward-looking but enjoying few external relations (see Plate 9.2a and
9.2b). The strengthening of internal solidarity in order to ease the circumstances
of isolation has thus ensured the development of a strong sense of community
which has been materialized in the discursive construction of the space in accordance with the cultural requirements of the inhabitants.

Moreover, the dense pattern of urban settlement on the slopes and tops of
inaccessible mountains has ensured a density of building. A stepped configuration
emerged so that 'one never knows whether one is standing on a floor or on a
roof, since the terrace or open space which contains the floor of one house forms
the roof of the one below', the roofs that are indeed the only open spaces of the
settlement (Hansen, 1960: 30) (see Plate 9.3a, 9.3b and 9.3c). In such a structure,
social life is usually lived in close relation to other houses and public spaces. This
ensured that the home is integrated into public life and facilitated the inviting of
neighbours to share leisure time in the open space in front of the home, on a
terrace which gives the inhabitants the ability to view the distant scenery.[6] In the

PLATE 9.2 (*cont'd*) (b) Amedy in Iraq

PLATE 9.3 The concept of neighbourhood in Kurdish society. A traditional settlement of (a) *Dowlob*, (b) *Awraman-e Takht* and (c) *Pâlangân*
Source: Hooshmand Alizadeh

(b)

(c)

PLATE 9.3 *(cont'd)*

words of Barth (1953: 105), who describes Kurdish cities as a 'rooftop society', the social gathering on terraces and roofs became the focus for socialization among the people.

While the open terrace space is nonetheless a part of the home it is also the only way for people to move around the city, and while the houses behind the open space have a boundary to define the territory of the family due to the intermingling functions, the terrace is considered an open space. Due to the physical structure of the city the openings of houses, especially those facing the terraced platform, are not fixed above eye level because in this context the passer-by is not a stranger but a neighbour who is intimate with the domain of the house and therefore is considered to be extension of the home and the family, so that 'The streets are not spaces that separate the public from the private, but outright extensions of the private: they can be occupied, appropriated and used, just like the electricity cables on lamp posts' (Gambetti, 2009).

This imparts a sense of fluidity between the inner and outer order. For example, Claudius James Rich (quoted in Barth, 1953: 103), who visited the city of Sulaymânia in Iraqi Kurdistan in 1820, 'was surprised at the regularities of the patterns of visiting, at the large groups that would congregate in the house of some or in the open, and spend their time smoking and drinking tea, while talking way into the small hours'. As he continues, this regular gathering to eat food occupies a very prominent place in communal life. In addition, the important status of hospitality among Kurdish people also lead to the institution of the guest rooms (*Mewân-khân* or *Dewâ-khân*) (Leach 1940) which traditionally played an important role in Kurdish culture (Hassanpour, 1996; Kreyenbroek, 1996). Based on the words of Bruinessen (1978: 83), the *Mewân-khâna* was a place of rest for travellers within which they were entertained, given tea and a good meal, and a bed for the night; 'here I first obtained an insight into the kindly hospitality I was to receive all through Kurdistan' (Harris, 1895: 454). The importance of *Dewân-khân* in the social life of the Kurdish people, as a symbol of hospitality, has importantly been maintained over time in the spatial structure of houses, where one room, usually on the second floor and located in the 'best' part of the house, is labelled *Dewâ-khân* and used mostly for the entertainment of guests (see Bois, 1966: 22).

Hospitality is one of the finest features in the Kurds character. It is, it is true, enjoined by their religion, and the same custom prevails amongst other Muhammadan areas. But the Kurd has carried it to a fine art (Hay, 1921: 49). Such close urban coexistence has moulded Kurdish culture to value social cohesion, hospitality and *Mewân-khâna* so that the people within each settlement act as an extended family to each other.

We return to a comparison of Islamic cities, through consideration of visual privacy in Kurdish urbanity, especially in the three cities of Sanandaj in Iran, Arbil in Iraqi Kurdistan and Diyarbakir in Turkey. Entrance doors are set opposite each other along narrow thoroughfares between homes, open to the view of passers-by, and windows overlook each other and the courtyard of other houses, which are bounded by low walls. This highlights significant differences between

(a)

(c)

(b)

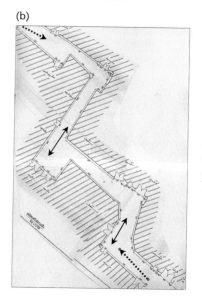

PLATE 9.4 (a), (b) and (c): Entrance doors opposite each other along the narrow thoroughfares and opposite to the view of passers-by in Sanandaj
Source: Hooshmand Alizadeh

Kurdish and other Islamic cities (see Plates 9.4a, 9.4b and 9.4c; 9.5a, 9.5b and 9.5c; and 9.6a, 9.6b and 9.6c). Moreover, entrance doors are usually left open during the day in order to offer a welcome to guests and to allow easy access for/to neighbours (see Plate 9.7a, 9.7b). Beyond the Kurdish context, the socio-cultural and physical structure of Islamic cities ensures that private domains are built in order to avoid unexpected overlooking. As such, the carpet pattern (see Plate 9.8a, 9.8b and 9.8c) of the urban form in Islamic cities (the existence of a uniform and absolute order in the layout of the city) generally allows restricted views and vistas to be maintained. This contrasts markedly with the topography of the terraced patterns which characterize Kurdish cities.

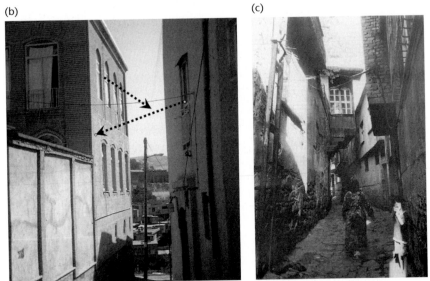

PLATE 9.5 Windows overlooking each other and the courtyard of adjacent houses: (a) and (b) the city of Sanandaj and (c) the city of Diyarbakir
Source: Hooshmand Alizadeh

PLATE 9.6 Low walls and the spatial element of *Bar-haiwân* facing the public: (a) the city of Arbil; (b) the city of Amedy;
Source: Hooshmand Alizadeh

(a)

(b)

(c)

PLATE 9.6 (*cont'd*) (c) the city of Sanandaj

(a)   (b)

PLATE 9.7 (a) and (b): Intimacy of the houses within the realm of public spaces in the city of Sanandaj
Source: Hooshmand Alizadeh

(a)

(b)

PLATE 9.8 Comparing urban forms: (a) the Kurdish city of Mardeen in south-east Turkey; (b) the Persian city of Yazd on the central Iranian plateau, with a carpet pattern; (c) the city of Sanandaj showing the terraced pattern of its built form with elaborated feature of *Bar-haiwân*

Source: Hooshmand Alizadeh

(c)

PLATE 9.8 (*cont'd*)

In a similar vein, while the basic plan of the houses in Kurdish cities – which usually has a tripartite structure, called *Se-bakhshi*, comprising a *Bar-haiwân* in front of the main living room flanked by two rooms[7] – can also be found elsewhere in Middle East such as in traditional houses in *Baghdad* (see Warren and Fethi, 1982), the point of difference relates primarily to the concept of privacy. In Baghdad, balconied windows which jut out from upper rooms have a special veil, called '*Shanashil*' – an elaborate overhanging screen, which prevents any view from outside, but which allows light into the room and inhabitants to see the outside (see Figure 9.1). Such spatial and physical examples highlight the significant differences between Kurdish and other Islamic cities (Memarian and Brown, 2003; Kheirabadi, 1991; Hakim, 1986; Warren and Fethi, 1982).

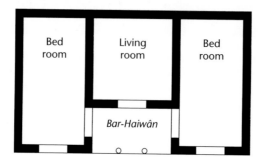

FIGURE 9.1 The basic plan of house design
Source: Alizadeh

Conclusion

While urban theory has generally disavowed the importance of studying privacy this chapter has championed the importance of the topic for understanding city life. In Western contexts where privacy has been considered in academic debates, issues such as territoriality and personal and community space have been discussed in behavioural studies in order to understand differences between public and private space. However, in Eastern cultures, and specifically those adhering to Islamic law, the concept of privacy has been considered as central to the nature of family life institutionalized by the holy *Koran* in order to ensure ideological and cultural stability (Saleh, 1998). For this reason, there has been significant amount of writing which directly emphasizes the concept of privacy in understanding urban form and everyday life. However, within this context, urban form in Kurdish cities particularly offers fertile ground for researchers who point to the specific cultural and geopolitical nature of Kurdistan urbanity. Such insights are important not only because they signpost the importance of studying Kurdish cities in order to contribute to the development of a cosmopolitan urban studies, but also highlights how a comparative approach to understanding the concept of privacy itself has much to offer the advancement of urban theory more broadly.

Notes

1 *Mahram*: People who are close, immediate family like brother, sister, father, mother, uncle, aunt, grandparents. Mahram people cannot marry each other (except husband and wife are Mahram). Thus, it is 'the legal term denoting a relationship by blood, marriage or sexual union which makes marriage between persons so related forbidden' (Khatib-Chahidi, 1981: 114). *Nâ-mahram*: Males and females who are not close family are Nâ-Mahram to each other.

2 This has mostly been observed within the houses of wealthy people or notables in the city of Sanandaj.

3 In this regard, there are many witnesses. The reader is referred to the work of Nikitine (1987: 228–9) and Bruinessen (2001: 95–107).

4 As Vasilyeva (2000: 13) indicates, Mâh Sharaf-Khânoum Kurdistani, who wrote the history of Ardalân's family – also known as the Kurdish poetess Mastura – was the only women-historiographer not only in Sennah/Sanandaj but in all the Near and Middle East till the end of the nineteenth century. She was from the city of Sanandaj belonging to the Wazieri family, the main supporters of Ardalân government (Alizadeh, 2007).

5 This is the year that Cyaxares, the grandson of Deioces (Díyako), the first king of the Medes' empire, occupied Nineveh and put an end to the brutality of the Assyrian empire in the lands under its occupation. Cyaxares, succeeded in uniting the many Median tribes into a single kingdom.

6 This concept was transferred/redefined into the structure of houses in the city in the form of a veranda in front of the main part of the house flanked by two rooms overlooking the courtyard in front and the distant landscape. This defines the common plan of the urban house type called *Sé-bakhshi* (triplex plan) in the city. In line with this interpretation of the life in Kurdish society, an Iraqi Kurd asked an Iranian folk musician Kurd, who was performing a programme in Iraqi Kurdistan for the people of Sulaymaniah (one of the great Kurdish city which can be recognized as twin of Sanandaj), if he was happy among the Kurds in Iraq. In response, he said it is a strange

question because wherever the Kurds are, there will be Bâwani Kurd. The word Bâwâni means the place of reference or safe haven. This concept is mostly applied to married women who move to a new place. For her, the house of her father is the main place of reference and refuge from any disputes. This notion is to somewhat similar to the view of Winnicott (1986, quoted in Menin, 2003: 5) who believes that 'cultural experience is located primarily, in the potential space between a child and the mother when experience has produced in the child a high degree of confidence in the mother, that she will not fail to be there if suddenly needed'.

7   An open veranda in front of the main part of the house facing the natural landscape, public domain and courtyard.

10

THE NETWORKED CITY

Popular modernizers and urban transformation in Morelia, Mexico, 1880–1955

Christina M. Jiménez

> Cultural conflict characterized nineteenth-century Latin America. On the one hand, the elites, increasingly enamoured with modernization first of an industrializing Europe and then of the United States, insisted on importing and imposing those foreign patterns on their fledgling nations. They became increasingly convinced that Europe and the United States offered solutions to the problems they perceived in their societies as well as a life-style to their liking. On the other hand, the vast majority of Latin Americans recognized the threat inherent in the wholesale importation of modernization and the capitalism accompanying it. They resisted, preferring their long-established living patterns to the more recent foreign novelties and fearful of their impact on their lives.
>
> *(Burns, 1980: 5)*

E. Bradford Burns' two-sided depiction of modernization in Latin America – elites' embrace of Western models of development countered by popular hostility to those approaches – long embodied the standard narrative of modernizing Latin America. This narrative portrays modernization as a project and process imposed by Eurocentric elites on local peoples despite their resistance (Esherick, 2000; Lees, 1985; Hahner, 1986). Of course, this narrative directly impacts our historical portrayals of cities 'beyond the West' since it positions national populations, in this case those of Latin America, as 'traditional', 'backward-looking' and highly local in their orientation. While accurately describing *certain* nineteenth- and twentieth-century contexts, when applied broadly, this narrative obscures more than it reveals about modern urban change in cities 'beyond the West'.

In Latin America from roughly 1850 to 1950, the negative consequences of many state-led modernizing projects led to widespread suspicion of such plans. In fact, part of the Mexican state's modern vision was a 'developmentalist agenda' shared by the regional and national elites, who argued for the need to control, discipline, educate, reform and 'civilize' ordinary Mexicans away from their trad- itional 'backward' customs (Knight, 1994: 396). Not surprisingly, in the Mexican countryside, peasant villages and indigenous communities often opposed large-scale modernization projects near their villages, such as the construction of intercity roads, railroads, irrigation works, dams, reservoirs and power plants. In an effort to defend their communities and their way of life, rural residents also resisted projects which they believed would compromise their local culture, political auto- nomy and access to vital resources, such as land, water or firewood.

In cities and towns, official state codes and municipal regulations sought to control people and urban spaces. Many urban dwellers resented regulations – such as those related to sanitation, public health, alcohol consumption and public behaviour – as invasions of individual choice, private space and family life (Lear, 1993, 1996; Tenorio-Trillo, 1996; Agostoni, 2003; Johns, 1997; Staples, 1994; Piccato, 1995; Beezley, 1987; Bliss, 2001; Meade, 1997; Martínez-Vergne, 1999). Historical scholarship has demonstrated how popular groups commonly resisted many state projects, including mandatory public education for children; birth and death registration; civil marriage; modern burial practices; mandated military ser- vice; a slew of urban sanitary regulations; mandatory medical inspections for the sick and for prostitutes; vaccinations; and required participation in state-sponsored public rituals such as parades, patriotic celebrations and state funerals (Coatsworth, 1981; Needell, 1987; Stern, 1999; Rockwell, 1994; Vaughan, 1982). Many of these state projects were part of a modernizing agenda. Hence, the narrative of popular rejection of modernization in Latin America is, in part, true.

Other historical perspectives shed light on another side of the story. Local studies demonstrate that popular groups were not necessarily opposed to the ideas or practices of modernity, but rather to the *method* by which the state often imposed changes, without the input of local people. In fact, counter to the standard narrative associated with Burns' two-sided description, local people often demanded that the state deliver the promise offered by the modernizing agenda, including health care, local educational opportunities, new infrastructure such as roads, and modern amenities including piped water, closed sewers and electricity. Ordinary Mexicans, in both rural and urban contexts, actively negotiated with the state in regard to the government's responsibility to fulfil their promises of modern life. 'Negotiation' is hence a key word used in the historical literature of Mexico since these interactions between the state and local people were cen- tral to the 'give and take' of political and rhetorical exchange (Joseph and Nugent, 1994; Fowler-Salamini and Vaughan, 1994). In these terms, many Mexicans actively politicked for modern projects *if* they could have a participatory role in the pro- cess of implementation. These ideas have also been discussed in the broader context of Latin America (Pineo and Baer, 1998; Mallon, 1995). In short, we must acknowledge the importance of local control and local participation, or the lack thereof, in order to understand the experiences of cities and peoples beyond the West.

Recasting the history of urbanization and modernization in Mexico as a struggle to *control* processes and resources can be enlightening. The generalization that non-Western, particularly indigenous popular groups rejected modernity ultimately romanticizes the traditional and rural orientation of Mexicans while also obscuring other dynamics that emerged around these processes. Moving away from the dichotomies so prevalent in the history of non-Western cities – 'modern-versus-traditional', 'urban-versus-rural', 'native-versus-foreign', 'popular groups-versus-the state' – creates space for other ways to understand the complex dynamics around modernizing agendas. By centring on *control*, we reframe popular efforts to resist, counter and condemn repressive state practices as a rejection of those *practices* rather than a rejection of the idea of modernization. Moreover, by highlighting popular agency in negotiating change, we can better understand the vital role ordinary Mexicans played in the transformation of civil society, the public sphere and popular politics in urban Latin America.

Beyond a simple binary, Latin American intellectual García Canclini proposes an alternative way of understanding modern change in Latin America through the notion of 'cultural hybridity'. García Canclini explains that in the Latin American context:

> The conflict between tradition and modernity does not appear as the crushing of the traditionalists by modernizers, nor as the direct and constant resistance of the popular sectors determined to make their traditions useful. The interaction is more sinuous and subtle: popular movements also are interested in modernizing and the hegemonic section in maintaining the traditional – or part of it – as a historical referent and contemporary symbolic resource ... Today, we conceive of Latin America as a more complex articulation of traditions and modernities (diverse and unequal), a heterogeneous continent consisting of countries in each of which coexist multiple logics of development.
>
> *(Canclini, 1995: 9)*

This chapter aims precisely to get at that type of hybrid modernity by considering the role of popular modernizers in carrying forward the modern transformation of urban public spaces. Historical dynamics of urban change in Latin America reinforce this reinterpretation of the popular appeal of modernity; the tendency of rural people to move from the countryside to cities partially indicates an embrace of what modernization and modern cities had to offer them.

In order to best understand the process of urban transformation in Latin America, we therefore need to contextualize the distinct historical experiences of Mexican and other Latin American cities. For example, in contrast to trends of modern urban change in the United States and Europe, throughout much of Latin America modern urban development took off much later, comparatively, in the second half of the nineteenth century (Hardoy, 1982: 28–9; Scott, 1982). From the 1850s on, many factors created the historical foundation for urbanization and modernization in Latin America, including post-colonial political consolidation of newly independent countries; the establishment of international trade and financial relations; the expansion of regional and national networks to facilitate travel, transport, and communication; the rebuilding and expansion of public infrastructure; the application of new technologies; and unprecedented population growth.

These changes fundamentally transformed Mexico, a predominantly rural country until the mid-twentieth century. Before being integrated into a system of cities, united by transport, communication and commercial networks from 1870 to 1910, Scott (1982: 14–15) describes Mexico as 'functionally divided into largely self-contained agrarian systems in which towns and cities served the limited commercial needs of their own rural areas and the needs of political and administrative control under successive forms of government'. Transport networks thus laid the foundation for urbanization. However, the application of liberal laws in the countryside from 1850 to 1910 and the export-orientation of agriculture under President Porfirio Díaz pushed rural people to cities, and *haciendas* (large landed estates characterized by mixed production); 'In the country as a whole 85 percent of communal villages and 90 percent of rural families were landless, and fully 50 percent of the rural population was tied to the hacienda system' (Scott, 1982: 33). The trend continued from 1910 to 1920, when the instability and violence of the Mexican Revolution further displaced people from the countryside (Knight, 1990; Hart, 1987; Carr, 1976, Camín, 1975; Wells, 1985; Gilly, 1971; Joseph, 1982). While certainly powerful forces were at play, the push and pull of the economy, liberal laws and revolution only partially explain rural to urban migration.

In Mexico, the city represented the promise of modernity and the opportunity for a different life. Many people were pulled to the city by the availability of modern infrastructural amenities, the increased access to public services, the ideas and rhetoric of liberalism, the desire for Western-oriented consumer goods, and the increased affordability and accessibility of popular entertainment. Opportunities in the city were created in part by lopsided government spending in urban areas. Through the 1920 and 1930s, post-revolutionary Mexican governments disproportionately invested in urban services and industries, drawing more Mexicans to towns and cities for jobs, modern amenities and public services (Quijano, 1975; Scott, 1982; Pineo and Baer, 1998). Rural to urban migration intensified with the initial economic surge associated with WWII, import substitution industrialization, and the phenomenal economic and industrial growth from the 1940s through 1970s known as the 'Mexican Miracle'. Expansion of government services, state bureaucracies, industry and the funnelling of public revenue to urban services all intensified the movement of people from rural to urban areas. By 2000, 75 per cent of the Mexican population lived in cities and towns (Goodwin, 2007: 7).

Notably, from the late nineteenth century onwards, the migration of people to cities in Mexico and other regions of Latin America quickly surpassed the city's ability to accommodate newcomers. Despite the development of Mexican cities and public urban investment between 1870 and 1940, for example, urban population growth significantly outpaced the capacities of the Mexican economy to provide jobs for people, partly due to capital-intensive nature of industrial growth in a capital-scarce economic situation as well as the unmatched growth in agriculture (Scott, 1982: 16). As a consequence, one of the defining features of urbanization in Mexico and other non-Western cities was the persistent inability of state governments to provide sufficient housing, employment and public services to urban residents. Since the late nineteenth century, these very common societal circumstances (unemployment, inadequate housing and the lack of public services) which have become emblematic in many industrializing non-Western cities generated common popular responses from urbanites (Violich, 1987: 5–37; Hardoy, 1975; Gilbert et al., 1982; Porter, 2003).

Notions of 'formal' and 'informal' have come to embody one of the central dynamics of Mexican (and Latin American) urban change – the inability of the formal economy or the state to accommodate the needs of the urban population, hence people's practice of taking matters into their own hands in order to forge their livelihood and other needs 'informally'. The lack of formal employment options and the persistence of underemployment and poverty generated a long history of informal, non-regulated and 'illegal' industries, from micro-enterprises, like food sellers in the street to clandestine production, as a normal part of the urban economy. Likewise, in response to the lack of affordable housing options, urban residents established makeshift settlements, squatter camps and urban slums as the population outpaced available and affordable housing (Auyero, 2000; Schneider, 1995; Illades and Rodríguez Kuri, 1996; Roniger and Waisman, 2000). Urban dwellers responded to the state's inability to provide infrastructural amenities (such as sewage systems, paved roads and clean piped water) and public services (like health care and schools) by proactively creating their own public services. Being beyond formal control of the state, resident-driven initiatives are often labelled as informal and illegal, casting a shadow over the local projects (Hardoy and Satterthwaite, 1989; Holston, 1999; Perlman, 1976; de Soto, 2002).

Ironically, resident-led initiatives and people's uses of city spaces are used to support negative representations of the failures of government in non-Western cities. Certainly, it is true that Latin American governments – ranging from national to municipal – are woefully guilty of failing to provide services and infrastructure to most of the population. However, emphasis is often placed on the failure of the state to order, control, fund and regulate these processes, not on their failure to provide them. In short, modernizing, non-Western cities are often presented as 'chaotic', in part, due to the non-standardized, informal nature of the building of public infrastructure in neighbourhoods and due to public access to urban spaces. Seen from another perspective, these trends also embody forms of popular *control* over the modernization of the city through neighbourhood-centred, resident-led efforts to improve and extend public services in outlying *barrios* (neighbourhoods) and popular *control* over urban spaces through the informal street vending economy. As discussed previously, the notion of control over processes and spaces is vital to understanding fully these dynamics.

Informal networking remains a vital form of survival, identity, culture and notions of community in contemporary Latin American cities; one study of a squatter settlement in Buenos Aires, Argentina, highlights how 'ongoing informal problem-solving networks [are] meant to ensure material survival and shared cultural representations' (Auyero, 2000: 14). Auyero continues, 'as residents improved their shantytown homes, 'the inhabitants constructed a series of neighbourhood institutions and a set of cultural understandings about life . . .' (ibid.: 14–15). Social theorists of the city, namely Castells, Gottdiener and Lefebvre, all emphasized the central role of the 'production of space' as a method of dominance. Conversely, they also suggested the importance of worker control over spaces in the city as pivotal to the articulation of a popular politics which could then serve to counter the continual pursuit of dominance of those spaces by capital and the state (Orum and Chen, 2003: 35–9). By stating their rights to claim, improve and expand shared urban spaces, like neighbourhood streets, plazas and parks, ordinary men and women produce their own vision of those shared urban community spaces.

The survival strategies of city dwellers and particularly the urban poor have been at the centre of research focused on developing, non-Western cities from the mid to late twentieth century to the present. By contrast, studies telling the history of popular strategies and informal networks before 1930 are few. Here I seek to address this imbalance and present a snapshot of such histories, centring on the dynamics of popular modernizing through informal networks and group associations in provincial Mexico from the 1880s on. In the remainder of this chapter, I tell the story of how residents of Morelia, Michoacán, actively pursued the promises of modern infrastructure and public spaces, namely plazas. The choice of Morelia, Michoacán, as the focus of the study is an attempt to move the discussion of these dynamics beyond national capitals and megacities in Latin America (see Bell and Jayne, 2006, 2009). As a state capital and provincial city with a long colonial past, Morelia (founded in 1531) represents an ideal city to illustrate how the story of state formation and modernization were negotiated and 'hybridized' – to draw on García Canclini (1995) – in the sub-national context, precisely where the nation was instantiated. Morelia experienced many of the internal dynamics impacting on other Mexican cities, such as the consequences of liberal laws and rural-to-urban migration (Rivera Reynaldos, 1996). Still, it was a small city. Through the mid to late nineteenth century, the urban population of the municipality grew from 23,835 in 1882 to nearly 60,000 in 1930, almost tripling in size over 50 years. Similarly, the population of the District of Morelia, a larger territorial unit, increased from 111,000 in 1882 to around 200,000 inhabitants by 1930, an average expansion for a Mexican provincial capital and its hinterland during this period (Sánchez Díaz, 1991; Meyer and Beezley, 2000: 425). Before and after the Revolution, Morelia served as an administrative, service and distribution hub for the diverse regional economy (which produced corn, chillies, beans, wheat, sugar, cotton, *aguardiente*, textiles and livestock). In fact, as a regional capital, Morelia was representative of the Mexican hinterland. Moreover, like other provincial capital cities, its position as a regional hub dramatically reinforced its identity as a modern urban centre in contrast to the surrounding 'backward' countryside.

In Mexico, similar to many regions of Latin America, the adoption of modern amenities from 1880 through the 1950s transformed urban lifestyles and public life. The building of modern amenities is commonly presented as the result of state efforts. However, popular groups also internally shaped the processes of state formation and modernization. By taking seriously popular desires for the benefits of modernity, we can rethink some of our assumptions about the modernization of Mexican cities, and other cities beyond the West. Urban residents of Morelia not only embraced but more often than not initiated efforts to modernize, commercialize and beautify the city through projects, such as the construction of sewers, public water fountains, drainage pipes, paved roads, street-lighting, urban renovation projects for plazas and gardens and efforts to expand advertising spaces (Jiménez, 2001, 2004). Private investment and popular initiatives in Morelia thus literally modernized, commercialized and opened to the 'public' significant portions of city space and urban infrastructure in the late nineteenth and early twentieth centuries.

Neighbourhood networking for modern amenities and urban public space

In 1879, 49 *vecinos* (neighbours) from two *barrios* of the outer edges of the city wrote to the city council explaining that although they had a fountain, several months had passed since it had been supplied with water. They explained in their letter:

> when the City Council arranged to build the fountain, named La Fuente de La Mulata Cordova, we all contributed to the project of our own free will [*franca voluntad*] with the amounts that our poverty permitted in order to carry out the completion of the fountain, knowing the fountain would be of such benefit as a necessity for the domestic relief of our families and the public in general.
>
> *(1879 AHMM c. 252, e. 121)*[1]

The problem was that three months after enjoying water in the fountain, the water service stopped because the faucet valve broke. The neighbours explained that they themselves collected the money from *vecinos* to fix the valve and have the water turned back on. However, after making the repair, there was only water service for three days. They explained that since that time three months had passed during which 'the rainy season helped us a lot and mitigated our thirst'. They were therefore asking that the city council restore water service to their fountain. Although there are not thorough minutes of this request, it appears that another fountain elsewhere in the city was granted the water originally supplying this fountain.

One year later in October of 1880, the neighbours of these same *barrios* wrote another letter, but this time to the Governor of Michoacán. They explained to the governor their initial plea to the city council, retelling the story of their generous contribution of a little less than 300 pesos to the construction of the fountain at the invitation of the *ayuntamiento* (city council). This original contribution was made, 'in the interest of immediately having the water which would provide domestic relief to their families'. They condemned the indifference displayed by the city council, writing: 'our just petition, therefore requires us, vecinos, to bother the higher attention of this Superior authority'. They proceed, requesting that if water was not going to be restored to the fountain then: 'please order it to be removed since in its state of idleness, it does not serve for any purpose other than covering the wickedness of crimes of incontinence committed in it'. In addition, they requested that the 'Municipal Body return to us the amount of money we contributed, being justified by the fact that it was for this improvement, from which we have received no benefit' (1880, AHMM, c. 252, e. 121).

This petition clearly illustrates how neighbourhood residents invested in local infrastructure and expected the municipal government to do its part to facilitate the success of the improvements. The letters also represented a condemnation of the lack of accountability and responsiveness of the city council during this period. In the second letter, the neighbours posed a challenge to the governor demanding accountability from that 'higher superiority'. Before concluding the letter, the neighbours explained that: 'the man in charge of the water piping declared in clear and precise words that the Fountain of the Mulata de Cordova does not have water because the vecinos of it do not give him money for brandy [*aguardiente*]'. They continued; 'we will do this if necessary, but we consider it a mere caprice that he does not give us the water we are asking for'. The residents were clearly bearing witness to the arbitrariness and personal abuses which plagued the city and which, presumably, the authorities were responsible for rectifying. Unfortunately, there is no documented response to this letter.

As illustrated by these petitions from the inhabitants around the Fountain of the Mulata Cordova, city residents responsibly responded to government requests to perform certain duties – such as contributing to the construction of the new fountain – as a way of justifying their own requests to the city council. Similarly, residents implicitly argued for new fountains and better water service by explicitly describing the benefits to public health and claiming their need for clean water in order to fulfil their daily duty of washing down their door stoops each morning, as specified by the Edict of Police. Countless examples of this type of rationale formed the basis of the political culture and moral economy of the city.

From the 1870s on, newcomers, migrants and long-established residents organized in neighbourhood-based networks to actively pursue common interests in their streets, often demanding improvements from the state. In the late nineteenth century, as the city's population increased, migrants from the countryside either squeezed into existing houses and tenements or set up shacks on the margins of the city. As urban real estate values – and presumably rents – in the city's downtown rose beyond the means of many recent migrants to the city around 1900, housing settlements along the periphery of the city proliferated (Uribe Salas, 1993: 8). Initially, these new residential settlements lacked paved roads, plazas and public services. After 1900, however, new housing colonization (sub-urbanization) projects gradually incorporated and transformed these makeshift peripheral settlements into established neighbourhoods of the city (Lear, 1993; 1996). Inhabitants of the expanding suburbs and makeshift settlements on the outskirts of the city pulled the modernizing and beautification projects of the centre into the city's periphery by pursuing their own agendas associated with urban modernity and notions of progress. By de-centring the typical focus on elite intentions and centrally-located spaces and by incorporating the experiences of residents in peripheral neighbourhoods, we get a very different vision of modernization in the city.

Residents thus came together, at first informally, through new forms and spaces of sociability in the city. Hilda Sabato suggests that in the spaces of a city people engaged in new 'modes of sociability', which is also a way of saying that people expanded their social networks (Sabato, 2001). Old and new forms of sociability, including mutual aid societies, voluntary associations, patriotic juntas, religious groups, peasant communities or political parties, created a foundation for these expanded urban networks. Beyond formal groups, urban residents living in close proximity to each other along a street or plaza interacted informally in these shared spaces. Statistical and descriptive evidence illustrates that neighbourhoods, markets, parks, plazas, *cantinas* and other public spaces brimmed with new and old residents. During their daily interactions people exchanged ideas and gossip, shared common complaints and stories of their experiences, and circulated information about life, work and politics in Morelia. People participated in urban social networks through their simple presence/residence in the city, not via claims of citizenship or social status. Popular collective engagement expanded notions of the public; through everyday forms of politicking and social networking, Morelia's working- and middle-class residents leveraged their collective identities in order to demand and often successfully gain public concessions and protections from municipal and state governments.

Social networks thus mobilized and strengthened urban identities, both of which were produced in relation to city space. A neighbourhood plaza, for example, was typically one of the first requests of outlying *barrios*. Plazas provided residents with a basic, yet essential, public gathering space off the muddy and dusty roads, a space where people could visit, sell, buy and recreate. The centrality of the plaza as a locus of community life, political expression and neighbourhood identity has been well established (Low, 2000). Urban plazas throughout Latin America also have been, and still are, viewed by urban lower classes as an important place of employment, particularly for sellers in the informal economy and workers in the service industries, including porters, shoe shiners and water carriers. According to the 1857 Constitution, men of proper age (18 years if married and 21 if single) who were Mexican nationals and who sought the status of 'citizen' (an exclusively-defined category which granted the right to vote and the right to engage in politics) had to prove that they had an 'honest occupation'. Significantly, the 1917 Constitution perpetuated this legal requirement of citizenship in Article 34. In Morelia, petitioners commonly invoked, implicitly and explicitly, their desire to sustain their families through the 'honest work' of public selling (Mexican Government, 1857). They also argued effectively for their dependence on the right to sell in public spaces in order to support their families, especially as the economic situation of the city worsened and viable alternatives dwindled. Residents may not have consciously realized that obtaining 'honest work' was part of the legal definition of citizenship, but they certainly were conscious of the fact that aspiring to this ideal was viewed positively by the city council. As with the discourse around public health and public beauty, honest work and sustaining one's family became rhetorical staples of Morelia's political culture.

Since plazas were foundational spaces for political, social and economic production for urban popular groups, outlaying neighbourhoods persistently organized to establish and to improve their local plazas and surrounding streets. The persistence of this request is illustrated briefly by two petitions sent to the city council nearly 45 years apart. In 1884, 11 residents of the San José parish community lobbied the municipal authorities to allow them to create a market so they could buy and sell in their neighbourhood plaza on the periphery instead of travelling to the more distant central markets. In their petition to Morelia's city council, these residents explained, 'With the desire to continue our struggle for acquiring the means of subsistence for the survival of our families, we present our case for establishing a market in the Plaza of San José to the attention of the City Council' (1884 AHMM c. 253, e. 232). They promised to provide cloth drapes to give the stalls a more pleasant appearance, but could not promise more since 'that is all our limited means will allow us' (ibid.). The city council agreed to their request. In 1928, over four decades later – and across the seemingly monumental chasm of the Mexican Revolution – a very similar petition reached the desk of Morelia's city council secretary. Echoing the decades-old concerns of city dwellers, these neighbours of Carmen Plaza requested permission to set up a market. They argued that a makeshift market would not only provide for their personal needs, but that it would also contribute to the greater good of society: 'It is unquestionable that for ourselves, for the surrounding neighbourhoods, for the traffic, and for the appropriate appearance of the city, our request [for the market] will be advantageous' (1928 AHMM c. 334, e. 56). In both 1884 and 1928, the city council of Morelia agreed to the petitioners' requests; it also stipulated, however, that residents were responsible for all costs related to the development of the market. Undaunted, neighbourhood residents established their markets, using their own labour and financial resources. These two examples, among hundreds of similar letters and petitions, typify the negotiations over the expansion of the city that occurred between municipal authorities and city residents during these decades. In these communications, Morelia's residents not only articulated their own visions of urban development and employed the rhetoric of modernization and development associated with the state and elites to achieve their own goals, but they also cultivated new (literal and figurative) spaces of political power from 1880 at least through the 1950s, and certainly beyond.

In contrast to small towns or Indian villages, the spatial context of large and small Mexican cities, like Morelia and other capital cities, enabled popular groups more opportunities to insert their claims into public dialogue because of the many shared spaces, like plazas, parks, fountains, paved streets, covered archways and sidewalks – all spaces which became emblematic of both civic and national identities. Notably, from 1880 to 1910, the rhetoric of liberalism, desire for Western-oriented consumer cultures, and popular cultures in Mexico under Porfirio Díaz allowed popular groups to forge urban public spaces for themselves. After 1910, the rhetoric and promises of the revolution blended with prior claims (rooted both in liberalism and other historical and customary group rights) to become a malleable and sophisticated language of contention for popular groups in Mexico.

Beyond providing a space of their own, neighbourhood plazas further united residents based on their selling, sharing, gathering, circulating and living in and around this common public space. The consolidation of neighbourhood identity occurred not only internally as residents came together in their shared identification with certain local spaces, but also through external recognition of the neighbourhood as a cohesive physical and communal space. While older neighbourhoods in the colonial footprint of the city were identified in relation to their parish or adjacent church, new neighbourhoods emerging in the peripheral areas of the city were often identified with their local neighbourhood plaza. Once established, an emerging neighbourhood could utilize its external recognition as a collective entity to petition authorities for other public services such as a public fountain or pavement along the main road connecting the city centre to the *barrio* plaza.

After 1900, many older neighbourhoods of Morelia were newly consolidated as official *colonias* (suburbs), often taking the names of nationalist or revolutionary heroes, (such as Colonia Cuauhtemoc, Colonia Vasco de Quiroga, or Colonia Issac Arriaga). Take the example of the old barrios of San Juan and Guadalupe, portions of which were incorporated into the newly formed Colonia Vasco de Quiroga in 1903. Officially founded by Rafael Elizamarrás, a pharmacist and land owner, who entered into a contract with the city council to subdivide and sell his ranch located next to the neighbourhoods of San Juan and Guadalupe (1903 AHMM c. 13, e. 1), these *barrios*, in and of themselves, had a long working-class history. During the first decades of the twentieth century, group and individual petitions from the Colonia Vasco de Quiroga commonly crossed the desk of the municipal secretary. Initially, many requests were from new plot owners for construction permits to build their houses, to install piping or sewers, to add or lower existing windows, and to add balconies (1908 AHMM c. 17, l. 2, e. 37, 39, 43, 52, 57, 65, 67, 68, 69 and 70). Requests for building permits often included a detailed design and description of the proposed construction or alterations (1909 AHMM c. 17, l. 2, e. 11). Residents' investment in the built infrastructure of this newly formed and expanded suburb, especially houses, seem to lay the foundation for a strong neighbourhood identity.

After 1916, however, collective petitions from the Colonia were received regularly by the city council. That year, for instance, 29 residents requested that the city council open three streets in their neighbourhood. They stated that 'even though we are mostly poor people, we are inspired by the progress of this neighbourhood'; thus they asked that these streets be created 'for the public good' (1916 AHMM c. 32, l. 2, e. 66). Again in 1919, 31 *vecinos* wrote to the municipal president requesting that they convene a municipal commission to investigate the 'awful state' of their public fountain. Notably, in this period following the proclamation of the 1917 Mexican Constitution, residents employed the rhetoric of citizenship and justice, in addition to the common complaints about public health threats. They wrote:

> since we are zealous for the fulfilment of our duties as citizens and desire to avoid the spread of epidemics which pose grave harm to all the social classes of this city, we do not doubt that the city council will promptly remedy this extremely dangerous situation. For such actions, [the city council] will receive special gratitude for giving us justice.
>
> *(1919 AHMM l. 110, e. 47)*

They proceeded to list all of the nasty 'indecencies' (*porquerias*) committed by the nearby troops:

> the soldiers bring their horses to the fountain to allow them to drink and to bathe them, converting the fountain into a dirty sewer drain [*atargea*] and a wash basin since the women of the same soldiers come to wash their filthy clothes along the edges of the fountain.
>
> *(ibid.)*

Working-class residents expressed concern for protecting public beauty, urban sanitation and order in their neighbourhoods, thus using values directly associated with the elite agenda to meet their local needs.

The consolidation of neighbourhoods, like the Colonia Vasco de Quiroga or the Colonia Socialista, and their incorporation into the official built environment of the city created not only new spaces for working-class and aspiring middle-class residents, but also new opportunities for political organization. In the wake of the Mexican Revolution, established neighbourhoods, like the Colonia Vasco de Quiroga, began to assert the collective rights to defend their neighbourhood spaces in new forms, namely through the formalized *junta de vecindad*, or neighbourhood associations.

From informal network to formal organization: the emergence of neighbourhood associations

The role of grassroots activism in the renovation of established neighbourhoods and in the development of the urban periphery has long been a focus of studies of post-1920 cities in Latin America. Scholars commonly treat urban neighbourhoods, *colonias populares* (working-class suburbs), and squatter settlements of the mid to late twentieth century as sources of popular activism, manifested through neighbourhood associations, from the 1920s to the 1980s (Cornelius, 1975; Jiménez, 1988; Eckstein, 1989; Foweraker and Craig, 1990; Escobar and Alvarez, 1992; Slater, 1985; Maffitt, 2003). This trend, however, clearly began before 1900. For instance, neighbourhood-based identities served as sources of empowerment and collective identity as early as the 1870s in Morelia, and likely even earlier (Anderson, 1976). In the immediate post-revolutionary period from the late 1910s through the 1920s, neighbourhoods formalized their collective claims to local spaces by forming official neighbourhood associations (Jiménez, 2004, 2007).

During these decades, informal neighbourhood-based networks transformed into permanent neighbourhood associations, complete with official titles and letterhead. Popular, informal collective action thus established the historical precedent for the subsequent corporatist structure of the post-revolutionary state, a trend evident in Morelia and likely in other Mexican cities as well. Neighbourhood associations also spearheaded requests for the construction of local markets, street repair, and plaza improvements in the 1920s and beyond (1928 AHMM c. 334, e. 56). In the 1920s, the gradual institutionalization of grassroots social networks into formal neighbourhood associations was part of the broader consolidation of the post-revolutionary state through its corporatist structure and expanding bureaucracy. Although much more needs to be learned about these early associations, their appearance during the late 1920s coincides with other social and political developments of that decade, including the polarization of classes, suburbanization and the consolidation of the Mexican state-federal power structure.

The first formalized neighbourhood associations in Morelia which wrote to the city council in 1927 and 1928 were not composed of members of the upper classes dwelling in the city centre, but rather originated in areas of the city where there was a mixture of people from different social groups (1927 AHMM c. 329, l. 1; 1928 AHMM c. 333, e. 2; 1928 AHMM c. 334, e. 56). These *vecinos* used neighbourhood associations to protect their immediate streets and community as well as to promote public works in nearby blocks. Through their associations, residents requested that the city council address several issues, ranging from the need for augmented water service and more street lamps, to requests to remove the animals and vendors from their parks and plazas, to a request for a new school and sports field (1926 AHMM c. 317, e. 6; 1927 AHMM c. 329, l. 1,

e. 10; 1930 AHMM c. 104, e. 4; 1931 AHMM, c. 117, e. 10; 1932 AHMM c. 125, e. 3). Conveying their sense of ownership over these neighbourhood places/spaces (which they had a direct hand in developing), collective petitioning enabled residents officially to lay claim to their rights as *vecinos* of Morelia.

One of the clearest examples of this sense of ownership over neighbourhood spaces was a 1927 petition from the Colonia Socialista to the city council. They explained how the *barrio* was 'victim of three things'. They utilized the language of victimhood to invoke the city council's duty and responsibility to protect them. First, they protested the fact that the nearby faucet where their animals regularly watered was 'in bad condition' and often too dirty for even the animals to drink. Second, that the small dam of 'corrupt water' which one neighbour used to water his gardens was leaking into the others householders' water supply. Lastly, they reported that 'some individuals have dedicated themselves to fabricating adobe bricks in a place designated for a market or garden'. The *vecinos* complained that this work left large potholes and dirty water which was very 'anti-hygienic for public health'. Therefore they asked that the city council order this brick manufacturing to stop (1927 AHMM c. 329, l. 1, e. 10). Even in the late 1920s, householders in outlying *barrios* continued to rely upon the city council to 'protect' them and the community's health from potentially threatening situations caused by poor infrastructure, unregulated industry and unsanitary private practices of certain *vecinos*. In response, the city council sent each of these complaints to a different state or municipal body, illustrating the regularization of state and local politics and the expansion of the bureaucracy. The commission on public health was to take care of the unsanitary dam, while the *encargado de orden* (deputy) of the *colonia* was to stop the fabrication of the abode bricks and fine the guilty parties 22 pesos each. Lastly, the overseer of public works was ordered to spend 4 pesos and 10 centavos to fix the pipe and faucet of the fountain. Through their particular associations, other neighbourhoods also organized to promote and protect their local interests (1928 AHMM c. 333, e. 2; 1928 AHMM c. 334, e. 56).

The Colonia Vasco de Quiroga embodied perhaps the clearest example of the formalization of neighbourhood activism and identity into associations. As discussed above, this *colonia*, established in 1903, had a long history of community activism and it remained active well into the 1950s. By the late 1940s, their neighbourhood association was further institutionalized and renamed 'La Unión de Colonos of the Colonia de Vasco de Quiroga'. Subsequently, the association's petitions were marked with the official Unión de Colonos stamp and printed on official Unión letterhead. This neighbourhood association even had it own Pro-Material Improvements Committee (*Comité Pro-Mejoras Materiales*). This official association, like others, became the official representative of neighbourhood needs as well as the primary conduit for resident–state dialogue. For these reasons, the municipal government maintained file records on its correspondence with associations. Vasco de Quiroga's Unión de Colonos had written a series of letters and petitions to the city council about a variety of issues including a collective complaint about the lack of water in the barrio (1947), a denouncement of the corruption of the

Encargado de Orden (1947), a request to improve street drainage (1948), and a request for permission to hold festivities in the barrio in celebration of Mother's Day (1949) (1947 AHMM c. 13, e. 1). The city council responded positively to several of these requests and others were passed on to other bureaucratic authorities. In many ways, the collective neighbourhood association thus came to represent both a platform to assert popular demands of government as well as a client in the corporatist structure of the post-revolutionary Mexican state.

As late as 1955, 36 members of the neighbourhood association of the Colonia Vasco de Quiroga, including both men and women, sent a letter addressed to the municipal president and carbon copied it to the Governor of the State of Michoacán and to the Manager of Coordinated Services of Health and Welfare (*Jefe de los Servicios Coordinados de Salubridad y Asistencia Pública*) (1955 AHMM c. 13, e. 1). They explained that it was 'materially impossible to travel along Zaragoza Street due to the immense quantity of mud that has accumulated, reaching the level of the sidewalk' and that the condition of the street was made even worse still by the passing trucks which created deep ruts in the mud. As a consequence, they lamented, 'Our families are not able to go out to do shopping for indispensable items. . . . On sunny days, [they continued] the stench of the rotten sludge, in the state of putrefaction, lingers into our homes'. The residents moreover apologized: 'we would be guilty of laziness, if we remained silent without appealing to your Honour who is known to be of the progressive spirit' and asked that the sludge truck be sent to clean up and level out the street. The letter included a mild warning and implicit nudge for the council to act to rectify this problem, suggesting that 'It would not be hard for an epidemic of a dangerous character suddenly to arrive, which would make us seriously lament this situation later' (1955 AHMM c. 13, e. 1), also requesting that the city council send an inspector to testify that the street was in grave need of paving. There is no documented response to this letter and it's unclear whether the city council responded to this clean up and paving request.

Conclusion

In this chapter, I have shown how residents of Morelia, Mexico came together around shared urban spaces (namely, plazas) to articulate a claim over those spaces and to assert their collective neighbourhood identities. Plazas served as figurative and literal negotiating spaces to demand fulfilment of the promises of modernity from the state, namely the provision of modern amenities, like piped water to outlaying suburbs. While research on the movements of the urban poor and organized urban groups in Mexico and other regions of Latin America has garnered significant attention, these mobilizations are often presented as the outcome of the political upheavals around the Mexican Revolution and Great Depression. Here I have presented an earlier history of these kinds of urban movements rooted in the politics of modernity and liberal discourses of late nineteenth-century Mexico. In doing so I have illustrated that in Morelia urban residents have been organizing – informally and formally – around collective urban-based identities since at least the late nineteenth century (Cornelius, 1975).

More broadly, I suggest that the history of modernization in Mexico, and perhaps other non-Western cities, tends to be embedded in a historical narrative which reinforces basic binaries, like traditional-versus-modern, native-versus-foreign, backward-versus-civilized. Moreover, the adoption of models rooted in the history of Western cities obscures as much as they reveal in the non-Western context. For instance, the state-dominated process of modernization is rooted in the model of Western cities – where the state had the financial capacity and bureaucratic structures to control and guide the expansion of state infrastructure. Rather, in this chapter, I have argued that the success of popular movements in securing long-term access to public spaces – namely streets, sidewalks and plazas – is a distinctly non-Western phenomenon. Historically, popular access to public spaces for selling, settling and improving was a significant concession given to the urban working-classes by savvy-minded political actors to appease both the need for jobs and housing beyond what the state could provide. In many US and European cities, the urban public space was closely regulated historically earlier due to the rise and expansion of the middle classes. Central areas of the city, in particular, became exclusionary spaces for consumer and leisure activities of the middle- and upper-classes. In contrast, in much of Latin America, the urban informal economy caters not only to the employment needs of working-class urbanities, but also to the consumer demand for inexpensive, non regulated, and non-taxed consumer goods. Similarly, I explain how resident-led improvements to the city public infrastructure are commonly presented as a failure of the modern state – and hence seen in a negative light rather than acknowledged as an example of popular initiative and agency in the often state-dominated process of modernization. In short, from the 1880s on, popular groups have actively participated in the production of their immediate neighbourhood spaces. In the process, they have created a place in urban politics and asserted their claim to the city.

Note

1   AHMM stands for the Archivo Histórico del Municipio de Morelia. The year indicates the year on the box/bundle/book/folder/document. Abbreviations used in text: c. = box [*caja*]; l. = book or bundle [*libro* or *legajo*]; e. = file [*expediente*].

PART III
Mobilities

11

DISTINCTLY DELHI

Affect and exclusion in a crowded city

Melissa Butcher

Introduction

Delhi is a physical city. It's push and shove; sliding through crowds. It's grit and sweat and light in the eyes reflected off new tower blocks and malls. It's fending off stares and squeezing between the interstitial spaces, the unexpected corridors of access that momentarily at least allow breathing space. It's red streaks of *paan*. It's the sound of *puja*, *namaaz* and the chanting of secular traffic. It's the scent of temple incense and hot chocolate at Café Coffee Day that recalls boyfriends past. It's the architecture that engenders nostalgia for other homes and other times. It's the smell of *nullahs* and meat that mark boundaries of purity and disgust. It's roundabouts and the judgements of others read off the surface of the body.

These responses, from interviews and journals kept by a group of young people from diverse backgrounds living and moving through the transforming city of Delhi, are part of a growing body of research describing the embodied experience of cities (see Rose *et al.*, 2010; Jiron, 2008; Degen, 2008; Watson, 2006; Wise, 2005). Despite the slippery task of finding consensus on terminology and definition, and the philosophical density of debates (see Pile, 2010; Anderson, 2009; Amin, 2008; Gunew, 2007; Terada, 2001), this work is further elaborating an urban geography of affect, as residents of cities embody their own maps that mark out routes of familiarity and avoidance, spaces of comfort and exclusion.

However, to date little of this work has paid attention to cities in the global south. The following examination of affect and distinction is therefore placed in Delhi as part of the project to 'post-colonialise knowledge production' (Robinson, forthcoming), and to think through cities as they are constructed and reconstructed in specific place-based processes and experiences. As Xiangming Chen (2009: 367) argues, global city literature has created an artificial dichotomy between global north and south, privileging economic and structural analysis over agency, missing everyday urban experiences as a result. Examining the plurality of cities and the interactions that give shape to urban life allows an exploration of not only the multiple, affective capacities generated in different city-scapes, but also the underlying frameworks of accumulated cultural knowledge that enable their circulation.

This focus on a cultural framework aims to reclaim some room for the role of subjectivity in the theorising of affect; putting flesh on arguments that affective responses stem from 'pre-cognitive templates' or 'tacit, neurological and sensory knowing' (Amin, 2008: 11). Rather than affective experience occurring 'beyond, around and alongside the formation of subjectivity' (Anderson, 2009: 77), I suggest that it is at the very heart. In particular, this chapter will focus on the role of affect in demarcating difference in a rapidly transforming, 'globalising' city where previously held cultural knowledge is challenged by the creation of new spaces and the circulation of new demands. Communal, class and gendered boundaries, for example, are maintained and reasserted in affective responses to others that delineate distinctions often based on judgements of civil and uncivil behaviour, that is, 'appropriate' comportment, movement, noise, smells and contact that generate comfort and predictability. These responses assert what and who is, and is not, permissible within Delhi's public spaces.

Affect and the city

> When I first came to Delhi it irritated me. The whole place seemed uncivilised. Men are sitting in seats reserved for women, someone is blowing horns at red lights, people are playing songs loudly on their mobiles and if they are not doing any of these then they are staring at you constantly. These things made me feel violated. I liked to point these out to them. It almost became a habit. Now, after three years it all seems normal. But it still irritates me. Still.

As Rabia flew into Delhi from her home in north-east India to begin her postgraduate studies, there was a palpable sense of excitement in her descriptions of the city lights below her. But the enchantment soon wore off as this later description above highlights. The city, its noise, its dirt, its incivility, 'violates' her body, chaffing at prior modes of thought and behaviour she has known and felt as correct, shaping them to the point where present circumstances become 'normal'. Rabia was one of 23 young people aged from 15 to 23 years old, from diverse socio-economic and cultural backgrounds, who took part in this study, using diaries, photographs and maps to document their movement through Delhi.[1] Many of their everyday experiences of the city's noise, crowds and perceptions of disorder, things being out of place – including themselves at times – were described in terms of embodied responses of both pleasure and discomfort, disgust and avoidance, association and adjustment.

The centrality of the corporeal experience in the formation and understanding of urban space is highlighted by Rose *et al.* (2010) and Latham and McCormack

(2004), who suggest that the city assembles both buildings and bodies as much as it is built by the latter. The body carries the capacity to affect, and be affected by, the city (Pile, 2010), entangled in relationships with both the human and non-human (see Conradson and McKay, 2007; Conradson and Latham, 2007; Ahmed, 2004). According to Amin and Thrift (2002: 28), 'the bulk of (the city's) activity' is constituted by the senses and habitual 'reflexes and automatisms', generated in everyday encounters and practices. Latham and McCormack (2004: 706) concur that it is necessary 'to take seriously the fact that much of what happens in the world happens before this happening is registered consciously in cognitive thinking'. Amin (2008: 5) goes further, arguing that the relationship between people and the built environment is 'productive of a material culture that forms a kind of pre-cognitive template for civic and political behaviour'. What he refers to as 'ethical practices' in public space are 'guided by routines of neurological response and material practice, rather than by acts of human will' (ibid.: 11).

However, I would like to argue that these responses are predicated on accumulated cultural knowledge that in turn informs subjective interpretation and judgement that has necessary, even if reflexive, involvement in material practice and urban encounter. The affective expression of disgust, for example, requires a presumption of what is not disgusting; an understanding of the other as 'different' or 'same' must be present. Rather than preceding subjectivity then, its shared meanings form the basis of the everyday practices and value recitation that inflect the use of space. This cultural frame is the 'mechanism', as Amin (2008: 9) refers to it, that guides our interactions with others and with the built environment, and on which affect, with its 'involuntary', habitual responses, is circulated as collective practice. The 'template' for affective responses to the city's entanglement of bodies, infrastructure, and different 'others', is cultural knowledge that is inherently relational rather than neurological. As Watson (2006: 157) has argued, urban public space is a repository of 'collective memory, cultural integration and environments for learning social relations'.

To contextualize these arguments, both extending our understanding of affect, and analysing the disjunctions and continuities in urban theory when applied to the global south, the following sections will focus on the city of Delhi through affective markers of 'civility', and the contemporary location of communal, socio-economic and gendered subjectivity in this transforming city-space. Delhi is being 'gentrified' through state intervention, and the ubiquitous public/private partner-ship. New infrastructure is built while other sites are designated as 'illegal' and the city is fragmented into 'deserving' and 'undeserving' localities (Butcher, 2009). Remodelling the built environment has seen pre-eminence given to an aesthetic of an imagined cosmopolitan 'global' city that overlays its new condominiums, shopping malls, public transport systems and green spaces (see Brosius, 2008). Distinctions are transposed onto this new infrastructure as boundaries of inclusion and exclusion are drawn and there is the physical removal of those that do not fit within Delhi's 2021 Master Plan for urban development (see DuPont's 2008 overview of slum clearances).

The construction and encoding of these boundaries is carried out not only by hegemonies of structural power centred on city authorities, but also by affective boundaries deployed and maintained through subjective understandings of belonging and spatial use by city inhabitants. As the city is demolished and rebuilt, inhabitants are redirected, evicted and rehoused along with their understanding of whom and what is and is not permissible within shifting public space. These boundaries are imbued with a sense of distinction and collective identity, expressed through affective markers such as 'civility', manners, the appropriate means to move through the city and behave in particular spaces.

Civility has become a discourse of a global 'middle class'[2] making new claims on how city space is to be used (see Anjaria's 2009 study on Mumbai; also Fyfe *et al.*, 2006). According to Patel (2009: 470), the middle class sponsor globalization and neo-liberal policies of redevelopment 'if only to maintain their social and spatial distance from the "other"'. The 'other', recognized through sensual appropriation (Tyler, 2006), represents a form of pollution and must be kept apart, a phenomenon not isolated to developing countries such as India (see Lawlor, 2005; Skeggs, 2005; and Tyler, 2006 for their work on 'chavs' and the construction of a middle-class identity in Britain).

Civility then becomes a representation of a particular form of urban space, and a means of ordering the city, as breaches encountered in interactions with urban others generate affective responses, such as disgust, that reinforce distinction (for example, see Herbert, 2008; Wise, 2005; Zukin, 1995). Phillips and Smith (2006), however, argue that incivility is more likely to be registered, not in the face of difference, but when the flow of the city is impeded, suggesting that it is movement and, importantly, expected, predictable behaviours that circumscribe comfortable, familiar spaces of belonging. This ability to flow through the city also stems from the cultural knowledge needed to avoid unpredictability; to avoid unsettling, insecure, ambiguity; to avoid collisions. People who are 'spatially untethered', or 'who appear unkempt and somewhat irrational, violate the shared norms that produce predictable behaviour' (Herbert, 2008: 659–60). To be 'spatially untethered' is also to be untethered from the cultural practices and knowledge that create that space. The out of place 'other' block the flow of the city in both their spatial and symbolic transgressions (Anjaria, 2009: 396). They do not conform to a sense of order demarcated in the correct use of space and the understanding of boundaries between public/private, moral/immoral, clean/dirty, in a city seeking 'modern', global, status. The resulting anxiety creates what Sibley (1995) has referred to as the attempt to 'purify' space by establishing clear boundaries and internal order based on dominant cultural values and practices. These territorial lines are made legible through 'collectively held ideas of the spaces where they occur' (Herbert, 2008: 661–2).

The emphasis on the desire for predictability takes on a particular salience when placed in the context of a rapidly changing city. Normal reflexes, based on knowing which way someone is more likely to move, the direction of traffic, the codes of expected civility in a particular space, no longer always apply. In London, Watson (2006) has noted that public spaces such as markets, areas destabilized by global mobility, economic and social change, have become 'unsettled spaces' which are now subject to attempts to order them; to tidy up the refuse, the smells and noise that may no longer fit within planning authorities' desire for imagined cosmopolitan or 'global living' (see Wilson, 2006; Brudell *et al.*, 2004; Drudy and Punch, 2000 among many others researching the impact of 'gentrification'). In a city such as Delhi where public spaces often have multiple uses, including dwelling, social and commercial ones, contestation over space use can become decidedly uncivil.

Uncivil boundaries of belonging and exclusion

A vernacular, affective map of the city is etched in the movement of young people as they travel through Delhi. Balbir showed me a photograph of a *masjid* (mosque) he travels past each day. He is a south Indian Hindu, from a low-income family, living in a *basti* area behind one of Delhi's most exclusive neighbourhoods.

> This is not telling that you don't come here but emotionally . . . Muslims only go to that [place] [. . .]. It's not written restrictions but it is a restriction to the other peoples.
>
> *MB: But would it be the same for your temple also?*
>
> Balbir: Yes, [around the] temple there is also restrictions.

Balbir described the unwritten regulations of, in this instance, communal belonging and exclusion, articulated by the senses, and 'felt' within the body; circumscribing sensual boundaries of discomfort and familiarity. Sight, smell, noise, touch and taste in both the physical and Bourdieuian sense of distinction, marked out for these young people spaces of civility and incivility, spaces where they felt comfortable and where they belonged.

Incivility appeared to be most correlated with particular crowds and their associated dirt and noise. For example, Selvaraj, a university student living in north Delhi, avoided Camp Market, in the northern suburbs of the city: 'I found it dirty, crowded. The first time I visited the place was in winter, dusk. So everything I saw was shady. Shady men, shady buildings, shady vehicles, shady shops.' Though unclear about the 'geographical demarcation of this area', there is a *nullah* (drain) running through Camp that 'stinks'. 'So when I can feel the smell, I know I'm nearing Kingsway Camp.' The 'reek' of open drains, public toilets, and meat markets delineated spaces to be avoided. For Leena, a high school student living in a wealthy suburb of south Delhi, the fish market was her least favourite place in Delhi. Smell delineated the clean and unclean, including people, marking out spaces of discomfort:

> There were people everywhere. Nothing smelt nice. I couldn't even pick up the scent of my own perfume. [. . .] The smell was unbearable. The sight of dead fish in frozen ice, their eyes still wide open as if they were staring

> right at you! [...] When I got out of the market I didn't want to go anywhere else. I just needed to go home and have a bath.

The uncivil crowd was often equated with disorder and both class and communal distinction, embodying the ambiguity and unpredictability of difference. Oditi, a Bengali Hindu, describes *Chandni Chowk*, the main market area of Old Delhi, as her least favourite place: 'It's just so congested, it's dirty and it's like an over-whelming Muslim culture that's there. Maybe I haven't grown up like that so that's one of the reasons'. There are specific sensory indications for Oditi of this area's difference: its 'layout', the 'kind of people', and 'the lifestyle of the people who reside there'. It is dominated by *Jama Masjid* (the main mosque), and the smells and sight of meat, raw and cooking, all indicate that this is not her place. However, Oditi links the need to 'grow up' in a place, that is, to accumulate knowledge, to acquire a level of familiarity, and with it, the comfort that comes from knowing the rules. Shveta, a high school student from an upper-income family living in north Delhi, has visited *Chandni Chowk* several times and has 'gotten quite used to the busy, dusty and dirty ambience'. But like Oditi, she demarcates the area as different on the basis of social distinction:

> It is mostly visited by the lower strata of society in Delhi. The market area is very congested and the lanes are very narrow. I saw how in crowded areas there is a difference in an individual's personal space from a more open area. In *Chandni Chowk* everyone was pushing.

Back in Camp Market, also in the north of the city, Ananya, a Bengali university student, felt 'awkward' in a crowd who collectively behave 'differently':

> Like the guys stare. They behave in a different way. The way they talk and the way they behave is different. They are not very friendly and look in a suspicious way.... I don't feel that safe, I don't know where to go so I feel I will end up ...

There is insecurity expressed in the trailing end of a sentence, in the ambiguity of this space and the incivility, not only of other people but, as Ananya describes, the inability to predict the flow of traffic. There are 'loads of rickshaws and buses and every-one coming in and like ... jumping onto you. Every time I feel that I will have an accident here because there's no traffic signal or any traffic like ... no rule'. For Ananya, a sense of predictable order and security is premised on the familiar frames of reference that the 'rules' embedded in cultural knowledge provide. Crowded, uneven sidewalks also breach rules, as do 'spatially untethered' beggars, transient workers and the mentally ill who generate awkward, sometimes volatile, encounters in many of their narratives, contributing to a sense of disorder and unpredictability.

Some crowds, however, are more predictable than others and as a result are more comfortable. While Ananya feels a level of distress at Camp Market, she enjoys the 'crowd' at the *chai* (tea) stall on the university campus, where she knows

PLATE 11.1 Connaught Place
Source: Ananya

she will meet friends and other students, that is, familiar faces. The crowd in Connaught Place (CP) (Plate 11.1), an upmarket central shopping precinct, adds to her pleasure 'roaming' there, as do memories of spending time with her former boyfriend in the coffee shops and cinema:

> It reminds me of my home [Kolkata]. You can go window shopping, you don't need to get in, and you can watch the crowd around the pavement. [...] It's open also, it's crowded but you get to meet many people at the same place. [...] I don't feel alone there. I feel that I have many people around me so I feel comfortable in that place.

CP is one of Vishaka's favourite places as well 'with its vast sense of space and beauty (it never seems overcrowded no matter what day and time)'. A college student from an upper income family in west Delhi, Vishaka also enjoys the nearby Janpath markets (Plate 11.2), where '[t]he familiar colours and smells and people made me feel good'. Selvaraj, a university student, enjoys the precinct for the very possibility of encounters with 'people from all sorts of backgrounds and many nationalities'. Originally from south India, Selvaraj has lived overseas (Japan) and represents in his 'rock star' dress and chin piercing a very cosmopolitan archetype of 'global youth'. These differing affective responses to crowds (comfortable association or disgusted repulsion) appear inflected by the individual experiences of these young people embedded in the cultural knowledge that informs their subjectivity; their own classed, communal and gendered distinctions.

PLATE 11.2 Janpath market
Source: Vishaka

Classed distinctions

Leena and her friends, Jaya and Shveta, all from upper income families and living in wealthier enclaves, explicitly delineate class in their distinction between civil and uncivil crowds at two very different shopping precincts: the public markets of Sarojini Nagar (Plate 11.3) and the new mega-malls opening in the wealthier southern suburbs of Delhi:

> Jaya: It's a different crowd [at Sarojini Nagar].
>
> Leena: Yeah, different crowd but . . .
>
> *MB: What's the crowd at Sarojini?*
>
> Leena: Um . . .
>
> Jaya: People like us.
>
> Leena: People like us but very rarely. I mean people like us but you see more of people like . . .
>
> Shveta: College going people.
>
> Leena: Yaar, and not good college going people.
>
> Shveta: Yaar.
>
> Leena: You know, not like Stephen's College going people but like . . .
>
> [Shveta and Jaya make suggestions]

PLATE 11.3   Sarojini Nagar
Source: Melissa Butcher

Leena: Like those very, what's the word, almost government kind of college
    going people.

MB: The smaller colleges, the state colleges?

[general yes] [...]

MB: And what sort of people go to Ambi mall?[3]

Leena: You know like nowadays . . .

Shveta: School kids . . .

Jaya: A lot of people just go, I mean, won't just be very higher class . . .

Leena: Exactly, that's what I was going to say.

Jaya: . . . probably won't end up buying much but they just . . .

Shveta: They just roam around.

The inability to shop, to consume, is a key point of difference between these young women and the 'lower classes' that use the new malls. They are perhaps representative of Anjaria's (2009) 'citizen-consumer' although should not be reduced to a single category. They echo Herbert's (2008: 661–2) argument that 'exclusion is a spatial practice in more ways than one – not solely a manifestation of spatialised power but a re-inscription of spatialised distinctions'. For Leena, in her favourite mall, Select City Walk (SCW) (Plate 11.4), she can 'see a lot of like-minded people there' that she can 'connect with'. Leena judges her ability to do this by observing what they buy, for example, demonstrating the same taste in clothes. It is also 'quiet'. Amin (2008: 10) has argued that social interaction 'rarely involves transgressing long-accumulated attitudes and practices towards the stranger' and the separation between these young women and others with whom they do not 'connect' is clearly demarcated. According to Leena, those from 'lower, maybe not lower but maybe people that have low jobs or whatever, like small jobs', behave differently as the space itself enforces a code of civility:

> [People of a different 'class'] acted differently here. Less noise, more manners! So I came to the conclusion it's not always the way you behave that puts boundaries around you and defines your space. Sometimes, the space you are in puts boundaries around you and defines your behaviour.

SCW, at the time of this study, was the newest and one of the most exclusive malls in Delhi with many Western brands represented, including Marks and Spencers, as well as branded Indian goods. It carries its own sense of identity, as 'Western', modern; a subjectivity consumed and expressed by this group. The young women agreed that 'hardly anyone' could be seen wearing a sari at SCW, and 'then they are like . . . aunties' (Leena, laughs).

These codes of consumption and affective distinctions exclude Balbir and Rabia; both university students but from very different backgrounds to Leena and her friends. Balbir's family is from south India and they live in a *basti* behind a wealthy suburb in south Delhi.[4] Rabia describes herself as from an 'average' middle class family. She dislikes places such as the upmarket cinema area of Vasant Vihar, which is a favourite haunt of Jaya, for example. She draws links between her discomfort and her cultural frames of reference:

PLATE 11.4 Select City Walk
Source: Melissa Butcher

Because I'm not from that background. I'm from a middle class family with average income. I was not grown up in a very rich way. I don't feel comfortable in very rich areas. [. . .]

MB: Why?

Appearance doesn't matter to me very much but sometimes [they] behave in different ways. They don't give us the space to say '[we] are welcome', through that gaze, that gaze makes me feel uncomfortable. Sometimes I look back in the same way they look at me. They scan me, from top to bottom, like X-ray gaze. [...] Their behaviour also. Loudly talking without thinking of others, they start music in the cell phones or talk loudly. You feel yourself different in some way. Sometimes it gives me a kind of empowerment also. I can see things in a different way. They are violating my private space.

While dressed in the uniform of global youth culture, jeans, and t-shirt, Rabia still feels out of place, and that those 'x-ray' eyes somehow know she is not from Delhi, that English is not her first language, that she cannot afford branded jeans, perhaps even that she is Muslim living in a predominantly Punjabi, Hindu and Sikh city. Balbir also notes affective restrictions on his entry into more upmarket areas such as Vasant Vihar, even though one of his relatives works as a driver there: 'because it is a posh area but they are looking at us [as] not their status. In between us, the status comes'.

A further dimension of distinction raised in the contested use of public spaces in Delhi centred on an ineluctable tension between the individual and the indistinguishable, unenumerated crowd, embodied in the unknown, unsettling, ambiguous 'anyone'. The fact that 'anyone can come in from any direction and you can get hurt' increased Ananya's dislike of Camp Market. Shveta avoided shopping in Noida because:

A lot of stuff has been happening [there]. Like I used to go there a lot because there were more malls there. [...] That time also used to be a little unsafe because you know anyone could just walk in. You know usually in malls, you know like middle class people or high class people go there. Over there just anyone walks in ... there's ... cheap crowd kind of people. [...]. It's really scary and a lot of stuff has been happening there.

The ability for 'anyone' to walk in increases unpredictability. Shveta expresses a fear of 'random' violence that others also acknowledge. Ananya no longer feels secure in Central Park after the bombing there in 2008. Balbir dislikes 'any place where anything happens suddenly wrong'. As a result he reflects a general preference for open, 'safe' spaces marked by familiarity, such as Rashtrapati Bhavan (the Presidential Palace): it 'feels secure', it is policed to ensure order. Within this open, yet controlled space, different others can be absorbed with minimum discomfort. He notes that if someone visits a temple or mosque 'you have to be Hindu or Muslim. The other person come (the stranger) is seen suspiciously. But in the park ... it is a common place for all, Hindu and Muslim. That's why I like that space'.

However, Rabia does not feel comfortable in Rashtrapati Bhavan. 'You feel dwarfed, it's so vast and well maintained. Sometimes I feel I am destroying the beauty of the place through my activity'. Preferences for ordered or disordered spaces at different times and places demonstrated the importance of personal disposition and a 'multiplicity of affect' (Rose *et al.*, 2010) engendered in the interactions between these young people and the spaces of the city they moved through. For example, Selvaraj enjoyed 'hanging out' in CP, but only with the 'right company: 'CP is a place which will make one feel real lonely if one is without the right company'. Balbir prefers the neatly manicured, depopulated avenues of south Delhi, as does Jaya, while Hemish, another student from north-east India, preferred the 'natural' state of Old Delhi with its jumble of dirty, crowded lanes, markets, *havelis*, shopfronts and overhead wiring.

There is a lawn behind Rabia's university department where she feels 'comfortable'. It is more isolated. 'I like it. Maybe becoz [*sic*] all the loud people are not here'. For Rabia, who recognizes that she is an 'introvert', spaces feel 'comfortable' when 'loud people' are absent but also when it is less ordered, like *Chandni Chowk*, an area unpalatable for Oditi but one of Rabia's favourite places despite the noise. Its crowded, narrow lanes are populated by salwar qamiz, saris, and the slacks-and-shirts style of Hindi film stars. Chaotic shops sell unbranded goods, and distinctions between class can be inverted, as much as the distinction between pavement and road disappears. The body can become anonymous in this crowd. However, in Chandni Chowk, a predominantly male space, judgements can shift from reiterating class distinction to that of gendered boundaries that must also be managed in Delhi's patriarchal public spaces.

Gendered distinctions

While the previous section has focused on the affective dimensions of class and communal distinctions in Delhi, there is no escaping the predominance of the male body in the city, a factor that inevitably led to a strong narrative of affective exclusion for the young women in the study. In their journals there are no stories of moving through the city at night unless they are inflected with fear and in-security (unlike in the writing of some of the young men).

Charu lives in a resettlement colony in the north-east of the city.[5] The young women living here have circumscribed mobility through customary communal monitoring. Charu's pleasure in 'roaming' on a rare excursion 'outside' the colony, to Rashtrapati Bhavan, became a reflection on the pleasure that open space can bring:

> It has lots of open space where you can walk. There's more security there, you have policemen. No-one can come in and go out just like that. [In the colony] you cannot even go out for a walk without someone commenting 'that girl is roaming around on her own'. So we don't have that space to roam around in, go for walks, which you can do in Rashtrapati Bhavan.

The tight, congested buildings of the resettlement colony enclose the young women that live there. Their narratives are heavily inflected with insecurity and frustration at the constraining gaze of the male and their families. For this group there is a greater sense of exclusion from the global, cosmopolitan city: 'I was feeling upset about such a life in which we can't go anywhere alone' (Adab). They live in ever-decreasing circles of mobility, avoiding *gullies* (lanes) and young men who 'say just about anything to all the girls who go that way' (Nomi), and enduring the physical bumps and verbal barbs that this area requires. They feel 'irritation', 'frustration' and 'fear' of boys and dogs and buffalos sharing the lanes (Tavishi), and 'anger' at the impertinence of others who 'misuse' public space (for example, queue-jumping in the ration shop, or bicycles blocking shop entrances). Many of the young women expressed a dislike of the settlement: '[We] don't like the ambience. Parents don't allow their daughters to go out to the shops, to roam around, and people are also jealous if someone does anything good. There are many restrictions' (Esha).

The 'ambience' of this quarter exemplifies Anderson's (2009: 77–8) description of 'affective qualities that emanate from but exceed the assembling of bodies'. It is a collective atmosphere that 'envelopes' and 'presses upon' these young women. However, the 'ambience' of this quarter is far from being ephemeral and unstable as Anderson argues. It is redolent, heavy, with 'tradition', the accumulated expectations of the place of the body of the woman in the home and the street. It shapes and manipulates that body as much as the ambience of mega-malls such as Select City Walk impact on the 'consumer-citizen' of Middle Class Delhi.

The sense of enclosure for women in the city in general is reflected in Rabia's comments that 'sometimes, again, I feel we all are in a harem'. These bodies enclosed are juxtaposed to Hemish's ability to adapt. Visibly an outsider in Delhi, he describes how he attempts to 'shape myself to be like a man from Hindi [north India] mainstream so that I can adjust myself from bus conductor to my classmates'. No matter how much they contort themselves, these young women are unlikely to leave these domains except through marriage. Even for the young women from upper-income families, comportment and space use is heavily determined by expectations of gendered behaviour. When Leena wore a short denim skirt to go to a bar at the five-star Ashoka Hotel, she encountered her own limits in the city:

> I immediately felt uncomfortable as what seemed like a hundred eyes seemed to look at me in amazement at the short skirt I was wearing even though it was at a hotel. At that moment my only wish was to miraculously disappear and reappear INSIDE *F Bar*. But unfortunately, I couldn't do that, I had to walk through the stares. However, the minute I reached *F Bar* I suddenly felt normal again. Like I fit in!! (Leena's emphasis)

The wish to disappear was also generated when contact with crowds became physically unbearable.

> Padma: I travelled a long time back [on the bus]. Just the *feel* of it, you know. The sweating [grimaces] and …

Leena: Stinky!

Padma: Yar. Like in those kind of places you feel . . . 'can you like not touch' or just . . .

Leena: . . . like hide in the corner.

MB: Leena, you're saying you try and hide in the corner?

Leena: Yar!

Padma: I want to make myself invisible! People fall on you . . .

Leena: Then the bus turns and everyone falls in the same direction!

Padma: . . . then you have to push and then they push each other and they fall on the other side! It's so . . .

For Padma and Leena, from well-off families living in south Delhi, bus travel is a novelty. With their own family drivers this form of transport, and its predominantly economically marginalized commuters, can be avoided. Unlike for many of the young women in the resettlement colony, where restrictions on their mobility and futures were a source of frustration, Padma, despite her access to private transport, voluntarily chose to remain close to home. She had only recently moved to Delhi, coming from a 'really small town' in the Punjab. At home, she says:

> I get my personal space. I get to be myself there. I don't have any restrictions that I have to behave in a particular way. I think I do in school and even when I'm outside my house . . . when I'm outside my comfort zone I have to behave in a certain manner so that nobody else, you know, thinks that it's odd or that it's out of the usual. [. . .]
>
> *MB: So you find Delhi a bit confronting sometimes?*
>
> Yeah. The kind of life in Delhi is not what I'm used to. In my home, my family is really disciplined kind of family. I think the whole environment of Delhi still it runs in a particular way which I'm not used to.

Padma repeats in her narrative the phrase 'I'm used/not used to it', reiterating that habitual practices stemming from accumulated cultural knowledge generate familiarity and define affective states of comfort/discomfort. As with Rabia, who prefers her isolated lawn on the university campus because 'I don't have anyone else with whom people can compare me', Padma avoids any threat of judgement that her body is out of place by remaining as much as possible within the familiar confines of her home and immediate neighbourhood. For Hemish, the sense of 'freedom' at not having to worry about 'bodily behaviour' is felt at the *chai* stall on university campus: 'We feel a sense of more freedom in that open environment where we haven't to be always careful about our bodily behaviour.' These comments raise the complexity of urban geography, however, in that the 'open' space of the campus is also enclosed, with security guards at the main entrance and affective barriers that would prevent, for example, the young women from the re-settlement colony accessing this space. Rabia notes the disdain with which the lower class/caste construction workers are treated by others within the campus.

Jaya also notes how her habitual patterns of movement through the city territorialized spaces of comfort. She travels regularly through Lutyen's Delhi, that planned part of the city which is generally wealthier, greener, more open and ordered:

> So for me Delhi is really clean and neat and nice. [...] you go through other parts of Delhi which are not what you're really used to seeing and I do find that uncomfortable because the part which I'm living in is really nice and posh and clean.
>
> MB: *What makes it uncomfortable for you do you think?*
>
> It's dirty and noisy and crowded.

These habitual forms of everyday practice appear to be guided by states of comfort and discomfort generated by, and in turn reinforcing, these young peoples' subjective understandings of where they belong, and their place in the order of Delhi. With this understanding embedded in their affective responses to the city and divergent others, their patterns of mobility enable the domestication of particular spaces, reinforcing their place within it and enabling the capacity to move through it. This corresponds with Amin's (2008: 12) claim that the movement of people in public space is less random and more 'guided by habit, [and] purposeful orientation'. While, there was a degree of 'roaming' – purposeless wandering in parks and malls – by most of the young people in this study, this was carried out in spaces designated as 'comfortable' or 'civilized'. A desire to explore new parts of the city was limited by these affective responses. However, these boundaries were not immutable and under certain conditions could be challenged.

Challenging distinctions

In this study, there was an obvious deployment of power in the bounding of space as cultural hierarchies were transposed onto the new surface of Delhi. But while the city was divided by precincts of inclusion and exclusion, these boundaries were not rigid. Spaces of inclusion could be created when there was spatial disorientation, and a concomitant shifting of power relations, even if only momentarily. For example, Esha, who lives in the resettlement colony in the north-east of Delhi, recalls an outing with her school friends to a nearby park:

> Then we went to the park and our teachers also had their meal with us. I thought to myself, that our teachers never have meals with us in school and also keep a distance with the students. I thought to myself that teachers

should be like this, like our *didi* [older sister] who moves around with us, plays around with us and mixes with us [...] I enjoyed myself so much!

The quality of a space, in this instance, a park away from the surveillance of the colony, allowed for new interactions. The boundaries of class distinction could also be challenged in spaces where the rules were unknown. Reservations about how to use the new Metro rail system were, in Balbir's case, initially inflected by his assessment of his place in Delhi as someone from a low income background:

> When I went to [university] first time by metro I afraid, because it was new to me. First, I thought that it is very luxurious to me and going on metro is prestigious to me also [...]. I heard a lot about the metro previously, that is, it is very fast, time consumption is less, highly stylish. [...] When I entered the metro station I didn't know from where it will leave and how I can get out of it. For three to four weeks of my journey on metro is very boring because I normally sit on the seat and look most of the time on the chart and hear the sound of the speakers for which station is next, and in between when I see the crowd I think they are superior to me and look like they know all information which I didn't. But this is wrong presumption of mine. Because, one gentleman came near to me and asked that 'from where I can catch the metro for Dwarka'. That time I realise most of them are like me and no need to be afraid about it because all are new to human, just we have to learn.

While the Metro was a system new to everyone in the city, and therefore incompetence in using its automatic turnstiles was, initially at least, classless, as previous sections have shown, barriers to accessing *established* spaces in the city were harder to cross. These boundaries, accumulated over time within the cultural framework of the city and its inhabitants, are thoroughly ingrained onto the mental maps of Delhi's residents.

Accumulating new knowledge has the potential to create more permeable boundaries and distinctions, open to negotiation, elaboration, redefinition and perhaps even dismantling. However, a more predominant strategy was simply to find other spaces of comfort. To avoid unsettling crowds, for example, there is a search for quiet spaces. For Vishaka, this is Statesman House (Plate 11.5), in CP, with its upmarket café that also enables a wide view of the streets below. 'It's a beautiful building', that gives her a 'soothing feeling' and makes her 'feel at home'. In the resettlement colony, rooftops become spaces of escape away from the surveillance of family and community (Plate 11.6). 'It's a very peaceful place. [We] can speak whatever [we] like, speak [our] minds' (Maha). However momentary, these 'peaceful' locations distinguish the capacity of urban inhabitants to find and occupy interstitial spaces in the city that provide a modicum of comfort in the face of exclusion and change.

PLATE 11.5 Statesman Building
Source: Vishaka

PLATE 11.6 Rooftops resettlement colony
Source: Nomi

Concluding thoughts

In the crowded public spaces of Delhi, in its streets and shopping precincts, or on its public transport, the body must be flexible. It must bend and twist and absorb the inevitable physical and sensory collisions. Or it can remove itself entirely if the discomfort generated in possessing different cultural frames of reference that guide reflexive and conscious movement and opinion is too great. It can be made invisible, hiding or removing itself. Such adjustments were evident in the movement through Delhi of the young people in this study, in their experience of the city, demarcating spaces of pleasure and discomfort, inclusion and exclusion, and affectively dissecting the city into the civil and uncivil. In particular, they reiterated boundaries of distinction as an affective response to the 'disorder' of things and people out of place, to a slippage in former distinctions (for instance, 'lower class' shoppers in the upper class malls), highlighting the intimate connection between the physical and affective bounding of space.

Distinction was drawn by and on the body, evident in appearance, consumption, responses to noise and smell. 'Proper' behaviour, civility, marked out boundaries of inclusion and exclusion created through adherence to the 'rules' and routines of public space generated in the recitation of culturally embedded everyday practices such as shopping. While stereotyping by some participants of others was obvious, these subjective judgements also provided a sense of order and comfort in a complex urban environment, and demonstrated a sense of self as the benchmark of determining difference. The 'rules and routines of ordering public space' (Amin, 2008: 14) were inculcated through ongoing, culturally embedded practice and social interaction with more familiar, comfortable others.

These reiterations of, and challenges to, boundaries of distinction, both physical and affective, question debates that remove the subjective from accounts of the affective experience of the city. The shared knowledge of 'correct' spatial organisation, activity and representations of civility, what and who is permissible or not in a particular space at a particular time, are cultural markers. The use and understanding of space informed by this accumulated knowledge was affectively circulated in the judgements of sights, smells, noise and touch of others, the crowds too close for comfort (Tyler 2006). This sensory experience was often the first indication that something was out of place, themselves or something/someone other.

As a result, the spaces of Delhi became part of a system of classification; containing rules and conventions that could differ according to gender, religious affiliation and class status (and also other factors not included in the scope of this paper, such as age). Deviations from cultural benchmarks, norms of order and predictability generated feelings of insecurity and exclusion, whereas habitual, expected, movement, smell and sounds maintained particular spaces as comfortable. Measures could be taken to manage or counteract the uncivil, the unpredictable or the unknown, for example, avoidance or the deployment of a 'suspicious' gaze against a body judged as out of place. From these findings it is possible to agree with Latham and McCormack (2004), for example, that the city is produced through human practice and contested by the experience and use of urban space, but this is also mediated by participants' subjective understanding of the constituents of that space. Distinctions, communal, socio-economic and gendered, were created in the embodied experience of Delhi, embedded in accumulated cultural knowledge and an understanding of their own subjective place in the city, and expressed in judgements often centred on a normative civility.

There is complexity in this process, however, as Rose *et al.* (2010) have argued. Memory, comparison, the re-appropriation of space, its multiple uses, in this instance in a regenerating city where new spaces provide an opportunity for a re-conceptualization of the rules, all compete for attention. While there was an assignation of meaning to space there were at times, and in specific contexts, challenges to that meaning and the position of boundaries. Spatial transformation and human agency has the potential to create fuzzy, flexible boundaries. The democratising potential of space is noted in particular locations, including rooftops, tea stalls and the Metro rail system. Despite these breathing spaces, the findings suggest that the 'cleaning' up of Delhi, the attempts to remove the sights, smells and sounds of poverty as part of the aesthetics of global living, reinforce affective distinctions as existing cultural hierarchies are transposed onto a gentrifying city.

Notes

1   The data was derived from field work, January to March 2009: first, collating data from 23 young people who 'mapped' their movement in the city over a period of one month; and second, collating thick description of specific sites identified by participants as popular points of interaction, including shopping precincts and the university campus. Research participants were contacted through existing relationships with NGOs and universities. Initial participants then recruited peers to take part. Their movement was recorded in journals, blog entries and street directories where appropriate. They also photographed places that had particular meaning to them which were later discussed with the author.

2   'Middle class' is used with the usual caveat that this is very broad terminology. However, as noted in the narratives of participants, this class distinction is often used to demarcate themselves and those 'above' and 'below'.

3   Ambi (Ambience) Mall is near their school and a popular 'hang out' spot. It is one of the largest malls in India.

4   There is a symbiotic relationship between these upper- and lower-income colonies. Residents of the bastis provide the manual labour (cleaners, cooks, nannies, drivers etc.) for the wealthier households.

5   During slum clearances in the 1970s, evictees were rehoused in designated 're-settlement colonies' in different parts of the city. The colony in east Delhi, where the group of young women in this study lived is one of the oldest of these. Other, informal, resettlement colonies tend to be densely populated and unplanned, resettled by migrants or squatters. They can also be termed a *basti*.

12

SHANGHAI BORDERLANDS

The rise of a new urbanity?

Deljana Iossifova

Introduction

China's concomitant demographic, economic and urban transitions have brought
about commodification of space in urban centres characterized by sociospatial
difference. This chapter looks at everyday life on the borderland between two
very different fragments of urban space in Shanghai: one new and flourishing,
representing the envisioned future; the other stigmatized and dilapidated, representing the unmemorable past. On the borderland, rural-to-urban migrants, urban
poor and a new generation of middle-class professionals coexist in space and time.
I look at the spaces that they appropriate and inhabit in the context of an ever-
present state, claiming that Shanghai's borderlands constitute the link between the
past and the future and might give rise to a new urbanity.

> New China rioted with gigantic building schemes, with barracks, schools,
> sports grounds and airports. The old China lived in narrow alleys, echoing
> with the chanted cries of the coolies and peddlers; it hung birds before its
> doors, smoked water pipes, bargained and haggled, ate and slept, played and
> smiled, and was happy.
>
> *(Baum, 1986)*

> A social transformation, to be truly revolutionary in character, must manifest
> a creative capacity in its effects on daily life, on language and on space.
>
> *(Lefebvre, 1991)*

The street in between the Village and the Compound was rather busy in April
2007. Around five o'clock in the morning, the local market janitor would usually
perform his morning exercises. Young workers would start unloading trucks stacked

with meat, vegetables, eggs and other products just minutes later, and on some days, a woman on a bicycle would deliver pigs, cut in half. Nearby, behind closed curtains, a group of players would perform the finishing moves in an all-night game of mah-jong. The middle-aged female cook at a soup restaurant would roll up her shutter, announcing the beginning of a new business day, and cross the street to collect the dishes of the previous night's dinner from the guards on duty at the gate of the Compound. Construction workers would then begin to pour out of their dormitory, carrying their orange helmets in one hand and their breakfast – usually a bag of soymilk and *youtiao*, freshly fried dough sticks covered with sugar – in the other. Street sweepers in blue uniforms would do their rounds to collect the garbage that had accumulated overnight. A couple of teenage boys, returning from their night shifts of selling Beijing duck in the streets of Shanghai, would store their carts and display boxes away, urinate against the fence of the Compound, and crawl into tiny spaces behind roller shutters along the street for a few hours of sleep. Men and women – some dressed-up in professional sports clothing, others wearing just shorts and flip-flops – would jog up and down the street.

By six-thirty, people would be up and busy in front of most shop-dwellings, brushing their teeth, gurgling and spitting. They would wash their hair in plastic tubs on the street, do their laundry and later hang it up to dry on clothes lines spanning between trees, lampposts and the fence of the Compound. Residents of the Village would empty and wash their chamber pots at the public toilet, which a sanitary worker in her blue uniform would arrive to clean a little later. Young women would gradually transform crammed dwellings into small but welcoming shops, placing their produce on display on large tables and taking over the narrow sidewalk. Residents of the Compound would buy breakfast at one of several food stalls along the street. Drivers would polish the outsides of their employers' automobiles, and taxis would start lining up in front of the Compound to pick up boys and girls with school bags and musical instruments. The young women in the shops along the street would hand over their children to neighbours to watch when leaving their shops to run errands. Men and women collecting used goods, old paper and bottles would ride their tricycles, smiling, ringing their bells and calling out for customers: '*Congtiao . . . diannao!*' – 'Air conditioners . . . computer!' The sound of somebody practising the saxophone in the Compound would fill the air, mixed with the noise of workers mowing the lawn.

In the 'urban', it has been argued, 'everything is calculable, quantifiable, programmable; everything, that is, except the drama that results from the co-presence and re-presentation of the elements calculated, quantified, and programmed' (Lefebvre, 2003). In Shanghai, ever since China embarked on its journey of 'Opening Up and Reforms', the 'drama' of co-presence and coexistence enfolds in the everyday, as long-term urban residents in decrepit housing prepare to vacate their homes in the city centre, about to give way to countless urban renewal projects; as members of a well-educated emerging middle class, the dutiful con-sumers in a new economy, take their place in newly-built, commodified inner-city housing; as migrants from the countryside arrive in hope of employment, trading

PLATE 12.1   The borderland between the old Village to the left and the new Compound to the right
Source: Deljana Iossifova

rural homes for shabby shacks and crowded dorms; as international students and expatriates, huddled together in gated compounds, experience the city from the air-conditioned interior of the taxis that take them to selected points on an ever-changing map. In the midst of processes of enormous sociospatial restructuring, people with different backgrounds rub shoulders in a city that keeps continuously changing (see Plate 12.1).

Almost a century earlier, grappling to understand the forceful changes resulting from the Industrial Revolution in the West, the Chicago School of Sociology developed ecological arguments about the city and the 'urban' (see, for instance, Park *et al.*, 1925), defining the city as a 'relatively large, dense, and permanent settlement of socially heterogeneous individuals'; proximity, density and diversity were seen as characteristic of the distinct ways of 'urban life' (Wirth, 1938). 'Diversity', generally regarded as a necessary condition for urbanity in the Western context, has remained a favourite subject of critical inquiry ever since (see, for instance, Sandercock, 1998; Fainstein, 2005). Some argue that physical proximity and everyday encounter between those who are different does not always translate into meaningful contact or a 'culture of recognition' (Valentine, 2008), while others see the experience of 'difference' as a prerequisite to the 'urban mindset' (Sennett, 1974) and understand co-presence, everyday social interaction and cultural confrontation in shared space as sources of social renewal, economic innovation and creativity (Amin and Graham, 1997).

Looking at diversity and the coexistence of ethnic groups – an ethnic group defined as a 'corporate group' which shares an identity 'based on some shared cultural traits' and 'finds itself in competition with other groups for wealth, power, opportunity, and recognition' (Eller, 2009). For example, Barth (1969) identified four alternative modes of coexistence within polyethnic societies: when ethnic groups occupy clearly distinct niches, they can co-exist in a stable condition of minimal interdependence when they compete minimally for resources; they can negotiate their border politics when they compete for resources; and they can reach close interdependence in political, economic and other fields when they provide services for each other. When, however, ethnic groups occupy the same niche, they either reach a state of accommodation through increasing interdependence, or one group is displaced by the other. In this chapter, I will show how some of the different groups and their (inter)actions in a selected case study neighbourhood in Shanghai[1] contribute to the formation and negotiation of multiple urban identities (and hence, to the definition of a new urbanity) on the very stage of Lefebvre's urban 'drama': the space in-between the new and old, the rich and poor, the wanted and unwanted – the borderland.

The sociospatial divide

The selected case study neighbourhood today contains urban fragments representative of different stages in urban development: the Compound, a gated high-rise development for the new middle class, and the Village, the remaining part of an old shantytown. The old shantytown came into being when, during the hundred or so years before 1949, thousands of immigrants arrived in Shanghai from nearby and faraway provinces. It was not unusual for people from the same towns and villages to cluster in certain parts of the city, their 'native place identity' leading to the assignment of a particular value to their respective neighbourhoods. Those who settled in the focus area, located on the north bank of Suzhou Creek, came largely from northern Jiangsu Province. Categorized by Shanghainese as '*Subei* people' – a label signifying lower quality and class – they found themselves sharing an ethnic-like identity that prohibited them from finding occupation beyond the lower ranks of industrial production or shelter outside the boundaries of their *Subei* neighbourhoods, which came to be feared and avoided by anyone who considered themselves to be of higher social standing (Honig, 1992).

Differentiation based on native place identity became less common when the household registration system (*hukou*) was introduced to restrict migration between rural and urban areas almost completely during the early years of socialist rule. Built on inequality from its very beginning, the *hukou* system has been described as a 'caste-like system of social stratification' between urban dwellers and the rural 'peasantry' (Potter and Potter, 1990), entitling urban *hukou* holders to the regulated supply of daily necessities (such as food and clothes), education, health services and housing in the municipality of their registration. The holders of a rural *hukou*, however, received none of these services, and they were not permitted to leave their villages without permission (Chan, 1996; Chan and Zhang, 1999). It was with the introduction of the *hukou* system that the immigrant population of the old shantytown became officially urban. Their new urban identity did not contribute to the improvement of their vastly overcrowded living conditions, as the Central Government, conceiving of cities as places of production, had little interest in the maintenance of urban infrastructure, including housing. Nonetheless, improvements took place gradually (bamboo shacks, for instance, were partly replaced by sturdy multi-storey dwellings), and by the late 1980s the shantytown had even acquired access to electricity and tab water.

With Opening-Up and Reforms, socialist state-owned factories and enterprises became increasingly less competitive and many of them had to fold (see, for instance, Wang *et al.*, 2005), not without consequences for the people in the focus area. Most of them had inherited from their parents their work place at the nearby factory and lost their jobs when it closed down in the early 1990s. Lacking the skills and training necessary to start their own businesses as propagated by the new maxim of 'getting rich first', many remained without a permanent job ever since, getting by on temporary jobs now and then and on the small allowances they received. Over the years, to accommodate their growing families, they, the locals, constructed additions to their homes wherever available space permitted, and, spending their abundant spare time together in tiny living rooms, backyards and alleys, they built friendships and close networks of mutual support with neighbours, who usually shared their fate (see Plate 12.2). Later on, when Shanghai

PLATE 12.2 Residents of the Village in their living room/kitchen/bedroom
Source: Deljana Iossifova

started its Suzhou Creek beautification programme in 1998, arousing a wave of shantytown demolition, resident displacement and state-sponsored, property-led gentrification (He, 2007; He and Wu, 2007), locals experienced the top-down sociospatial transformation and commodification of space first hand – the factory, their former working place, was demolished, and in its place appeared the new, 'modern' Compound. Surrounded by fences and secured by gates, the Compound was slowly populated with a younger generation of well-educated professionals – the urban newcomers to the area, their 'collective form of social identity' defined by privileged access to the real-estate market and the relatively new experience of home ownership (Tomba, 2004).

They were not the only newcomers to the area. With the relaxation of migration restrictions, a so-called 'floating population' (*liudong renkou*) of rural-to-urban migrants began to re-emerge in China's cities. The 'rural migrants' are generally portrayed as peasants and temporary guests in the city, seeking to make money but rooted in (and bound to return to) the countryside. Because of the urbanization processes taking place back 'home', however, many of the recent rural newcomers in the focus area stated that they had never worked in 'farming'; they did not depend on the state for its services. Better educated than the previous generations, they were often not employed in the 'typical' industry or construction sectors and frequently operated their own businesses in the city. Furthermore, in their narratives about their reasons for coming to Shanghai, they often spoke of the image of 'the sparkling city' conveyed to them through magazines and television – a city they were genuinely interested in, one that they wanted to be part of and one they wanted to explore. In the case study area, the rural newcomers settled mainly in vacated housing in the Village, resulting from the reduced willingness of commodity housing owners in the Compound to rent to them, but also from ever-new policies introduced by the Municipality to prevent them from so doing.[2] Consequently, the living conditions of rural newcomers in the Village were much worse than the living conditions of locals, but they paid higher rents per square metre in comparison. Rural newcomers looked up to and associated certain spaces, places and lifestyles with the 'urban' and 'other'. As a rural newcomer resident in the Village for over ten years put it:

> There are so many migrants around in the Village now; because of the many migrants, people lack the sense of belonging. Look, people in the Compound [...] are very well educated. And if you happen to live with them, you will be better off yourself. In the Village, there is no chance for people to progress!
> (*Interview with CZQ, September 2008*)

The psychology of place

In outlining the recent developments in the case study area, the previous section focused on the ways in which certain social groups are assigned particular identities and their respective place in the city because of various, mostly political, motivations. It is interesting in this context to look at Graumann's (1983) 'multiple identities' model, which builds on the understanding that a person, a place, or a thing may have more than one identity, and that the various social identity formation and maintenance processes take place simultaneously – sometimes complementing and sometimes contradicting one another – making a certain minimum of interaction between 'self' and the 'other' necessary for the maintenance of individual and inter-group stability. Graumann positions identity as the product of constant negotiation between the following three modes of identification: identification of, being identified and identification with. 'Identification with' refers to the role models we choose as things and places become representative of our values. The second mode, 'being identified', refers to the ways in which the individual or object becomes subject to (sometimes historically handed down) typifications. 'Identification of' refers to the experience of sameness and the feeling of familiarity. It is a process of appropriation, particularly in regard to language, as it involves the assignment of pre-existing categories (such as names) to objects or people.[3] It is useful to keep in mind the model of 'multiple identities' when looking at the different groups and their actions in the focus area.

Having appropriated it physically and psychologically over a lifetime, most of the locals, for instance – born and raised in the focus area – had developed strong ties with the neighbourhood and its people. However, the validity of their 'identification of' their neighbourhood was scrutinized by recent and ongoing changes. Many of their former neighbours and co-workers who had succeeded in adapting to the new economy had gradually left the old Village to move into new, 'modern' commodity residential areas, just like the one across the street that had taken the place of the old factory. Lacking the skills to do so for themselves, the locals left behind depended on the Government to decide upon their future, that is by selling the land on which their homes were built and resettling them to 'better' and 'newer' apartments at the edges of the city. Waiting in their impoverished neighbourhood (the negative perception of which was exacerbated by the possibility of direct comparison with the new-built Compound), they watched rural newcomers move into the now vacant homes of their former neighbours and friends in the Village. The sociospatial transformation brought about feelings of discontinuity and insecurity, and locals felt increasingly alienated from their once familiar neighbourhood, excluded from a better off society, and restrained to the 'ghetto' in their residential choice. In this context, it is interesting to consider the findings of Dixon and Durrheim (2004), who studied place-related identity processes in post-Apartheid South Africa and found that the disruption of place may become a way to 'justify collective resistance to social change' and to sustain ideologies of segregation. Similarly, the appearance of increasing numbers of rural newcomers reinforced the locals' feeling of decreasing social control, which used to be particularly pronounced in the Village owing to the large amount of time that residents spent in their neighbourhood because of unemployment and mobility restrictions. With time, locals had established the norms of 'appropriate' behaviour in the Village, and they found that rural newcomers did not readily adhere to these norms. They clearly identified rural newcomers as the trigger of environmental and social decline, and referred with nostalgia to the 'good times' before Opening Up and Reforms.

In the Compound across the street, urban newcomers tried very hard to distinguish themselves from those occupying the Village. Torn between their sympathy with life in the Village, the 'other', and a hostility emerging from the desire to make use of and assert their new identity and power as the dominant group in the city, urban newcomers strived to 'behave' according to their newly acquired role as the new 'urban elite'. In interviews, for instance, they stated frequently that the services and products offered across the street were 'obviously too cheap to be any good' (contradicting their statements, however, they could often be seen buying groceries and small everyday goods from hawkers on the street). Equally, space for them had acquired the status of commodity, and in contrast to locals who appropriated space predominantly through locomotion, 'doing', and personalization in an expression of both the lack of accessible private space and the struggle to survive in an increasingly commercialized environment (see Graumann, 1976, for a detailed elaboration on the different modes of appropriation), urban newcomers appropriated space in the first instance through purchase. The consumption of space as commodity can be read as the result of selectivity, and hence as an expression of categorization: choosing one place or location over another. Choosing a gated residential compound in a particular location is the expression of previous 'identification of' – marking a process during which space becomes a status symbol (see Plate 12.3). Duncan (1985) has argued that individuals in individualistic societies express their identity through material objects, whereas home is not seen as a symbol of social status in collective societies. Hence, the new role of space as a status symbol for urban newcomers indicates the transition from a 'traditional' (or collective) to 'individualistic' society.

Better city, better life

In 2005, President Hu Jintao introduced a new vision of an 'orderly society' in line with the concept of 'scientific development'. The leadership, acknowledging threats to social stability like the growing gap between rich and poor, an inadequate social security system, and increasing unemployment, seemed headed toward a shift from economic development to a new paradigm: the doctrine of a 'harmonious socialist society' (see Bo, 2005; Department of International Organizations et al., 2005; Fan, 2006). The main goals included maintaining rapid economic growth, establishing the rule of law and an adequate social security system, strengthening the xiaokang (middle-class-oriented) society, and, in particular, improving the 'morals' of the population. Consecutively, Hu launched the 'eight honours and disgraces', a 'social engineering campaign' (Suessmuth-Dyckerhoff et al., 2008) promoting a set of values, including 'social morality' and 'cultural harmony', and calling upon, for instance, the virtue of being 'disciplined and law-abiding', rather than 'chaotic and lawless' (Yan, 2006). 'Civil society', so Short (2006) argues, 'emerges from the

PLATE 12.3 Internal paths in the Compound
Source: Deljana Iossifova

practices established in the shared space of the city'. While Hu's campaign was mainly targeted at combating corruption on all levels of government, it also impacted upon the street. Power relations become explicit in space through the language of architecture and urbanism; they become physically perceptible in the urban 'landscape of power' (Zukin, 1993). It appears that when a city is striving to portray itself as global (and especially so in view of an event like the World Expo 2010, bearing the motto of 'Better City, Better Life'), some practices are far less desirable and desired than others.

Shanghai seemed determined to erase any possible trace of poverty or disorder within the city proper. Local customs, like wearing pyjamas in public (see Plate 12.4), drying blankets on the street and spitting, for instance, began to be portrayed as backwardly or rural, and they were distinctly ascribed to the residents of 'old residential areas' by the Government and its media (Lu, 2009). In the focus area, residents and homeowners in the Compound had started to complain loudly about the 'chaotic' conditions in their neighbourhood: the fence around the Compound, they claimed, was constantly 'abused' by the residents of the Village, in that they converted it to suit purposes other than that for which it was intended (for instance, it was utilized to serve as the community drying rack); unwanted individuals from whom the fence and guards were intended to offer protection persistently succeeded in 'invading' the Compound (see Iossifova, 2009a). Moreover, vendors on the street were accused of causing traffic jams, pollution and noise. The *jiedao* (subdistrict level government) decided to take measures in an effort to pacify the

PLATE 12.4 A scene from the past: rural newcomer wearing his pyjamas in public
Source: Deljana Iossifova

Compound dwellers. Within only one week in the summer of 2008, a two-metre-high concrete fence appeared in front of the very dwellings and shops along the street that tenants rented from the *jiedao*, without adjustments to the rents to reflect the worsening conditions. Standing in front of shop fronts and housing facades, the fence was praised by *jiedao* representatives and residents of the Compound for successfully preventing shopkeepers from occupying parts of the street; surprisingly, even some of the tenants behind it found that it succeeded in 'disguising' the decrepit appearance of the Village. The changes triggered by the micro-transformation of the borderland mainly affected rural newcomers, with largely negative consequences in transforming terms of continuity and the right and ability to negotiate livelihoods in the city (see Iossifova, 2009b for a detailed elaboration).

Partly because of the cultural bias against spatial mobility, traditionally regarded as the source of disorder and instability (see Zhang, 2002), 'urbanizing' efforts also included a ban on street vendors from public spaces in the city. Street vendors, hawkers and entrepreneurs who provided cheap services in comparison to competitors in local supermarkets but did not pay rent or taxes became the subject of a daily pursuit by the authorities (see, for instance, Cai and Liang, 2009; Yang, 2008). Consequently, the long-established presence of street vendors, hawkers and service providers on the street between the Village and the Compound in the case study area was not welcomed any longer. As the tenant of a hardware store put it: 'Now they (the migrants) are not allowed to do business any more due to the World Expo 2010. So, my business is going down – many of my customers were from the countryside' (interview with ZLH, July 2009).

The *chengguan* (city management), for instance, had taken to patrolling in the mornings and evenings, appearing in convoys of three to four trucks carrying ten officers each. Their visits ranged from the routine patrols up and down the street to extensive raids, during which they searched shops and housing along the street for hiding hawkers. They were said to be very violent at times, and whereas they were not entitled to arrest and detain street vendors, *chengguan* officers were in a position to confiscate their products. Hawkers had quickly grown familiar with the *chengguan* routine and had established strategies (like hiding their goods in the shops along the street for the duration of the *chengguan*'s patrol) which enabled them to disappear from the street in less than 30 seconds. Correspondingly, they re-appeared to occupy their preferred spots just minutes after the *chengguan*'s departure. The *chengguan* officers, on the other hand, were fully aware of the futility of their efforts and even sympathized with the harassed street vendors – but stated that they were just doing their jobs.

In October 2008, residents of the Village still emptied their chamber pots at the public toilet, the saxophone player in the Compound still practised in the morning, and '*Congtiao . . . diannao!*' still could be heard occasionally. After an initial phase of conformity and adaptation, the people on the borderland began to work around imposed rules and regulations to reinvent their lives and livelihoods, regardless of the moral efforts and physical barriers instituted to prevent them from so doing. Shopkeepers behind the fence had gradually begun to remove concrete bars to facilitate better access to and more light inside their businesses (see Plate 12.5). In early 2009, young women could once again be seen on the street with large plastic tubs, doing laundry, washing their hair, brushing their teeth and bathing their children. A bicycle repair shop – its tenants busy repairing bikes and mopeds around the clock – had taken the place of the former Beijing duck store. Everyday life on the borderland was about to return to a slightly modified version of what it had been before the construction of the fence.

The borderland

Following Lefebvre's (1991) argument that social transformation must manifest effects on daily life, language, *and* space, we may assume that the ongoing transformation of contemporary Shanghai is 'truly revolutionary' in character. Graumann's (1983) 'multiple identities' model can help to illustrate and understand the different aspects that contribute to the formation of an urban identity for the city as an entity. The *desired* identity of the city (its 'identification with'), its buildings, and its residents is carefully drafted and propagated by those who manage it: Shanghai wants to be the global city[4] and glittering metropolis. Outsiders who have absorbed the image of the successfully branded city ('being identified') as well as resident groups, not unlike the locals, urban and rural newcomers outlined in this chapter,

PLATE 12.5 The fence in front of a shop on the street between the Village and the Compound, and a street vendor selling fish to customers
Source: Deljana Iossifova

have to negotiate the mismatch between these envisioned or imposed identities and the reality that they experience in their everyday lives through appropriation of space and the building of alliances and shared identities – the 'identification of'. Negotiation and re-negotiation between the different groups involved form the foundation of an emerging 'multiple identity' made up of the different layers created and appropriated by diverse actors. It is precisely their interactions and linkages that can lead to conflict as much as to the emergence of a 'new' urban culture and experience – a new urbanity; the pre-eminent site for this transformation is the borderland, the 'in-between' that carries 'concrete contradiction' (Lefebvre, 2003).

Urbanity, then, can be thought of as the condition of ongoing negotiation among co-present actors as well as among actors and their environments. The state and its citizens engage in the iterative transformation of society and space: the state, imposing an urban identity envisioned from above, is continuously challenged by the counter-tactics of citizens and their everyday lives. The co-presence of gated communities and decrepit neighbourhoods in immediate proximity in Shanghai makes existing social hierarchies readily perceivable. We find several of Barth's (1969) conditions of coexistence between the different groups: the locals and rural newcomers occupy the same niche (the Village), but they do not necessarily compete for resources and hence can achieve a stable condition of close interdependency. Similarly, whereas they occupy distinct niches, the 'Villagers' and the residents of the Compound compete minimally for resources, as they have very different needs. Rather, they can reach a stable condition because they can coexist in close interdependency through the services that they provide for one another (rural newcomers selling food to urban newcomers, for instance, or using the urban newcomers' park area within the Compound for recreation). However, the group dynamics between the distinct groups may take a dangerous turn in the future, owing largely to an urban identity politics propagated by a state that clearly favours its emerging urban elite. Where political ideology has such a strong impact on identity formation processes, social engineering should be taken seriously. Here, the role of the borderland – the space in-between, facilitating everyday encounter – in contributing to the process of positive appropriation of the 'other' through the continuous re-negotiation of learned values in everyday practice is of particular importance.

In Shanghai, coexistence of extremes in immediate proximity make possible an alternative urbanity, which invents its own rules and specificities, allows for informality at all scales, and permits high degrees of interaction between unlikely partners in a process of appropriation. The borderland in this chapter – that exists between the Village and the Compound – has been shown to reaffirm difference, rendering the street simultaneously a site of encounter and a representation of the social practice of spatial differentiation (Newman, 2003; Newman and Paasi, 1998; van Houtum and van Naerssen, 2002). In between gentrification, very real, and ghettoization, real or perceived, emerging coexistence requires constant negotiation between the individual members of the different socioeconomic groups involved. Co-adaptation among various groups of citizens (the locals, rural newcomers and urban newcomers, for instance) appears as they experience the conscious and unconscious transformation of values and belief systems and the changes in their everyday culture in response to the immediate presence and continued exposure to the respective 'other'. This co-adaptation is further manifest in space, as the existing built environment is transformed to respond to a new popular consumer culture, and as design intentions are ignored and long-established traditions and customs find their place in new environments. Shanghai's borderlands offer a unique chance for the emergence of a new urban culture across the different scales and times, as confrontation with the 'other', rather than an exception to be feared and avoided, becomes the normal condition of everyday life.

Notes

1   The study is based on a total of ten months of fieldwork in Shanghai between October 2006 and July 2009. The research strategy allowed for as much flexibility as possible in the selection and application of research methods. Photography helped to establish important contacts with residents and extended periods of open observation as well as over 60 interviews revealed the particularities of everyday life, the rituals of residents and visitors, and their interactions. In this article, abbreviations are used for the names of people and places.
2   The division of new-built apartments into smaller units and their rent-out to migrants, for instance, was prohibited in order to 'protect tenants' safety' (Yan, 2007).
3   It is through this kind of appropriation that 'space' becomes 'place' (Graumann, 1976).

4   Shanghai, as 'one of the world's most rapidly growing, emblematic twenty-first century cities' (see http://www.aaschool.ac.uk/STUDY/VISITING/shanghai.php), has recently been the city of choice and subject of investigation for summer schools, architectural and urban design units, field trips and courses in diverse disciplines at universities all over the world.

13

CONTEMPORARY URBAN CULTURE IN LATIN AMERICA

Everyday life in Santiago, Chile

Jorge Inzulza-Contardo

Introduction

Patterns of physical and socio-economic change have been observed in many large metropolitan urban areas in Latin America since the 1990s through the proliferation of urban policies to regenerate inner-city neighbourhoods. This chapter focuses on Santiago, Chile, which has undergone a 'tertiarization' of the economy including investment in new offices and housing projects, which has rapidly modified the urban landscape of a city previously characterized by its human-scale skyline. More specifically, I consider the impact of urban re-development on a historic area of the city. Drawing on empirical data from Bellavista which is one of the oldest inner-city neighbourhoods of Santiago, I show how developing an understanding of collective memory and local identity has much to offer Latin American urban planning theory as well as providing important insights which challenge 'Western' theorization of gentrification.

Gentrification and urban culture

Gentrification was initially observed in Islington, London, and Soho, New York, in the 1960s. Since that time gentrification has become associated with the displacement of working-class people (Glass, 1964; Pacione, 1990; Zukin, 1988), an increase of urban land values (Wyly and Hammel, 2001; Ulusoy, 1998), and the emergence of post-industrial urban production and consumption cultures (Smith and Herod, 1991; Smith, 1996; Thrall, 1987; Ley, 1996). More recently, theorists have developed a number of key indicators to highlight the changing nature of gentrification as a globalizing process that has spread to urban areas throughout the world (Hackworth and Smith, 2001; Lees *et al.*, 2008; Atkinson and Bridge, 2005).

Firstly then, the starting-point of gentrification from the 1960s onwards has become associated with neighbourhoods in the US (Soho, New York) and Western Europe (working-class quarters in Paris; Islington and the Docklands, London). Through investment from public/private partnerships in urban regeneration initiatives and the aligned housing investments of gentrifiers (artists and young professionals) improvements in the urban fabric of historic neighbourhoods began to unfold. Following such initial investment a second wave of gentrification in the 1980s, underpinned by new economic and cultural processes at global and national scales, began to attract middle-class residents back into the centre of cities, and along with the intensification of public–private partnership an 'aggressive entrepreneurial spirit' led to the 'anchoring of gentrification' (Lees et al., 2008: 175). A third wave of gentrification in neighbourhoods such as Brooklyn, in New York City, and La Condesa, Mexico City, has been described as a process of 'a generalised strategy of capital accumulation, extending and intensifying the processes seen in the second wave' (Cameron and Coaffee, 2005: 2). Moreover, this process 'is [a] purer expression of the economic conditions and processes that made reinvestment in disinvested inner areas so alluring for investors' (Hackworth and Smith: 468, 1167, in Lees et al., 2008: xviii). In most of the cases, gentrification is now generally being facilitated by corporate developers rather than through state initiatives. Finally, a fourth wave of gentrification is described by Atkinson and Bridge (2005: 1) as 'new urban colonialism' [and that gentrification is] ... no longer confined to western cities. Processes of neighbourhood change and colonisation represented by an increasing concentration of the new middle classes can be found in Shanghai as well as Sydney, or Seattle'.

Despite an understanding of the proliferation of gentrification throughout the world there has been a tendency for writers to apply theories developed in North America and Europe to cities elsewhere in the world with little critical reflection on the limits to such theoretical work. In the remainder of this chapter I outline how the case of Santiago helps to interrogate conceptualization of gentrification and in doing so I show how a Latin America case study can offer new insights that can advance research agendas focused on developing cosmopolitan accounts of urban gentrification.

The legacy of Latin American historic neighbourhoods

There is a growing understanding that models of urban development that have emerged from study of cities in Europe and North America have been applied to cities in 'developing countries' in a manner that ensures that those cities are always judged as 'backwards'. Such accounts have, of course, tended to ignore the specific development trajectories and internal differentiation of urban territories and societies in Latin America (Panadero, 2001). For example, Table 13.1 highlights general trends where pre-industrial architecture and symbolic values of quarters inherited from medieval, renaissance, baroque and classicism found in Islington, London, and Le Marais, Paris, have been restored through gentrification of housing from multiple-occupancy to family homes. In contrast, however, large cities in North and Latin American cities can be characterized through the emergence of high-rise buildings and suburban development (White, 1984; Yarwood, 1974). However, in many US cities such urban change has gone hand in hand with gentrification of nineteenth-century quarters with old industrial buildings being converted into home-studios, for example the loft living associated with Soho, New York. Similarly, in cities in Latin American a mixed heritage of both European and US style urban areas can be found in the inner cities. However, the eclectic mix of European architecture and a post-Hispanic gridded urban fabric characterized by local materials adobe, wood and brick has been observed as now existing next to the high-rise glass and steel buildings of international companies and gated communities in the last decade.

Moreover, socio-spatial segregation has only been a defining feature of Latin America urban landscape since the 1980s, through 'a dramatic spatial separation between the residential areas of bigger income groups and the residential areas of the poor people, revealing other differences such as ethnical and racial ones' (Lungo and Baires, 2001: 6). Spanish terms such as '*aristocratización*' used by Morse and Hardoy, (1992) and more recently '*la reconquista urbana*' (Sargatal, 2000) or '*elitización*' used by García (2001) highlight how gentrification in Latin American and Hispanic contexts is a process that has only recently began to rapidly change

TABLE 13.1 European, North American and Latin American inner cities in the 1990s

European features	North American features	Latin American features
More compact centre	Less compact centre (diffused)	Less compact centre (diffused)
Irregular urban fabric	Gridded and irregular urban fabric	Gridded urban fabric
Mainly terraced buildings	Mainly terraced buildings	Terraced buildings
Mainly middle rise buildings	Middle rise buildings	Mainly low rise buildings
	High-rise buildings	High-rise buildings
Old quarters (before 19C)	Old quarters (before 19C)	Old quarters
	Suburban areas	(around 20C over)
		Ghettos communities
		Gated communities
More pre-industrial architecture:	More post-industrial architecture:	Post-industrial architecture:
Medieval	Eclecticism	Post-Colonialism
Renaissance	Modernism	Eclecticism
Baroque	Industrial buildings	Modernism
Colonialism		
Conversions of	Conversions of	Conversions of
Victorian and	Industrial buildings (lofts)	Industrial buildings
Georgian buildings		and colonial houses
New studio-flats	New studio-flats	New studio-flats

Source: Based on White (1984); Yarwood (1974); Zukin (1988); De Mattos (2002).

the physical and social patterns of city centres. However, the '*reconquista urbana*' (urban recapture) described by Sargatal (2000) offers an important insight into gentrification in the Latin American context, as a way to reverse the depopulation of inner cities, but not necessarily with the returning of middle classes or old inner-dwellers to the core of cities; but 'as a subset of wider issues covered by terms like … rehabilitation, urban renewal and revitalization' (Rubino, 2005: 226). Jones and Varley (1999: 1,548) add that 'gentrification [for the Latin Amercian context] does not [necessarily] require the occupation of renovated properties by a new residential population', as noted in the urban renewal experiences associated with first and second waves found mainly in Europe and the US, 'but involves the rehabilitation of deteriorated properties [with other activities such as commerce and office] and in many cases, the replacement of existing housing areas with commercial activities'. It is to how these broad processes have unfolded in a specific historic neighbourhood that the chapter now turns.

Everyday life in Bellavista, Santiago

Bellavista, and more specifically the La Chimba quarter, has been described as one of the first suburban areas of Santiago, emerging as early as the sixteenth century (Winchester *et al.*, 2001). By the end of nineteenth century the growth of the León XIII quarter and the catholic Gran Santiago area (Hidalgo and Cáceres, 2003) ensured that Bellavista became one of the most important and diverse residential neighbourhoods in the city. Today, the area is constituted by two municipalities, Providencia (east side) and Recoleta (west side), with Pío Nono Street as the administrative boundary which is combined with the physical boundaries of the San Cristóbal hill (including the metropolitan zoo) and Mapocho river. As Figure 13.1 shows the geographical location when combined with the distinct mix of building typologies gives Bellavista a clear social, cultural and spatial identity. In particular, the neighbourhood is considered to be a key cultural hub in the city, being populated by writers, musicians, painters, architects and designers characterized by peaceful streets which are seen as a refuge for ideas, thinking and creativity.

Table 13.2 summarizes the main two main kinds of residents of Bellavista – existing residents who have lived in the quarter for at least 20 years (and in many cases easily for more than 40 years); and 'new residents' (young urban professionals). The 'older residents' group is estimated to contain 1,276 people, and is made up of artisans, painters, sculptors, architects, and so on and who tend to be owner-occupiers of terraced houses built at the end of the nineteenth century. The main reason for this social group in selecting Bellavista as a place to live was the affordability of housing, and access to cultural activities (theatres, galleries, schools) and the surrounding natural environment. In contrast, the 'newer residents' are usually aged 31 to 45 years of age; tenants and/or owner-occupier with mortgages; are from middle-class family; have a university degree; and are employed in the tertiary sector (marketing, computing, graphic design, architecture-drawing,

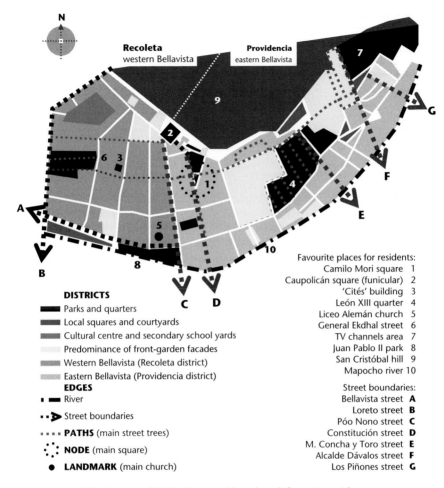

N

Recoleta
western Bellavista

Providencia
eastern Bellavista

7

9

G

2

6 3

1

4

F

A

E

5

10

8

B

C D

DISTRICTS
▬ Parks and quarters
▬ Local squares and courtyards
▬ Cultural centre and secondary school yards
▬ Predominance of front-garden facades
▬ Western Bellavista (Recoleta district)
▬ Eastern Bellavista (Providencia district)
EDGES
▪ ▬ River
▪ ▪ ▶ Street boundaries
▪ ▪ ▪ ▪ **PATHS** (main street trees)
⸫ **NODE** (main square)
● **LANDMARK** (main church)

Favourite places for residents:
Camilo Mori square 1
Caupolicán square (funicular) 2
'Cités' building 3
León XIII quarter 4
Liceo Alemán church 5
General Ekdhal street 6
TV channels area 7
Juan Pablo II park 8
San Cristóbal hill 9
Mapocho river 10

Street boundaries:
Bellavista street **A**
Loreto street **B**
Póo Nono street **C**
Constitución street **D**
M. Concha y Toro street **E**
Alcalde Dávalos street **F**
Los Piñones street **G**

FIGURE 13.1 The image of Bellavista neighbourhood from its residents
Source: Based on author's fieldwork (2007)

etc.). It is estimated that at least 1,620 new residents (12.6 per cent more than old residents) have located in western Bellavista (Recoleta district) since 2002. Interviews with newer residents highlight that they decided to move to Bellavista because of its proximity to their workplaces in central Santiago and respondents made no secret of their intentions to move to live in houses in outer Santiago in due course.

There were also significant differences amongst older and newer residents with regards to their perceptions of 'neighbourhood', with a 'friendly environment' valued more by the older residents (29.8 per cent) than by the newer residents (15.9 per cent). However, the 'diversity' of Bellavista and its 'cosmopolitan culture' were valued by 20.5 per cent of both older and newer interviewees with all appreciating the mix of social – groups and nationalities living in this area. As a

TABLE 13.2 Old and new residents in Bellavista, samples 2004

Profile	Old residents (living 20 years or over)	New residents (YUPs) (living since 2000)
Resident profile family composition	*Amount of residents*: estimated 1,276 residents 521 residents (66 or + years old) 755 residents (46–65 years old)	*Amount of residents*: estimated 1,620 residents 904 residents (31–45 years old mainly) 716 residents (from flats, estimated 2 people × 358 flats)
	Family composition: 2–5 people per household (married couple with children, or sons & daughters) 1 person per household (single, widow)	*Family composition*: 1–3 people per household (young couples with or without children) 1 person per household (single, woman householder)
Housing type and tenure	*Building typology*: Type A: terraced building Type B: garden-city building Type C: mainly 'Cités' Range 60–400 sq/m built area	*Building typology*: Type A: terraced building Type B: garden-city building Range 60–400 sq/m built area Type C: mainly gated communities formed by middle and high-rise building Range 35–80 sq/m built area (flat)
	Housing tenure: Tenants Mainly owner-occupied (without debt like mortgage)	*Housing tenure*: Tenants (poss. finance lease) Both owner-occupied (with and without mortgage)
Education & Employment	*Education degree*: higher education (professional degree, technical, artist, paint, writer, etc.) lower education from 2nd and 3rd sector (technical, worker)	*Education degree*: mainly higher education from 3rd sector (professional degree, postgraduate, manager, technical, graphic design, press, some workers, etc.)
	Employment: paid employees, self-employed, some managers, unemployed and house-keeping	*Employment*: paid employees, self-employed, some managers and unemployed

Source: based on (1) resident profile from 60 semi-structured interviews of Bellavista residents; (2) Municipalidad de Recoleta (2005); (3) Municipalidad de Providencia (2007); and (4) Census of Population and Housing 2002.

Canadian female journalist who has lived in Providencia district since 1980 pointed out 'the whole Chilean people can be found here … allowing an important human patrimony'. In addition, a 33-year-old female architect who is a new resident in Bellavista suggested that it was important to 'have in the same place something very heterogeneous'. In a similar vein, a 34-year-old female social assistant who is a new resident highlighted that 'Bellavista is democracy . . . poor and rich people are living next to each other . . . and this means a tolerant neighbourhood'. Two women who live in the Providencia district, a 75-year-old older resident and a 33-year-old actress new resident, both pointed out that 'this neighbourhood should be considered as the main patrimonial-area of Santiago' because '[Bellavista] is a very good urban area and a very good natural area at the same time' and 'it is the main cultural centre of Santiago'.

However, older residents, especially from Recoleta, described the 'negative' aspects of recent waves on newcomers to Bellavista with 23.5 per cent considering that their area had becoming 'unsafe'. For example, a 74-year-old artist argued that 'the houses have changed the residential land-use and this neighbourhood' and '[western Bellavista] is more a commercial area . . . with restaurants and industries so nowadays Bellavista is more annoying and risky'. A second respondent, a 73-year-old architect, claimed that Bellavista 'is dying . . . there are ugly stores and high rise buildings . . . [Bellavista] is a gloomy neighbourhood' (see Plates 13.1a and 13.1b).

Moreover, older residents in the Recoleta district argued that Bellavista has become associated with antisocial behaviour (alcoholism and violence) and a generally dirty environment. A 75-year-old male cobbler suggested that 'his neighbourhood [Bellavista-Recoleta] is too quiet during the day', and a 59-year-old woman hairdresser, who has lived there since 1983, also recognized this situation, adding that as 'this area [Bellavista-Recoleta] is not quiet and safe especially during the nights [she] normally [arrives] early at home so [she doesn't] have to go outside later'. The displacement of residential by commercial land use was constantly criticized by Bellavista residents as ruining the historical feel of the area (see Plates 13.2a and 13.2b).

In summary, the re-scaling of Bellavista in order to service the high-rise commercial and housing needs of the post-industrial core of the city has led to the alteration of the physical form of the neighbourhood which is ensuring the obscuring of traditionally valued vistas and views, the loss of open and green areas, and the destruction of street culture. In these terms, the physical, social and cultural heritage of Bellavista characterized by everyday life in the traditional terraced houses of Providencia and Recoleta districts valued by older residents is being gradually swept aside by newcomers who wish to live in new middle and high rise buildings. Furthermore, the character of the area is further being irrevocable altered through the conversion of one- to two-storey houses into commercial spaces or through their demolition and replacement by high-rise offices and housing.

PLATE 13.1 (a) and (b): Images of high-rise buildings located since 2003 in western Bellavista

Source: Jorge Inzulza-Contardo

PLATE 13.2 (a) and (b): The historic feel of Bellavista
Source: Jorge Inzulza–Contardo

Conclusion

Gentrification in Santiago is unfolding in a manner that replicates some of the broad characteristics of urban change in Western cities with regards to social and spatial segregation (Sabatini, 1997; Portes, 1989; The World Bank, 2000; Rodriguez and Winchester, 2001); the emergence of gated communities and the building of high-rise buildings (Rojas, 2002); and the increasing dominance of globalized consumption cultures that marginalize local traditions and identities (Jones and Ward, 2004). However, I have shown in this chapter that there are nonetheless significant points of departure which highlight how gentrification in Santiago is unfolding in a manner that diverges from theoretical models developed in relation to Western cities. For example, Bellavista has 'jumped' the first and second waves of gentrification observed in Western cities which is defined in terms of the displacement of the working-classes from historic neighbourhoods due to the return of the middle-class to the city centre. Indeed, gentrification since the 1990s in Bellavista can best be described in terms of '*la reconquista urbana*' (Sargatal, 2000) or '*renovación urbana*' which characterizes a process of the replacement of one-, two- or three-storey terraced houses by middle and high-rise mixed-use commercial and residential buildings which are fundamentally altering the social fabric of the neighbourhood. The case study of Bellavista thus importantly highlights how '*gentrificación*' in the Latin American context needs significant academic and policy attention (Jones and Varley, 1999). Such comments notwithstanding it is vital that theoretical work is undertaken that does not simply define urban change in Latin American cities with reference to practices and processes associated with case study research and theoretical debates that have been developed in relation to Western cities.

14

URBAN (IM)MOBILITY

Public encounters in Dubai

Yasser Elsheshtawy

'My City. My Metro', is a slogan introduced by the advertising firm Saatchi and Saatchi to entice Dubai's residents, local and expatriate, into accepting the Metro as a mode of transportation. The statement is meant to evoke some sort of communal feeling – a sense of 'we are all in this together', promoting a sense of social cohesion that has so far escaped the multitude of nationalities, ethnic groups, and social classes making up 'cosmopolitan' Dubai. Yet given the fact that less than 9 per cent of the city's residents actually use public transportation and taking into consideration the inhospitable climate that dominates the city throughout much of the year, it remains to be seen whether these rosy pronouncements will bear any relation to reality. Given the city's fractured urban form – in many ways echoing its demographic – it is indeed questionable whether a light rail transit system on its own can miraculously transform this incoherent urban fabric into a harmonious whole.

The situation in Dubai is seemingly unique in that the city is primarily known for its 'exclusivity'. One of its major landmarks, the Burj Al Arab, is inaccessible and can only be entered through a gate. Similarly, many of its attractions, it seems, derive their allure from this remoteness – the gated residences of the Palm Jumeirah, for instance, and even the Burj Khalifa, the tallest building in the world, can only be seen from a distance, or by purchasing a ticket to reach its observation tower. Yet while this inaccessibility is typical at the level of individual buildings, it acquires a more serious dimension if we take into account movement across the city and its various zones. For the city's morphology and its street system preclude any meaningful pedestrian movement. Confined to certain zones, outdoor activities are relegated to either highly regulated public parks or to street corners and empty building lots. Particularly for its low-income population who do not have the necessary mobility devices, public interactions are confined to informal outdoor settings. Unlike their mobile counterparts who can traverse city streets with relative ease in their air-conditioned cars to reach a shopping mall, a public beach, or a park, they have to endure long waits at bus stops, lengthy queues and multiple exchanges on buses.

In this chapter, I investigate Dubai's public settings and the extent to which its infrastructure has contributed to the creation of ethnic zones and the intensification of existing societal divisions. I argue that this is precisely because of controlled mobility even though the city has an extensive bus network and recently launched a light rail system. I set this within a theoretical framework that identifies the 'mobility turn' in the social sciences and explore the role of infrastructural networks in shaping the contemporary city. Moreover, I review the city's urban development to explain its current fractured morphology. My focus then shifts to its informal public spaces and how they are used by its low-income migrant community. In these spaces, which are invariably connected to a major bus stop, activities are intensified, and they act as a node responding to low-income residents' mode(s) of mobility. They are, in fact, quintessential transnational spaces. Moreover, the mere existence of these gathering points constitutes an act of resistance, a tactic employed by the 'less mobile' to overcome the strategies employed by city planners (de Certeau, 1984). Having set the stage, I investigate the role of Dubai Metro and its potential role, or lack thereof, of correcting these exclusionary trends or further exacerbating its splintered structure. I conclude by discussing both the uniqueness *and* the commonness of the Dubai case, and the extent to which it may offer potential lessons for urban theory.

The mobility turn

'In recent years there has been an increased focus in the social sciences on mobility, both in its spatial and social aspects. Largely prompted by advances in communication technology, and increased migration to cities, many scholars have begun talking about a new 'mobility paradigm' or a 'mobility turn' (Creswell, 2006; Sheller and Urry, 2003; Blunt, 2007). Research has focused on transportation and communication infrastructures, migration, and citizenship as well as looking at the politics of mobility and the 'material contexts within which they are embedded' (Blunt, 2007: 684). What has been of particular importance is the degree to which the increased movement of people has challenged conventional boundaries of nation states, notions of citizenships, and a re-constitution of conventional geographies, and this has had particular implications for the conceptualization of urban spaces.

One particular strand of research on transnational urbanism, and more specifically, the examination of the relationship between places of migration (Jackson *et al.*, 2004; Ley, 2004) has been to address issues generated by the proliferation of transnational experience and connections. According to Conradson and Latham, transnationalism 'has become increasingly recognized as an analytically nuanced way in which to think about contemporary forms of global mobility' (2005: 227); it has in fact been constituted as a counterpoint to studies that have focused on the 'hypermobility' of sophisticated cosmopolitan travellers. Realizing that they constitute a small portion of humanity, a shift has happened whereby researchers are looking at 'grounded attachments, geographies of belonging, and practices of citizenship' (Blunt, 2007: 687).[1] Closely associated to this are studies of diaspora, notably a sense of feeling at home in the host country while retaining significant ties to one's place of origin. These studies have challenged conventional theories of mobilities that distinguished between sedentary and nomadic modes of travel and movement (Creswell, 2006) by blurring the distinctions between them and also showing that people can navigate easily between the two. Additionally, as some have pointed out, it is not just new forms of mobility that are important but also the relationships which are engendered as a result, and the capacity to form networks with people who are not necessarily proximate – what Urry (2007) calls 'network capital'. Among the elements which he identifies as necessary for this is the presence of 'appropriate, safe and secure meeting places' a capacity that is dependent on 'affordances of the environment'.[2]

Thus an exploration of the affordances of the built environment, in terms of its ability to foster modes of interaction among city inhabitants, can illuminate the degree to which various modes of mobility have led to social inequality, for instance. Moreover, looking at the 'everyday practices inherent to transnational mobility' can highlight the degree to which these are processed through a complex interweaving of individuals, social networks and places (Conradson and Latham, 2005: 228). What is of particular interest to me in the context of this chapter is the extent to which these new mobilities have impacted upon people's use of, and interaction with, urban public spaces. I am further interested in the extent to which the city of Dubai differs from or conforms to these practices and the implications this has for urban theory.

Certain processes concerned with these issues have been emphasized by social theorists. For instance, Tim Creswell notes that mobile people are seen as a 'threat' that needs to be contained; hence planning for them involves legibility, order, hygiene and sedentary values (2006: 42). Moreover such practices are apt to result in inequality, underlining John Urry's assertion that the existence of multiple mobility systems has the effect of 'producing substantial inequalities between places and between people' (2007: 51). He also observes that a lack of citizenship and new kinds of social exclusion result from 'a combination of distance, inadequate transport and limited ways of communicating' (ibid.: 190).

The case of Dubai confirms these tendencies but it also challenges them by suggesting that a new mode of citizenship circumvents the limitations imposed by the built environment. By this I mean that the expatriate inhabitants of Dubai – with no realistic chance at obtaining citizenship or permanent residency[3] – have nevertheless carved out a transnational urban space within the city in spite of the existence of controlled mobility in Dubai, as I will subsequently discuss. While the presence of multiple modes of mobility – cars, taxis, buses, the Metro – may appear to integrate the 'splintered' city, it has nevertheless led to increased inequality. Before I examine the case of Dubai in more detail, in the following section I discuss the degree to which these new mobilities have led to what some have called a 'fractured' metropolis.

Fractured urbanism: the contemporary metropolis and its infrastructure

The modernist city came to consider the car as the pre-eminent symbol of freedom and mobility. Cities were thus designed accordingly by building freeways that in many instances cut through established neighbourhoods, in the process demolishing them and uprooting residents (Gans, 1962). Le Corbusier's vision of a 'City of Tomorrow' composed of freestanding towers within an idyllic landscape may not have been realized as originally intended in the centre of Paris but found its application all throughout the globe (Le Corbusier, 1925). Indeed, his book/ manifesto, *Towards a New Architecture*, uses the image of the car as the ultimate symbol of a new society (Le Corbusier, 1931). He particularly focuses on Italian architect Matte Trucco's Fiat factory in Turin, which used its rooftop as a racing track, as the definitive symbol of synthesizing cars, buildings and people. The story of how this vision transformed cities throughout the world is well known. Instead of idyllic landscapes surrounding majestic towers we ended up with slums and crime-ridden housing estates (Wolfe, 1981), what Urry (2007) calls 'places of absence'; and rather than enriching our public realm people retreated into their own enclaves, protecting themselves from the intrusion of others, to the extent that some cities have come to be known as 'city of walls' (Caldeira, 2000). Of course the prevalence of the car and the focus on its infrastructure cannot be solely blamed for these social ills, but their (continuous) impact cannot be discounted.

Urban architecture and design have become a function of movement. Efficiency and speed are the main determinants of a successful urban fabric. Any interruption to this flow is detrimental to a healthy city and thus needs to be dealt with effectively. Such has been the premise of city officials and transportation engineers. For them, people need to be moved quickly from one place to the next. Yet, as many urban sociologists and scholars have pointed out, there is a clear distinction that needs to be made between *urbanization* and *automobilization*. The former is concerned with the city as a locale that is moved across on foot – the act of strolling becomes a way of discovering and interacting with the city. Modern life is thus located within neighbourhoods and rooted in places (Sheller and Urry, 2000). The *Situationists*, who aimed to dismantle the modernist city by uncovering its hidden spaces and evoking a psychological association between people and places, used the act of drifting, an aimless wandering among city streets, to create an alternative geography that runs counter to the dictates of transportation planners and is thoroughly grounded in the experience of the pedestrian (Sadler, 1999).

Automobilization, or an excess of mobility, leads to a dismantling of the concept of place, and the extension of a person's geography beyond the limits of the city. A new sense of time and place is thus created – freedom for some, but for others an unfortunate loss of a sense of place, community and social cohesion. Terms such as 'unbundling' territory (Sassen, 1996) or 'space of flows' (Castells, 2000) are used to evoke and conceptualize this new spatiality. Indeed as Sheller and Urry

(2000) argue, automobility creates a split between various parts of the city that would otherwise have been accessible on foot.

Moreover, advanced transportation structures tend to bypass areas within and around cities, thus leading to geographies of social exclusion and ghettoization. Moving swiftly through the city people lose the ability to 'perceive local detail, to talk to strangers, to learn of local ways of life, to stop and sense each different place' and as a result, 'the sights, sounds, tastes, temperatures and smells of the city are reduced to the two dimensional view through the car windscreen' (Sheller and Urry, 2000: 210). Such visualizations have become an entrenched part of our popular culture to the extent that they have been evoked in novels and films – such as Ramin Bahrani's description of car mechanics living in the marginal spaces of Queens, New York, in his 2007 movie *Chop Shop*. As the narrative unfolds there is a juxtaposition between an impoverished and deteriorating neighbourhood, a highway that runs overhead, and the Manhattan skyline in the background, suggesting a distant and inaccessible dream.

According to Graham and Marvin, this new 'landscape infrastructure' contains embedded normative visions and social bias. Moreover these processes cause an 'unbundling' of infrastructure networks which helps in 'sustaining the fragmentation of the social and material fabric of cities' (Graham and Marvin, 2001: 15), leading to a phenomena they have termed *splintering urbanism*. Thus, the very tools of modernist planning become devices to further produce inequality: streets are constructed for the sole use of vehicular traffic; there is an absence of sidewalks; shopping centres become *internalized* spaces; and the prevalence of 'spatial voids' isolate buildings by surrounding them with large parking lots. Given that the entire structure of the modern metropolis is geared towards car movement, Graham and Marvin note that 'mass mobility does not generate mass accessibility' (ibid.: 118). In fact according to Hamilton and Hoyle (1999: 20), visibility does not guarantee accessibility: '(T)he shops maybe in full view across the road from the place where you live, but if there is a three-lane dual carriageway in between, and the nearest footbridge is half a mile away, the shops are pretty inaccessible to you'.

A significant outcome is the existence of an 'other' or 'hidden' city. These interstitial urban zones have been described as 'off-line spaces' (Aurigi and Graham, 1997) and 'lag-time places' (Boyer, 1996). Such varying spatial conditions sometimes overlap and intersect causing conflict. Christine Boyer distinguishes between a 'figured city' and a 'disfigured city'. The latter are the 'abandoned segments' that surround and interpenetrate the figured city. Remaining 'unimageable and forgotten', the disfigured city is largely 'invisible and excluded'.

What are the social implications for these divisions? For some it represents another form of poverty – a poverty of connections (Demos, 1997). Here a person's or a group's ability to extend their influence in time and space is limited. They are thus condemned to locally based ties and connections; they are prevented from connecting socially and technologically to the spaces of the modern metropolis. Mitchell (1997) observes that living in these marginalized spaces clearly shows that they are confined by time–space barriers, rather than being liberated from them. Such readings rather suggest a passive citizenry which simply accepts the limitations imposed upon them without any active resistance. As I will show in my readings of these marginal spaces in Dubai, there is, in spite of severe mobility limitations imposed by the built environment, a form of resistance as well as connections that go beyond the local and connection to larger global spaces. In fact recent scholarship in transnational urbanism has shown that many migrants engage in a process of transforming their limiting circumstances (Peter-Smith, 2000). Moreover, as Graham and Marvin (2001: 308) correctly observe, different cities face varying circumstances, each is 'embedded within a different economic, cultural, social and geopolitical context and history'. Thus an alternative narrative is being constructed, which attempts to circumvent the dystopian portrayals of the likes of Mike Davis (1990) and Michael Sorkin (1992), but also attempts to move beyond the idealization of traditional neighbourhood life (Jane Jacobs, 1961; Richard Sennett, 1974; Michael Sorkin, 2009; and many others).

For rapidly urbanizing cities such as those in the Gulf, the expansion of infrastructure is closely equated with modernization. Bhabha (1994: 84) keenly observes that they are an 'assertion of an embryonic national identity in the form of airports, four-lane highways and power stations'. I examine in the following section the degree to which this has been manifested in Dubai, looking at how road networks have shaped the city and led to its contemporary form, influenced by the desire to 'modernize' and establish a 'material representation' of the city's progress.

Dubai's urban development: fragmented morphology

'In 1960, British architect John Harris was asked by the ruler to develop Dubai's first Masterplan (Plate 14.1). The condition of the city at that time was quite underdeveloped, having no paved roads, no utility networks, and no modern port facilities. Water was only available from cans brought into town by donkeys. Travelling to Dubai from London took several days in unreliable piston-engine planes with overnight stops. Communication was also difficult. There were few telephones and cables were sent by radio. The Masterplan developed by Harris aimed at rectifying this by addressing some fundamentals: a map, a road system and directions for growth. This initial plan would guide Dubai's development and be modified with the discovery of oil in 1966.

Developments followed the call in Harris's Masterplan for the provision of a road system; zoning of the town into areas marked for industry, commerce and public buildings; areas for new residential quarters; and the creation of a new town centre. These rather modest goals were in line with the emirate's limited financial resources (oil had not yet been discovered in sufficient quantities). In 1971, due to the city's expansion and increased economic resources a new Masterplan by Harris was introduced. The plan recommended the construction of a tunnel running beneath the Creek connecting Bur Dubai and Deira (the Shindagha Tunnel) and the construction of two bridges (Maktoum and Garhoud); in addition, the building of Port Rashid was also envisioned (Gabriel, 1987).

PLATE 14.1 Dubai's first Masterplan by British architect John Harris
Source: Courtesy of Harris Architects

The late 1970s and early 1980s can be characterized as a period of rapid expansion (AlShafeei, 1997). Of particular note was the emergence of the city's growth corridor along Sheikh Zayed Road towards Jebel Ali. Dubbed the 'new Dubai', this area emerged as the new commercial and financial centre of the city. Numerous projects were constructed along this stretch of highway and the skyline of the city changed. These rapid developments are a result of increased resources and an attempt to provide alternative sources for revenue. Yet the main problem caused by these new axes of growth has been fragmentation and the emergence of a city composed of disjointed archipelagos or islands. Furthermore, the speed with which some of these projects emerged necessitated an approach that would not be based on a 'rigid' Masterplan, hence the development of the Dubai Structural Plan in 1995, the main aim of which was to be flexible enough to accommodate any changes. Conceptually the Structural Plan is based on a series of nodes and axes of growth which, for the most part, account for the city's form as it appears today (also see Elsheshtawy, 2010; Dubai Municipality, 1995).

As a result of these various development plans with their primary focus on the efficiency of car movement, one of the first impressions of the city is its fragmentary nature. Dubai is composed of multiple, disconnected centres which are separated by multi-lane highways. This precludes any meaningful pedestrian circulation or for that matter, the kind of conventional urban fabric that is only found in the 'traditional' areas of Bur Dubai/Shindagha and Deira. This *tabula rasa* type development has resulted in large gaps or patches between developments, generally consisting of vast expanses of sand which need to be filled. Thus the general feeling of the city in its present state is that of a construction site – it is still a work in progress – which has been further exacerbated by the financial crisis. Furthermore, lacking the high population density that would sustain a continuous rate of building, many areas appear empty without a sign of life. Such neighbourhoods lack a sense of community; they have a transitory feeling.

To get a clearer sense of the city's urban morphology, Plate 14.2 shows a figure-ground analysis of the city as a whole (without the Palm Island). The analysis clearly reveals its disjointed appearance and the lack of integration between its various parts. And while some areas display a typical, traditional urban fabric, such as the spaces surrounding the Shindagha district as well as Deira, the appearance of a traditional morphology – narrow and twisted alleyways – is deceptive since the majority of the buildings are four- to five-storey concrete structures built in the 1960s and 1970s. Others simply exist as isolated islands or archipelagos. Buildings are set within vast spaces, surrounded by parking. While these diagrams may suggest a deterministic quality, an impoverished public realm and a limited mobility imposed on its low-income populace, a closer examination reveals that within these vast anonymous spaces, an alternative order has emerged, slipping between the cracks of this carefully controlled metropolis.

Sites of encounter: Dubai's public spaces

As I indicated in the previous section, the city's urban development has produced an urban fabric that is composed of isolated communities that are not related to each other. Crossing from one area to another is difficult as the city is crisscrossed by multi-lane highways which separate districts from each other by creating barriers. As a result the ability to move freely by foot appears restricted. Officially sanctioned public spaces can be found in the city's multitude of shopping malls, public beaches, public parks, outdoor cafes and a newly opened pedestrian mall in the upscale Jumeirah Beach complex. However, to describe these settings as public is debatable. Though no one is explicitly excluded and they are theor- etically accessible for the entire population, various factors come into play to ensure that only certain groups do actually use these places, including the distance from work and accommodation, the presence of security guards, and charging for entry. And while sociologists have debated whether there is such a thing as a truly public space accessible to all and whether this is even desirable or possible (Madanipur, 1998), my concern here is not so much with whether these spaces are accessible or not, or with their degree of accessibility, but with the ability of less mobile residents to reclaim parts of the city as their own. In this sense, such groups may circumvent the official version of what constitutes publicness in the establishment of an alternative order, a transnational space in the city's streets, at street corners, next to its bus stops, and in abandoned parking lots that emerges as a result of their lack of mobility (Plate 14.3).

PLATE 14.2 Ground analysis of Dubai (without the Palm Island)
Source: Yasser Elsheshtawy

PLATE 14.3 Transnational spaces in Dubai?
Source: Yasser Elsheshtawy

In my investigation of these settings since 2003, I have been able to identify a series of nodes where I have observed an intensification of gatherings of low-income labourers and service workers hailing mostly from South Asia (Elsheshtawy, 2004). These settings are located in the city's central areas of Deira and Bur Dubai where one can find the highest density of residents. Given the support of surrounding settings – retail outlets, money exchange operators, ethnic restaurants, a Hindu worshipping temple, as well as transportation nodes – they contribute to the proliferation of such spaces, akin to what Amos Rapoport (1990) calls a *system of settings*. These sites serve a variety of functions but perhaps the two most important (based on observations and interviews) are communication and maintaining transnational connections. In terms of communication, they provide opportunities for meeting friends, exchanging news and generally serve as a venue for reinforcing a sense of identity *vis-à-vis* a city which does not allow them to engage more evident public displays of community formation. And they aid in maintaining transnational connections made visible by the ubiquitous presence of money changers, the proliferation of ethnic eateries and Internet cafes, and the informal selling of mobile phone airtime – these latter two being recent technologies of mobility that enable people to communicate and interact thus enhancing their 'network capital', or 'motility' (Kaufmann *et al.*, 2004).

These are of course places located in the city centre, next to well-known landmarks and in close proximity to major traffic interchanges. Being in them is relatively safe and incidents of serious crime or harassment are rare. The city is very much part of the fabric of these spaces and its close proximity, adds to its allure and attraction. But, there are other spaces, less known and thus a bit more dangerous – or where there is at least a perception of danger. Among these is the district of Satwa (its hidden alleyways rather than the main commercial artery), the Al Quoz Labour camps located in parallel to Sheikh Zayed Road, and International City, a new residential development placed at the city's outskirts next to Dragon Mart, one of the largest Chinese wholesale centres outside of mainland China (Simpfendorfer, 2009).

What is so striking about these places is that they are relatively close to the city its gleaming skyscrapers, clearly visible. Yet precisely because of this proximity there is a general sense of helplessness among its inhabitants because they know that they cannot access these exclusive places. The sense of 'being so near but yet so far' is evident. As a result these districts appear isolated and cut-off from the city's spectacular spaces. Unlike the settings in the city's centre however, this detachment has led to a general perception among Dubai's local population that crime is rampant, ranging from drug and liquor consumption to prostitution and in-fighting among gangs. These are sometimes imagined but in other instances such perceptions are supported by facts and incidents reported in the press. As Graham and Marvin (2001) point out, such places, because they have been bypassed by infrastructure networks, become breeding ground for various forms of illegalities. The absence of the state and forms of social regulation, an anonymous architecture that does not foster any sense of attachment, a lack of any effective urban planning guidelines, and the absence of any participatory mechanisms in urban governance are all contributory factors to the social breakdown in these areas. This is even more surprising given the intensive level of control exerted over various other areas of the city. International City in particular has become a prime example for what is in store for the city if current policies are implemented. It has effectively become a dumping ground for the city's low-income workers, has no direct connection to any transportation link, and is primarily accessed from the Emirates Roads highway. Intended as a site for medium-income residents and their families, it has since operations began become a popular spot for the accommodation of 'bachelor' workers who reside in large numbers in apartments designed for families. A thriving prostitution scene has emerged which has attracted the attention of the local press.

Of particular note is that this policy of sanitizing the city proper and effectively moving its less desirable elements to the urban fringes, in the desert, has been temporarily halted given the financial crisis. Satwa, for example, was on its way to being completely demolished and replaced by a high-end luxurious development called 'Jumeirah Gardens'. Plans have also been drawn up to relocate the massive Al Quoz Industrial area, mostly home to small industries and warehouses but also containing numerous labour camps and residential areas for workers, in addition to a mall specifically built 'by the workers for the workers'.

(Im)mobility in Dubai

One of the main problems facing Dubai is traffic, which involves endemic queues, air pollution, noise and accidents. Many of its major thoroughfares are in a constant state of gridlock, particularly the central area in Deira and Bur Dubai. Based on Dubai Municipality estimates some numbers are quite staggering: 470,000 vehicles are registered in Dubai and additional vehicles arrive from the nearby emirates of Sharjah and Abu Dhabi, constituting almost 1 million vehicles on the road (based on two trips per day), which is roughly equivalent to Dubai's population.

Prior to the financial crisis it was estimated that by 2020 the city's population would reach 4 million, and the number of car trips 13.1 million per day (Dubai Statistics Center, n.d.). The rate of growth in the number of cars is rising at an average of 10 per cent per year, which far exceeds the world average of 2–3 per cent. Such rates can be attributed to the relative affordability of cars and cheap fuel. Furthermore, public transport – while increasingly being expanded – is not as ubiquitous as it should be. To combat these problems and to develop a transport strategy, Dubai Government formed the Road and Transportation Authority (RTA). One of the major objectives of the RTA is to integrate the various transportation modes currently in operation, including bus, water-vessels and light rail (which is under construction). Another strategy is to restrict the numbers of people driving on certain roads through the implementation of a toll system – also known as *Salik*. Initially installed along the city's main artery, Sheikh Zayed Road, it is currently being expanded and will eventually cover other parts of the city. Other measures include dedicating special lanes for buses and taxis.

Dubai is the site of an entrenched car culture, with activities and movements revolving around the automobile. Even small trips are carried out by car and most buildings are surrounded by vast car parks to facilitate easy access. Some retail centres such as the Deira City Centre are primarily accessible by car. Pedestrian entries exist but are hardly visible or prominent. Instead the entire complex is surrounded by huge car parks that have been developed over the years to enable entry to its various additions. Recognizing these serious limitations and their implication on sustainability, authorities in Dubai have embarked on a drive to limit movement on roads, as noted earlier, but also to develop an integrated transit system with the Dubai Metro at its centre. Given that about 9 per cent of its population use public transportation, it remains to be seen whether they will succeed in changing the dependence on cars (Ahmed, 2010a). Otherwise, the Metro may simply end up being an expensive mass transit system whose main aim is to transport tourists from one shopping centre to another, acting as a sightseeing device, in addition to catering to the city's white-collar workforce at the expense of its labourers and other low-income residents who are most in need of an effective transit system and lack any other alternative.

Dubai Metro: spectacle or mass-transit system?

From its outset, the opening and operation of the Metro was meant to be a spectacular affair very much in line with the city's image. Its construction is seen by officials as another milestone in the city's glorious march toward modernity, international recognition and legitimacy. Even its opening date was cast in almost mystical terms, with great emphasis placed on the corresponding numbers, 9.09.09 at precisely 9 pm, 9 minutes and 9 seconds. There is no doubt though that the actual construction of the Metro is quite a remarkable feat. It is considered the first light rail network in the region of the Gulf Cooperation Council (GCC) and represents the pinnacle in advanced driverless technology. Built at a cost of $7.7 billion it contains a VIP section in addition to a women's only carriage as well as WiFi access across the entire network. It was built in less than four years although at the time of writing only the first of its two lines is partially operational. Remaining stations on the red line which runs from Rashidiyya near Sharjah/Emirates Road towards Jebel Ali near the border with Abu Dhabi are expected to open in 2010, whereas the green line, which intersects with the red line at Union station in the city's centre, is expected to open in 2011/12.

The lines are mostly above ground on specially constructed viaducts, but they go underground in the city centre due to a high concentration of residents and buildings. It traverses the creek and emerges again (red line) near the Burjuman Shopping Centre in Bur Dubai from where it continues its journey along Sheikh Zayed Road (Plate 14.4). It is here that the visibility of the Metro is maximized – the image of the futuristic carriages moving swiftly along the street next to the city's most spectacular high-rises no doubt suggests an image of modernity that officials were not going to neglect by going underground. According to Mattar

PLATE 14.4 Dubai Metro
Source: Yasser Elsheshtawy

Al-Tayer (2009), chairman of RTA: 'RTA was keen to make the Dubai Metro a global icon of transportation systems' and 'the launch of the Dubai Metro ... will make Dubai a world class destination.' DTZ, a real estate advisory firm, suggests that:

> the image the Metro portrays of Dubai should not be underestimated either. It represents the first and only urban transport system in the GCC and we consider that this will have a positive impact, attracting inward investment and globally mobile occupiers.
>
> *(DTZ, 2009)*

In accordance with these visions its stations are gleaming examples of a futuristic transit node, fully air-conditioned with plenty of escalators. Crossing roads is achieved via specially designed sky bridges. The stations are clearly visible, their entryways in some instances blocking views towards buildings, or they occupy entire street corners. Entrances of traditional light rail networks or underground systems tend to be marked discreetly by a sign and a staircase. Yet in Dubai entries are acquiring a monumental dimension, through the construction of large 'entrance pods' with little attempt at subtlety. Even with this focus on grandeur and spectacle there is still the possibility once all stations are operational that the Metro will indeed cater to the city's various social classes and that it will somehow unify the disparate districts constituting its urban fabric. In order to achieve this, the RTA has embarked on a comprehensive strategy linking its various transport modes as part of an Integrated Public Transport Plan.

Dubai's public transit system consists of an elaborate network of water boats, buses and the recently opened Metro. The water boats (also known as *abra*) traverse the creek at a cost of about $0.3 linking the city's two sides, Bur Dubai and Deira. These are situated in the historic core and are encountered by most tourists and visitors, although they are mainly used by the city's low income residents. The buses are more extensive and have recently been modernized. According to the RTA (Ahmed, 2009b) they operate around 1,800 buses, which serve most areas in the city. In order to integrate bus operations with the metro they introduced what has been called feeder buses, taking metro passengers to various destinations in proximity to respective stations. The effect of these remains to be seen once all stations are operational. In addition to feeder buses, the RTA has begun construction on a surface tram line called Sufouh, linking the 'upmarket' area of New Dubai along Internet City, Media City and Jumeirah Beach Residence amongst other prestigious areas. It will also link with the Palm Island monorail. The settings linked by this network are considered to be the most luxurious in the city, containing the Burj Al Arab, as well as the Madinat Jumeirah complex. In addition it will facilitate access to the high-end Jumeirah Beach Walk an open-air pedestrian mall. This seems to be reinforcing the idea of exclusivity. According to officials, '(T)he aim of the tram project is to encourage people in the upmarket areas to use an alternative mode of transport instead of private cars' (Ahmed, 2010b).

Assessing the Metro

Given that the Metro is still in its opening phase any attempts at assessment are ultimately speculative. Yet I would like to examine its operation to see the degree to which it falls under the theoretical framework that I have set out in this chapter – namely the extent to which the Metro will enhance residents' mobility or contribute to further isolate its already disparate communities. Will it merely serve as an image-making spectacle, reinforcing the city's reputation as an exclusive setting for a highly mobile, technologically advanced workforce?

One of the main arguments made in support of mass transit systems in general is that they are a key ingredient for a sustainable city. Yet one of the requirements for the success of such systems is that they serve a vital and large central business district (CBD). It is only when a city has a large and dense CBD that a large number of people who happen to live on a particular corridor need to travel in the same direction (Mohan, 2005). For example, Tokyo, New York, Paris and London each have CBDs with more than 750,000 jobs. Tokyo has one of the world's largest, with approximately 2.3 million jobs, with an employment density of approximately 58,600 per square kilometre, and almost all of central Hong Kong is a business district (Wendall Cox, 2004). In Dubai, the total number of the white-collar workforce is about 220,000 and they are not concentrated in a single CBD but are dispersed along various specialized free zone areas through-out the city (DTZ, 2009). This is not conducive to a successful rail operation. According to Mohan (2005), when business districts are dispersed and incomes relatively low as in Asian cities (compared to cities in high-income countries), the situation worsens for rail-based high-capacity transit systems. Shanghai City has about 82 kilometres of metro and light-rail lines, but rail transport only accounts for 2 per cent of the local traffic volume. Mexico City (population 10 million) has 201 kilometres of metro rail and it is the cheapest in the world, but it carries only 14 per cent of trips.

Given Dubai's polycentric and fragmented urban form a rail-based network may not have been the best alternative for its transit system. Other options were never explored or considered; for instance the much cheaper and efficient Bus Rapid Transit (BRT) systems successfully implemented in South America. BRT systems can achieve very high coverage at low investment costs. As they are road based they can go near homes and destinations and cover most of the city, as has been planned for 80 per cent of Bogota residents, for example. This would not be possible techni-cally or financially with rail systems. When road systems are modified for BRT, it results in complete urban renewal (Mohan, 2005). Moreover such systems can easily adapt to changing urban forms, which would be particularly suitable for a city such as Dubai given its rapidly changing cityscape. Ultimately, BRT systems mostly serve the poor and are effective in reaching outlying, low-income districts.

A key factor in determining the extent to which Dubai Metro will be successful is the share of public transportation noted earlier. RTA hopes to increase this percentage to 30 per cent by 2020 (from an estimated 5–9 per cent), an ambitious

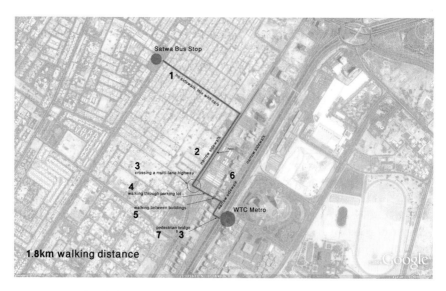

PLATE 14.5 Inhospitable spaces for pedestrians in Dubai
Source: Yasser Elsheshtawy

goal if compared to other major cities in the world. Another factor relates to acceptable walking distances, usually considered to be around 500 metres (or temporally, a 5 to 10 minute walk). Given Dubai's hot desert climate, that distance may have to be shortened. In addition to climate there is also the issue of land use which substantially impacts upon a pedestrian's perception of distance. Thus, if one is surrounded by varying retail activities, people, street vendors and performers, the effect of distance is substantially less than walking through a desolate landscape. Consider walking from the Satwa Bus station to the Emirates Tower station, which falls within the 500 metre walking radius. Analysing the actual path which a pedestrian takes reveals that s/he needs to walk across an inhospitable terrain for about 1.8 kilometres (Plate 14.5).

What exacerbates this situation further is the absence of any informal transportation network that may alleviate these problems. Remarkably, attempts at car-pooling are severely restricted by the RTA and can only be considered after submitting an application. Yet as of 2009 out of more than 8,000 submitted applications, only 200 permits were issued (Ahmed, 2009a). In addition, there are no dedicated pool lanes. Yet those who are most affected by this are the workers who, in their labour camps located at the city's fringes, are bypassed by the public bus network and have to resort to the use of taxis who overcharge them (Ahmed, 2007).

In spite of these misgivings, the Metro currently seems to be popular with users, although anecdotal evidence suggests that more than half of its use falls under 'fun', including the practices of tourists and sightseers, and only 21 per cent use it for work (Kurian, 2010). Land use values along the Metro stations have not increased, and though this may be related to the financial crisis, it may be an indicator that white-collar workers have as yet not given up on moving in the comfort of their cars. Some observations suggest that those who use the Metro live on the city's outskirts (McGinley, 2010). Recognizing that the success of the Metro's operations is not so much dependent on high-tech hardware but relies on upgrading the areas surrounding its stations, RTA has embarked on an extensive study involving TOD or transit-oriented-development, inspired by world-wide practices. This may prove to be a significant factor in moving the city towards a more sustainable urban form.

Conclusion: riding the Metro

Speeding along the elevated tracks of the metro the city seems to be somehow making sense. Passing its various landmarks – high-rise buildings, shopping malls, low-rise districts – as if they are frames in a movie viewed in close succession, creates an animated reality that does not appear if one is driving in a car or moving on foot. A sense of illusion is created, as if the whole city can be experienced through this high-tech moving device. Yet, the ride does reveal hidden spaces which would not have been perceived otherwise. The industrial district of Quoz, for example, with its backyards located next to warehouses, or the low-income residential area of Karama, revealing striking images of back streets and close-ups of balconies containing clotheslines. The lack of people in these spaces is striking, suggesting lifelessness (Plate 14.6).

Such images contrast sharply with the gleaming interior of the metro stations. Similar to lobbies in five-star hotels, they are highly regulated. Both private and

PLATE 14.6 Metro on the raised viaduct, heading towards the city centre, Dubai
Source: Yasser Elsheshtawy

government security personnel patrol its corridors and entryways ensuring that a carefully crafted script of movement is followed. Suspicious characters are checked for identification; a woman daring to drink a bottle of water is scolded for breaking the no-food/drink policy. Eateries and cafeterias have not opened yet further adding to the sense of a transient space that does not encourage lingering. The situation is no different once one emerges from these luxurious, cavernous spaces. Rigga station, located along one of the city's commercial thoroughfares, has not been transformed since the building of the station. The lack of any signs of spontaneous activities, gathering point or presence of vendors is evident. Once passengers (without a car) exit the Nakheel Harbour station along the Sheikh Zayed Highway, they move to the other side, along the road, to catch a 'feeder' bus. Standing in the relentless sun there is no sign of activity around them; watching them is like observing a set of hapless travellers stranded on the roadside waiting to be rescued by a bus, transporting them to yet another lifeless spot. The distance between the city's districts seems to have been intensified and made more visible, further separating its various social groups and seeming to produce a manifestation of the fractured and splintered city alluded to above.

Yet, in spite of this bleak picture, the city's migrants, as I have tried to show in this chapter, are creating their own order irrespective of a seeming state of limited mobility. For their network capital has been enhanced, and rather than looking at their movement from sedentary or nomadic perspectives, the distinctions are blurred. They are in perpetual movement, moving across the city from its distant labour camps, descending on its inner spaces. Through this mobility they manage to maintain ties between themselves, but also to their homelands. In this way they are able to sustain vital networks, and create a form of diaspora that is unprecedented, since ultimately, no matter how long their stay, they will have to leave. This reveals that though the physical setting may play a role in keeping people apart, the ingenuity of human behaviour can always manage to overcome such obstacles. The downside is the perpetuation of homogeneous districts and enclaves. People arrive in a place, but then remain there, unable to move beyond the confines of their gathering points. As Abdoumaliq Simone (2010b) reminds us, movement is an essential prerequisite for heterogeneity, a lesson that Dubai should heed if it aims to create a truly mobile, and socially sustainable, city.

Notes

1   For specific examples see Katharyne Mitchell (1997). Also see Michael Peter-Smith (2000) whose research on cities in South Asia on the everyday practices of city inhabitants inspired a wide range of studies.
2   Kaufmann, Bergman and Joye (2004) identify what they term as motility, or mobility as capital, which aims at exploring the relationship between spatial and social mobility.
3   See Elsheshtawy (2008) for a detailed discussion on citizenship in Dubai.

PART IV

Imaginaries

15

REALITY TOURS

Experiencing the 'real thing' in Rio de Janeiro's favelas[1]

Beatriz Jaguaribe and Scott Salmon

According to popular stereotypes of travel lore, the average Western tourist contemplating a vacation in Rio de Janeiro is led to anticipate a sojourn in a modern metropolis dividing sun-drenched days between the white sands of famed beaches and the picture-postcard venues of its iconic landmarks, while balmy nights can be spent drinking caipirinhas and swaying to tropical rhythms in chic nightclubs amidst the friendly company of 'exotic' locals. Sun, sand, sex – and samba – a Brazilian version of the tried and true travel agency formula. Given the currency of such tourist clichés, it is surprising that growing numbers of Rio's international visitors are embarking on an improbable and seemingly incongruous tourist pursuit: 'reality tours' through one of the city's teeming shantytowns or favelas.

Rio's favelas are mostly unplanned neighbourhoods of self-built dwellings which have sprung up on unoccupied land in the affluent southern zones – where they typically occupy the city's vertiginous hillsides – and in the sprawling industrial suburbs of the north and west. They are inhabited largely by mestizo people, many of them migrants from Brazil's rural hinterlands, who often don't possess legal ownership of their land or dwelling and whose scant access to education, sanitation and health is exacerbated by their beleaguered position in the crosshairs of what was, until very recently, an increasingly bloody war between often corrupt police forces and murderous drug lords. Life in the favela can be difficult and dangerous; they are an unlikely setting for a pleasurable tourist excursion.

In the global context of an inexorable planetary urbanization the proliferation of favelas and slums outside the West continues to question the sustainability of metropolitan living. Mike Davis provocatively warns us that the largest urban agglomerations of the twenty-first century will be located in 'third world' metropolitan centres where '(I)nstead of cities of light soaring toward heaven, much of the twenty-first-century urban world squats in squalor, surrounded by pollution, excrement, and decay' (Davis, 2006: 19). Rather than representing an anachronistic throwback to pre-modern conditions or transitional spaces outside the vocabulary of modernity, favelas and slums are an integral part of the uneven processes of capitalist urban modernization and pose a crucial challenge to its future.

Yet even in Rio de Janeiro, where an estimated 1.3 million people live in more than 600 favelas, the very meaning of the favela is changing and the growing popularity of favela reality tours is one indication of this. The favela has emerged as a bankable cultural brand. A wide variety of disparate products can now be bought and consumed under the rubric 'favela'. Aside from samba, funk, Carnival and the images and narratives of the favela fabricated by the media and the artistic imagination, the word 'favela' has come to connote a certain kind of cultural authenticity or alternative fashion statement. In fact the favela brand has become an international cultural phenomenon. Trendy restaurants in London, Paris (*Favela Chic*), Tokyo (*Favela*), New York's hip Williamsburg neighbourhood (*Miss Favela*), and Sydney (*Favela*) are all stylistically and gastronomically organized around a favela motif. Likewise a 'favela chic' has emerged in fashion circles where favela cooperatives such as the Coopa-Roca have become a successful fashion enterprise and the rubber flip-flop 'Havaianas', the emblematic footwear of the poor Brazilian, has been reborn as a designer item fetching outlandish prices in the boutiques of the world's fashion centres. Capitalizing on a growing recognition of the favela trademark, the Brazilian government chose to sponsor a 'Favelité Exhibition' at the Luxembourg Métro station during 'The Year of Brazil in France' in 2005 (Freire-Medeiros, 2009a). The walls of the subway were adorned with giant images of the Morro da Providência, Rio de Janeiro's oldest favela. Conversely, during the 'Year of France in Brazil' in 2009, the French photographer J.R. transformed images of the faces (and enormously expressive eyes) of women whose children or grandchildren had been gunned down in the battles of the drug trade into a giant installation projected onto the façades of shacks in the Morro da Providência. A version of the installation was also shown at the gallery of the Casa França-Brasil in downtown Rio. As a meaning-laden piece of cultural iconography, the favela, it seems, has arrived.

At the same time, the favela has attained a renewed prominence in civic debates surrounding Rio's urban future and its putative position in the global urban imaginary. Problems of poverty, crime and injustice have once again captured the public agenda. Of course, these issues are nothing new to Rio: it is one of the world's most unequal and violent cities, where the rich and poor live side-by-side yet in apparently distinct universes. But they have recently attained a new and pressing urgency as security forces in November of 2010 stormed one of the city's most notorious favelas, the complex of the Morro do Alemão in the northern zone of the city and evicted hundreds of heavily armed gang members, many of whom fled in advance of the arrival of the police forces. With the 2014 World Cup and 2016 Olympics now on the horizon, authorities are engaging in two simultaneous battles to transform the favelas: implementing 'pacification' schemes to control crime in the slums and investing billions of dollars remodelling the favelas as part of a wide assemblage of urbanization initiatives undertaken by both federal and municipal authorities. The future of Rio now seems inextricably linked to the fate of the favela.

Favelas have long occupied an ambiguous location in Brazil's urban imaginary. In the case of Rio de Janeiro, favelas were the paradoxical by-product of urban reform in the city (Abreu, 1997). During the modernizing reforms of the early twentieth century, poor inhabitants ousted from their tenements in downtown Rio resorted to building makeshift shacks near their work places. In fact, while conforming to the ideals of modernization, the urban reforms that prevailed throughout the twentieth century were mostly exclusionary and helped fuel the growth of favela enclaves in the city as a shortage of public housing coincided with the influx of poor rural immigrants searching for better conditions in the city. Despite the link between Rio's pattern of uneven urban modernization and the rise of favelas, chroniclers of the early twentieth century tended to perceive them as atavistic remnants of a pre-modern past. But by the 1930s favelas had emerged to occupy a prominent but ambivalent position in the national consciousness – as the locus of extreme urban poverty but also as wellsprings of popular culture – even as programmes of favela removal were still being implemented. By the 1980s the modernist perception of the favela as the 'national popular community' had been stained by the rampant violence of the drug trade that generated new and unprecedented patterns of social insecurity and violence. At the beginning of the twenty-first century, in the wake of the re-democratization processes of Brazilian society, favelas have once again been recast in the public consciousness, this time as the crucible within which the future viability of the city of Rio de Janeiro must be forged.

The favela lies at the heart of one of the most pressing and perplexing questions in Rio and indeed much of the increasingly urban world outside the West: how to transform, develop and integrate sprawling, often crime-ridden slums into the fabric of the city? In this chapter we explore aspects of the paradox of the favela at the beginning of the twenty-first century by examining how local realities intersect with the global imaginary in the increasingly popular phenomenon of Rio's favela reality tours. Framed by discourses of political agency, favelas are celebrated as a locus of cultural authenticity in the form of 'organic community'. They represent the 'real' Rio as emblems of popular culture, community agency and cultural inventiveness in the midst of adverse circumstances. At the same time, the favelas are disputed terrains where intractable problems of social justice, violence, citizenship, distribution of goods and services, and the vitality of cultural inventions are being negotiated by a plethora of social agents.

Locating Rio's favela tours

Quite apart from buildings, people, technologies, and social practices, cities are also fundamentally constructs of the imagination.[2] These imaginaries are manifest in media products, artistic endeavours and collective and personalized readings. As such all cities are contested terrains as different social actors, policies and economic forces engage in an active dispute of the fabrication of social reality (Bender, 2007). In the case of the favelas of Rio de Janeiro, we argue that they emblematize a particularly charged site of conflicting imaginaries. Rio's favelas have become intensely overwritten and exposed because they highlight contradictory imaginaries about urban modernities and cultural identities – and because so many of them are located in the midst of the city's upper-middle-class neighbourhoods. As heavily contested terrains favelas have gained an unprecedented visibility, not only linked to their formidable expansion but also triggered by new democratic expectations concerning social agendas, shaped by the ambiguity and anxieties of the middle classes towards the favela, and by the conflicting and juxtaposed representations of the favela as both the terrain of a vibrant popular culture and also the site of drug violence, poverty and crime.

The symbolic repertoires of Rio's favelas reflect both local and global expectations and they signal shifting perspectives on modernity, urban life and favelas themselves. Through an exploration of the meaning of favela tours, we seek to explore how global media-scapes, tourist consumption, specific kinds of urban poverty and violence are entwined and conflate the demand for the 'real' thing in the tourist experience. Contradictory representations of the favelas emerge within the specific contexts of the 'reality tours' that cater to foreign and local tourists by purporting to sell the experience of a brief sojourn in the favela as the 'real thing' in counterpoint to the standardized marketing of Rio de Janeiro as Cidade Maravilhosa (the Marvellous City).

The first favela tour was started in Rio de Janeiro in the early 1990s by Marcelo Armstrong, a young Brazilian who returned to his hometown after a globe-trotting stint in the international tourist industry. It was during one of Armstrong's pioneering Favela Tour excursions that the Englishman Chris Way envisioned the idea of presenting the slum of Dhavari in Mumbai as part of a reality tour package. According to this rendition this was how the concept of the 'reality tour' began its journey from local experiment to global phenomenon.[3] By 2005, at least seven favela tours were catering to tourists in the city, with Favela Tour; Exotic Tours; Jeep Tour; Be a Local Tour; Indiana Jungle Tours being the most established operations. Almost all of these tours centred the tourist experience on the favela of the Rocinha, although Marcelo Armstong's Favela Tour also included a stop in the small neighbouring community of Vila Canoas. In 2006, the Rocinha, popularly recognized as Rio's largest favela, was included as one of the city's official tourist attractions by the municipal tourist agency Riotur (Freire-Medeiros, 2009b). Recently the Jeep Tour has also included the favela of Dona Marta in its itinerary. Embedded in the neighbourhood of Botafogo, with panoramic views of the bay, Dona Marta – formerly notorious as violent and lawless – has become a major showcase of the new urbanization programmes and policies of public security exemplified by the occupation of the favelas by the Unidades de Polícia Pacificadora (UPPs), the 'pacifying units of the police'.[4]

A number of other favela communities have also sought to insert themselves into the tourist itinerary. These include Chapéu Mangueira and Morro da Babilônia in Copacabana, Pavão-Pavãozinho located between Copacabana and Ipanema, the historic Morro da Providência in downtown Rio that has been envisaged by state authorities as a kind of open air museum, and Morro dos Prazeres in Santa Teresa that attempted to organize a community-based favela tour that was eventually disrupted by local conflicts (Freire-Medeiros, 2007). Aside from 'reality tours', there has also been a recent boom of favela hostels, bed and breakfasts, restaurants and cultural centres. In the favela Tavares Bastos, the English expatriate Bob Nadkarni has set up 'The Maze', a music bar and an inn explicitly catering to foreign visitors. Located in the bohemian neighbourhood of Santa Teresa, the Pousada Favelinha advertises on its website that tourists will have the unforgettable experience of seeing 'at close range how a great part of Rio's population lives and also feel the receptive friendliness of the favela dwellers' (Pousada Favelinha, n.d.). It is, however, revealing that although the tourist boom in Rio's favelas was greatly fuelled by the global success of the film *City of God* (2002), there are still no organized tours to Cidade de Deus itself. Until recently, the immense and extremely dangerous favelas of the northern zone and the western zone of the city such as Complexo do Alemão, Borel, Favela da Maré were conspicuously absent from tourist sites. But the occupation of the Complexo do Alemão by the pacifying units of the police (UPPs) on 28 November 2010 and a series of urban reforms are dramatically altering the lives of many favelados or favela dwellers. Despite the recent intervention of the UPPs in a number of favelas in the northern zone of the city, Rio's reality tours remain primarily centered on in a relatively small number of favelas of the southern zone that have not only been somewhat 'pacified' but also offer spectacular views together with a vibrant popular culture.

Three basic premises underlie our reading of Rio's reality tours. The first relates to how imaginaries about the favela reflect conflicting agendas of modernity and modernization in Rio de Janeiro. The journalistic and artistic representations of the favelas in the early years of the twentieth century and their representation by foreign travellers and modernist artists in the 1930s to 1950s mirror shifting cultural perceptions that veer from a view of the favela as an atavistic element to be eliminated by Europeanized modernization and progress to an apprehension of the favela as a source of aesthetic invention and popular culture (Jaguaribe and Hetherington, 2004). In contemporary terms, the outpouring of films, reportage, public photographs and novels that have the favela as a central theme reveal an unfolding battle of representations. A crucial aspect here is that the favela is not only being represented by middle-class artists, writers and journalists but has also become an iconic site for the emergence of the authorship of favela dwellers themselves. Through NGO workshops in photography and film, through academic studies, through new forms of music and political activism, new readings of the favela by its inhabitants signal the force of the 'favela rising'. Almost all of these representations in film, fiction, reportage and documentary make use of a variety of realist aesthetic codes. The use of realism as a form of representation is not merely an aesthetic option but it also taps into the naturalized perception of realism as a legitimating discourse that renders social reality legible. The relevance of realist aesthetics is linked to their naturalized legitimacy as a direct encoding of reality itself and not merely as a 'representation' of reality.[5] The second contention that shapes our analysis and is directly connected to the previous argument about modernity and realism is that contemporary favela tours are geared toward the selling of the '*real thing*' as a form of direct contact with alternative realities. The promotion of the 'real thing' in the favela tour makes use of the spectacularization of realist aesthetics. The realist perception of the favela is fomented by media images in the daily newspapers and by widely viewed films such as *City of God* (2002), *Elite Squad 1* (2007), *Elite Squad 2* (2010), *News from a Particular War* (1999), *Falcão: Boys from the Drug Trade* (2006) and *Favela Rising* (2005), among a host of others. Journalistic reportage, testimonial narratives and literary novels also provide framing structures that shape amorphous and multi-faceted realities in the making. In this sense, artistic and media productions as well as the favela tours themselves are consumed as spectacles of reality. Finally, our last premise argues that the visibility of the favelas has lately become a motif of *branding* for the city of Rio de Janeiro. Faced with their inexorable presence – there are over 700 hundred favelas in Rio de Janeiro alone – municipal authorities, NGOs and grassroots associations of neighbourhood coalitions have all attempted to insert a more positive and domesticated perspective of favela culture as a means to promote Rio de Janeiro in the highly competitive process of establishing the city as a place of positive symbolic resonance in the global imaginary.

Rendering favela reality: exoticism, modernism and self-representation

In Brazil, favelas have a long history that is practically concomitant with the formation of the modern city. Indeed, the first favela enclaves in Rio de Janeiro trace their beginnings to the 1890s when soldiers returning from the suppression of the messianic uprising of Canudos in Northeastern Bahia camped on the hill of the Providência located in the downtown area. They were never granted the government housing they were promised and termed the hillside where they lived as a 'favela' in a reference to the vegetation that covered the hills of Canudos. During the urban reform programs of Pereira Passos from 1902 to 1906, the downtown areas of colonial Rio de Janeiro comprising of crumbling colonial mansions converted into proletarian tenements were razed to the ground. The consequent lack of popular housing spurred the expansion of proletarian suburbs but also the creation of favela settlements in downtown Rio.

At the beginning of the twentieth century, the journalists and writers who ventured into the favelas framed their excursions as travel narratives where they were cast in the role of the emissaries of the 'civilized Rio', depicting the lives of the 'excluded' for their middle-class audience (Jaguaribe and Hetherington, 2004). Likewise Alba Zaluar and Marcos Alvito (1999: 12) suggest that favela enclaves served as 'an inverted mirror in the construction of a civilized urban identity'. In a chronicle published by the writer and journalist João do Rio

(1881–1921), *The Free Camps of Misery*, the favela of the Morro Santo Antonio is not only characterized as a zone of scarcity, samba and promiscuity but also portrayed as a backward enclave, a place standing outside of modernity:

> I imagined that I was arriving from a long journey from the other end of the earth, from a promenade through the camp of sordid merriment, through the unconscious horror of a singing misery spun by the vision of those hovels and the faces of vigorous people who wallowed in poverty instead of working, people constructing right in the center of a great city an unusual camp of indolence freed from all law.
>
> *(Martins, 2005: 58–9: 83)*

Despite enjoying a camaraderie with a few samba musicians, João do Rio ends his chronicle observing that those spaces of poverty were surely rife with smallpox and he rushes towards the normative city as if fleeing from the menace of contagion. This dichotomy between the 'modern' city and the 'backward' favela permeates the writings of authors from this period (Zaluar and Alvito, 1999). Thus, the Parnassian poet Olavo Bilac (1865–1918) could characterize the favelas as 'a city outside the city' (ibid.: 12). In the chronicle *Fora da Vida* he pontificates that the washer woman who never left the favela enclave was 'so morally distant from us, so concretely separated from our lives, reclusive in space and time she was living in the past century, in the depths of China' (ibid.: 11). By contrast Lima Barreto, the foremost critic of the discriminatory practices of the Old Republic observes in regard to the modernizing reforms of Carlos Sampaio in 1922, 'One can see that the major concern of the present governor of Rio de Janeiro is to divide it into two cities: one will be European and the other will be Indigenous' (ibid.: 12).

The chroniclers of urban life in Rio at the beginning of the twentieth century sought to encapsulate the favelas within the vocabularies of classification, metaphor and moral judgment as places that seemed incompatible with the modernizing desires of the city's elites. The polarized presentation of 'civilized Rio' and the favela as the blighted terrain of the marginalized poor would be maintained for many decades. The dichotomy encapsulated by the hills (*morros*) versus the asphalt opposition signaled that the favela was outside the jurisdiction of the modern. By casting into metaphor the favela as the place of anachronistic debris, cancerous residue and dangerous terrain, the chroniclers of the Rio Belle Époque were repeating the usual refrain about the 'dangerous classes' without weaving the connections between unequal patterns of modernization and the emergence of the favela itself (Zaluar and Alvito, 1999: 9).

Like the backlands of the sertão, the favela seemed to present an intractable obstacle to the positivist motto of 'Order and Progress' emblazoned across the Brazilian republican flag. But whereas the sertão and the saga of the messianic uprising of Canudos were given heroic narrative in Euclides da Cunha's (1902) famous book, *Os Sertões*, the favelas of Rio de Janeiro were accorded no such status (Valladares, 2005). Embedded as it was in the scientific prejudices and prerogatives of his time, da Cunha's master work nevertheless denounced the genocide perpetuated in Canudos by the Brazilian republican army. But the customary neglect of the city's poor by the municipal authorities, the razing of popular housing during the reforms of Pereira Passos in the early twentieth century, the rapid expansion of favelas during the 1930s and '40s and the subsequent removal policies of the 1940s, '60s, and '70s were naturalized as necessary actions that would make Rio an ordered and livable city. The desire for progress notwithstanding, the razing of tenements during Passos' reforms did not even contemplate the provision of any alternative housing for the poor.[6]

It was during the 1930s and '40s, as debates around conceptions of a national identity were intensifying, that a novel premise emerged: the notion of a possible compatibility between modernity and mestizo culture. Just as the language of civic discourse was being infused with often competing vocabularies of modernity and nationalism, it inspired new, albeit contradictory contemplations of the meaning of the favela. Celebratory nationalistic interpretations of the favela as the organic cultural hearth of samba co-existed with pejorative depictions of favelas as zones of scarcity, poverty and lethargy. This apparent ambivalence was nowhere more evident than during the Estado Novo ('New State') of President Getúlio Vargas (1937–45). Samba schools were given government sponsorship and compelled to organize the Carnival parade around national motifs, the National Radio broadcast samba to enthusiastic audiences, and Carmen Miranda, the white Baiana, was exported as a folkloric national symbol to the United States. At the same time, the 1940s remodeling of Rio in accordance with modern parameters led to the proliferation of monumental public edifices that mirrored modern architectural vocabularies and entailed the construction of the massive Avenida Presidente Vargas that tore down housing for the poor and demolished pre-existing historic monuments. Concomitant with the elevation of Rio de Janeiro's popular culture as the national culture, the nation's lettered elites were largely betting on a progressive whitening of Brazil and seeking to propagate the presence of the state as a means to the eradication of the poverty and underdevelopment exemplified by the favela enclaves (Skidmore, 1974). Yet in the invention of the modern Brazil based on its mestizo identity, the favela also emerged in the representations of Brazil produced for foreign eyes.

At different moments during the decades from the 1920s to the 1960s, a serious of illustrious travellers, including Marinetti, Levi-Strauss, Le Corbusier, Blaise Cendrars, Albert Camus, Orson Welles and Marcel Camus, ventured into Rio's favelas and produced narratives, photographs and architectural designs inspired by their experiences (Jaguaribe and Hetherington, 2004). In his two trips to Brazil (1929 and 1936), Le Corbusier, upon viewing the city from atop the favela hills, was spurred to imagine an immense viaduct-like building, a gigantic functionalist worm slicing off the entrails of Rio de Janeiro's southern zone. The sinuous viaduct-building would emerge from the tropical landscape as a functionalist utopia amid the chaos of a city overridden by makeshift favelas. Albert Camus

(1991), during a brief visit in 1949, found most of the local elite and the intellectuals to be pompously boring and colonialized. His sojourn in the samba community and experience of the Afro-Brazilian religious macumba ceremonies were portrayed as an energizing respite from the fakery of the tropical bourgeoisie (Jaguaribe and Hetherington, 2004). But it was through the work of Frenchman Marcel Camus that the favelas of Rio de Janeiro were to achieve an unprecedented international reputation. Camus' film *Black Orpheus* (1958) won international acclaim and depicted the favela as a vital, music-filled, lyrical and folkloric terrain.[7] According to the dictates of modernist architecture and planning, the chaos of Rio's favelas had to be replaced by the functional orderliness of the *parques proletários* espoused by the Vargas regime. But in the visual arts, music and modernist literature, the favelas possessed local colour and the vitality of popular cultural invention. Samba was extolled as an authentic expression of popular culture by key modernist intellectuals like Brazil's foremost classical composer, Villa Lobos, who achieved a prominent role during Vargas' Estado Novo (Vianna, 1995). In this sense, the modernist excursion to the favela anticipated the appreciation of an authentic cultural experience, something that would later be repackaged as the 'real thing' by the selling of the favela tours.

By the early decades of the twenty-first century, Rio's favelas have received a more positive appraisal derived from more democratic social agendas, the critique of social inequality and the positive value attributed to 'third-world agency' that also relies on a diffused anthropological appraisal of culture. The notion of 'culture', previously associated with the domains of the lettered elites, modernizing technology, or the grandeur of past legacies, has shifted to include visions of a world shaped by a diversity of ethos and symbolic meanings that incorporate cultural diversity, class struggle and multiple social identities. From this perspective, the hierarchies between 'high' and 'low' culture evaporate. Furthermore, the overwhelming presence of media culture insures that cultural manifestations are bought and sold through diverse marketing strategies. In the wake of the valorization of popular culture, the favela is recast as a site where local communities seek to overcome scarcity, violence and poverty by means of an inventive cultural hybridity that adapts and maintains community networks in the face of adversity. This is a space where the precariousness of urban existence is mitigated by images and narratives of resistance, inventiveness and agency. However, the fact that the drug trade in Rio de Janeiro is primarily conducted there means that favelas are also seen as crime ridden, violent and lawless spaces. Enmeshed in the web spun by neighbourhood associations, international NGOs, the globalized drug trade and state interventions, the favelas are zones of porous negotiation. The positive rendition of the favela by NGOs, neighbourhood associations, and the tremendous impact of favela cultures upon sectors of middle-class youth do not guarantee, however, its entrance into the domains of the privileged city. Until 2010, real estate properties adjacent to violent favelas suffered massive devaluation. But the recent action of the UPPs' pacifying police units have meant that property values – even those proximate to favelas – have experienced a considerable upsurge in value. Symbolically significant locales such as Copacabana and Botafogo have had their real estate markets warmed up by the direct intervention of the UPPs. At the time of writing, it was impossible to evaluate whether these policing activities would have a lasting impact and whether they would meaningfully contribute to the wellbeing of favela residents themselves. But it was clear from the moment of their implementation that such policies serve dual agendas: they are geared toward addressing serious public security concerns and respond to dire social need, but they also conform to an urban marketing strategy – made yet more pressing by the awarding of the 2016 Olympic Games to the city – that seeks to recuperate Rio's image in the global tourist marketplace.

From the eradication policies of the beginning of the twentieth century, the removal strategies of the 1960s to '70s, to the urbanizing programs of the later twentieth and early twenty-first century, the relationship between the favela and urbanized Rio has undergone a radical shift. Indeed, until the late 1990s, the prevalent rubric that defined Rio was of a 'divided city' (Ventura, 1994). But although this dichotomy prevailed due to its synthetic power it was always an over-simplification of the negotiated, unequal yet porous relations between the favelas and the city of the middle classes. Favelas have not only come to stay but they represent a crucial feature of the modernizing impulse itself. Both populist political rhetoric and the just appraisal of the rights of the favela dwellers have ended their indiscriminate removal. The expansion of favelas and the rampant violence generated both by the drug trade and its repression by police forces have demonstrated not only the failure of a repressive modernizing project but also the complex web of entanglement between corrupt police forces, corrupt politicians and the drug trade (Arias, 2006). The increasing social democratization of Rio de Janeiro – despite the routinely unchecked abuse of civil rights – the political agency of the favela, and the partial success of some urbanization projects reveal that favelas are, now more than ever, a crucial part of the city fabric.

In several key respects the transformation of favelas into tourist sites both contributes and responds to their increasing visibility, and reality tours form part of the attempt to incorporate and domesticate terrains that were, until recently, considered outside the jurisdiction of the state. The uncertainties that besiege the favela made manifest by the ambivalent juxtapositions between images of the 'national popular' and the spectres of marginality, and between the celebration of cultural hybridity and the fear of the 'favela rising', are now intensified by its very media visibility. It is significant that the clichés, the cultural innovations and new agendas surrounding the favelas are usually couched in the registers of realism. Realist registers with their claim to accurately represent 'reality' are called upon to reveal the 'flesh of the world' by intensifying perception through photographic, journalistic and prose tropes that portray social realities. The point is that the naturalization of these registers posits them not as 'representations of reality' but as a direct manifestation of reality itself and in this sense they provide an entertaining 'pedagogy of reality' (Jaguaribe, 2007b).

Packaging the favela as spectacle

Although the favela has been exhaustively interpreted from both popular and scholarly viewpoints and is the subject of a plethora of reports, policy and planning projects, it is only fairly recently that it has become the focus of tourist exploration. While the visitors and travelers that came to Rio de Janeiro in the '30s and '50s made brief forays into the national popular and exotic favela, these modernist incursions were largely individual enterprises that lacked the organized contours and carefully scripted elements of contemporary tourism. Officially the Rocinha has been considered an urban district since 1993 but it is still perceived as a favela by both dwellers and outsiders. In fact, in media and everyday discourse, the Rocinha is routinely described as the largest favela in South America. In 2006, the Rocinha was officially recognized as a tourist site by the municipal tourist agency Riotur. The former incompatibility between the favela and tourism has shifted completely and favelas are now enthusiastically advertised as a unique facet of the city; just as the favela tour is sold as an entrée into the realm of the 'real thing'.

The contemporary favela is recognized as a symbolic source of cultural invention that fabricates artistic expressions, forms of fashion, and modes of socialization. Yet the term favela encompasses such an urban variety that it no longer has a singular vocabulary that can account for its diversity. If there are many aspects to the lives of the poor and a diversity of cultures of poverty, there are also many forms of social critique that have increasingly surfaced within the urban imaginary. The global impact of the media, neighbourhood associations, NGOs and the multiple agendas of identity politics provide a wide range of social interpretations. Tourists who formerly came to Rio intending to visit the usual postcard sites and relish the local are now also being called to engage in the favela tours – and now in those favelas that have been occupied by the UPPs. The favela tour is still shaped by the presentation of the 'real thing' but now adds a narrative of favela life 'before' and 'after' state intervention policies. Thus, the current thrust is not centered solely on the 'spectacle of poverty' and less in the 'thrills of outlaw violence' but rather in the vision of a community on the rise, of a new Rio that is shedding the ingrained dichotomy of being a 'divided city' in order to advertise its potential future as an 'integrated metropolis'. Nevertheless, despite the recent urban reforms undertaken by the PAC and the costly repaving of one of the streets of the Rocinha flanked with colourful buildings inspired, according to the Secretaria Estadual de Obras do Rio de Janeiro (State Secretary of Public Works in Rio de Janeiro), by the touristic Caminito in Buenos Aires, the tourist still can view raw sewage, makeshift buildings and the material signs of persistent poverty.[8]

What is it that makes urban poverty attractive to the tourist gaze? It is possible to reconcile the implicit voyeurism of tourist activity with a genuine apprehension and appreciation of social reality? The voyeuristic appeal of poverty is hardly new. As Phillips (2003) points out, as far back as 1884, 13 years before the first record of a Brazilian favela, the *Oxford English Dictionary* defined the verb 'to slum' as the tendency to 'visit slums for charitable or philanthropic purposes or out of curiosity'. 'Especially,' the entry continued, tellingly, 'as a fashionable pursuit.' Well over a century later, more and more people from across the world are now making the pilgrimage to Brazil's shantytowns, to the reality spectacle of the favela. Tom Phillips (2008), a British writer living in Rio sees it this way:

> There are around 600 redbrick favelas in Rio de Janeiro, home to around a million of its poorest people. The slums are considered no-go zones by most Brazilians. Foreigners, however, have always shown more interest in Rio's impoverished underbelly. Since the 1980s, when guidebooks suggested cunning ways to sneak a peek at the favelas without actually going in, 'poorism' has been a growing trend. Within a few years tour companies began offering visitors the chance to talk to locals, visit social projects and buy art from 'authentic' (i.e. dirt poor) Brazilians. Today a new, less savoury generation of 'poor guides' has sprung up. For around £30 a head, they offer to transport thrill-seeking foreigners into a real-life version of City of God, the acclaimed film about Rio's gang culture.

Rio's favela tours are catering to a growing urge to experience the 'real thing' as a form of experiential interaction in a world of globalization. As Bianca Freire-Medeiros (2009b: 580) suggests, favelas are inscribed as part of the package that frames reality tours. Of course, the experience of the 'real thing' does not necessarily mean a nuanced engagement with a social reality; it can also entail purchasing intense thrills by a sensationalist viewing of otherness. Indeed there is an inherent ambiguity in reality tours which seek to appeal to tourists in search of social awareness, while simultaneously catering desires to plunge into adrenaline drenched 'extreme tourism'. In practice this contradiction is partially resolved by market segmentation through the varying approaches and discursive framings employed by the different tours operating in Rio. For example, tours such as Favela Tour, Exotic Tours and Be a Local are largely conducted on foot and the tour guides (occasionally young favela residents) place heavy discursive emphasis on authenticity, organic connection to place and the presentation of the favela as a 'everyday community' (albeit an exotic or folkloric one). On others, such as the Jeep Tour and Indiana Jungle Tour, tourists are driven through the favela safari-style, seated in open-backed jeeps, stopping only at predetermined – and pre-scripted – locations where an 'arrangement' has been made with a local resident or store-owner. Likewise, the discursive presentation of the favela offered by the jeep-based tours is more extreme than that of the pedestrian tours, emphasizing the exotic, sensational and dangerous elements of the favela spectacle. On one such tour in 2008, as the jeep sped towards the Rocinha, our guide turned to his gaggle of tourists and announced melodramatically that we were about to enter 'no-man's land' where the 'State of Rio has no jurisdiction' and the 'drug lords rule'. He followed this up with alarmist warnings about the need to refrain from photography and to 'not look the drug dealers in the eye'.

Aside from the particularly sensationalist neocolonial vocabulary of these more 'extreme' tours, most of the reality tours operating in Rio de Janeiro attempt to depict places of poverty, exclusion and violence as 'authentic communities'. In this sense, reality tours would seem to endorse MacCannell's (1976) influential argument that tourism is a search for authenticity in a modern world devoid of enchantment. In the case of Rio's favela tours, this search for 'authenticity' is tempered by a desire to experience 'the real thing' safely ensconced within the familiar, protective confines of the organized and guided tour. Although the seductive ingredients of exoticism still permeate the marketing and discursive framing of favela tours, this exotic appeal is quite distinct from the early packaging of travel tours to pristine or 'barbaric', faraway regions. The exotic element here is not framed by a perception of the local native as a radical otherness; neither is it couched in a picturesque rendition of the favelados as 'primitives' or rural yokels transplanted to the big city. The exotic element is the selling of daily ordinary reality as an experience of the extraordinary not because this reality is unusual but because it is not the reality of the tourist. What the favela tours promise is that the tourist will have access to an experience of a reality that is unfamiliar but nonetheless intensely 'real'. Reality tours presuppose that foreign tourists live cocooned in media-saturated environments that curtail direct access to other worlds. The attraction of the favela is thus underpinned by the confluence of contradictory imaginaries. Favelas are depicted as authentic communities inhabited by poor, hard-working and cheerful people who creatively attempt to survive amid the poverty and violence that afflict them in their gaze-arresting topographic scenarios. This enticement, however, is not just premised on agendas of political agency, on the desire to contribute to the economic wellbeing of impoverished communities, nor is it just framed by a voyeuristic desire to see 'real' poverty without personal consequence. Favela tours also offer the excitement of venturing into the violent terrains of the drug trade depicted in myriad films, photographs, reportage and narratives. In a world of standardized package tourism, favela tours ignite the desire for an 'authentic experience' that is, nevertheless, heavily mediated – and validated – by previously tested symbolic repertoires. Within these overlapping representations, favela dwellers themselves generate modes of self-fashioning that are also influenced by the global media, cultural identities and local practices. The tourists that venture into the favelas generally have an 'anticipation of the experience' (Urry, 2002) framed by their own prior readings of popular narratives and images (Jaguaribe and Hetherington, 2004). Yet despite this voyeurism and aesthetization, favela spaces still cannot be readily absorbed as 'theme parks' because they are unstable, contradictory and porous.

The spectacle of the favela, pre-packaged for the tourist gaze, is now routinely offered as an excursion that can be purchased at the hotel desk alongside trips to Corcovado, the Sugar-Loaf and/or other sites of interest and amusement. If previously it seemed that Rio could be redeemed by the favela if only it was understood as a reality rather than a myth – and this message was certainly encoded in the selling of the reality of the favela tour (Jaguaribe and Hetherington, 2004)

– more recently, the wholesale intervention of the state suggests a vision of the 'real' wherein civil society, favela inhabitants and the agents of the state are reworking the conditions of life in the favela itself. Despite differing emphases in their approaches (and clientele), all of the tours insist on the presentation of the favela as the 'real thing'. While the Indiana Jungle Tour and the Jeep Tour provide some information about the Rocinha, they do not stress the socio-political aspects of favela reality because their aim is to offer tourists a taste of excitement and adventure which can be sold as a thrilling tourist experience alongside a hang-gliding excursion or a hiking expedition on the jungle trails of the Floresta da Tijuca. But selling the favela as part of a quasi-anthropological, social and cultural experience is the central premise of Marcelo Armstrong's Favela Tour, Rejane Reis' Exotic Tours and the Be a Local Tour. A sampling of the rhetoric that packages the tours can be found on their respective internet sites.[9] Favela Tour, the oldest enterprise organized by Marcelo Armstrong, extols its product as 'an illuminating experience if you look for an insider view of Brazil'. The Favela Tour,

> introduces you to another Rio, within Rio city: the favela. Picturesque from a distance, once closer they reveal their complex architecture, developing commerce and friendly people. Don't be shy, you are welcome there, and local people support your visit. It you really want to understand Brazil, don't leave Rio without having done the Favela Tours.
>
> *(Favela Tour, n.d.)*

Just as in any industry, rivalries have emerged and competing claims of authenticity are hotly contested. As the Exotic Tours website insists,

> Exotic Tours was the first Tourism Operator to do sightseeing inside Rocinha Favela and work with residents. Visit the largest slum in South América with local guides from the Rocinha's Tourism Workshop. Rejane Reis, owner of Exotic Tours, is the pioneer in tourism inside Rocinha and now she is the coordinator of the workshop which train local youngsters as guides to take you on this unique tour. Your visit will help our sustainable project as well as create work opportunities within the community. Be assured this is an absolutely safe tour as the locals welcome tourists and visitors alike. You may also purchase locally handcrafted items while on tour, in Rio de Janeiro, Brazil.
>
> *(Exotic Tours, n.d.)*

The third prominent competitor, Be a Local, a more recent entry into the market, promotes its tours by reference to their more direct and personal approach, targeted at the younger, youth hostel, segment of Rio's tourist population:

> Be a Local tours started operating in January 2003 with the idea of showing tourists a local viewpoint in Rio de Janeiro. The founders, Luiz and Marco

have a team who offer tours of Rocinha Favela, trips to Maracaña stadium for local football matches, and tips to legendary Favela parties. Motivated by complaints that tourists were seeing the Favelas (townships) from the back of a Jeep, they chose a more immersive approach, taking tours through the Favelas, meeting people, and shattering any preconceptions that outsiders may have about what goes on.

(Be a Local, n.d.)

All the operators of these now well-established tours, interviewed for this chapter, insisted that their tour operations benefited in different ways the communities that host them. Apart from providing local artists, store-owners and children with an opportunity to sell their wares to tourists, such benefits ranged from providing direct financial support for a community school (Favela Tour), providing training and employment for local children to work as guides (Exotic Tours), and the support of a children's daycare center (Be a Local). However, the exact amounts of these transfers – either as net figures or as percentages of profits earned – proved hard to determine. Similarly, all the tour operators definitively rejected the familiar charge that favela tours cater primarily to the 'fashionable voyeur' or thrill-seeker. Here, Rejane Reis' (Exotic Tours) response was typical:[10]

The overseas visitor wants to visit a favela because of its culture. We have families visiting who bring their children so that they can see another side of the world, the reality of life. We don't go to favelas to show people poverty.

In contrast, Tom Phillips (2008) offers a disturbing portrait of familiar tropes being invoked in a decidedly less professional approach to favela tourism elsewhere in the city:

Some years ago I was approached by a dishevelled-looking North American, who described himself as an 'alternative tour guide'. He claimed he was taking a young English couple to meet some gangsters. Would I like to come? The next day, we were led up a steep concrete staircase towards the top of a shantytown in Copacabana. Halfway up the hillside we stopped to chat with a pair of traffickers and for the guide to fill his nostrils with cocaine. I asked the guide if he didn't think the tours, which included the opportunity to pose for photos holding the traffickers' weapons, a tad over the top. He shook his head wildly, a thick white ring now etched around his nostril. It was important, he pontificated, for tourists to know the 'real Rio'.

The packaging of Rio's favelas for external consumption can best be comprehended by their reliance on a plausibly authentic display of the 'real place'. In other words, favela tours purport to offer what mainstream tourism does not. In a media-saturated world characterized by the proliferation of increasing homogenized landscapes of standardized hotel chains, fast food franchises, indistinguishable shopping outlets, and entertainment districts, reality tours offer a counterpoint to the dulling effect of late modernity's 'non-places'. Marc Augé (1995) designates urban spaces that have been overly-inscribed by a functionality geared towards the promotion of an instrumental rationality devoid of singularity 'non-places'. Despite the various differences between non-places in different cultural settings, they are all subsumed to an instrumental logic of efficient productivity, functionality and branding. In this sense the 'non-places' of late modernity can be seen as modes of entering the contemporary world through widespread and popularized vocabularies of use. After all, whether in a shopping mall in Paris or Bogotá or a Peruvian or Swedish airport, passengers and consumers know, abstractly at least, what kind of behaviour is expected of them and the kind of services to which they are entitled. Whether such expectations are gratified or whether people comply with the norms is a different issue. In contrast, the entrance into the domain of another person's poverty entails a degree of unexpectedness, a confusion of interpretations where even the voyeuristic element does not altogether negate the allure of spaces outside the jurisdiction of the 'non-place'. Yet once it is framed as a tour, the visit to the shantytown still relies on an overarching inscription. It is symptomatic, however, that having inscribed and prepackaged so many places, mass tourism needs to find new outlets of excitement and veers towards enhancing experiences of the quotidian with an defamiliarizing element. Reality tours, therefore, position themselves in direct opposition to the themed landscapes (and 'riskless risk') of the developed world's 'fantasy cities' (Hannigan, 1999).

So why are tourists drawn to favela tours? A definitive answer is impossible but anecdotal evidence and examination of the internet's copious assortment of blogs, photologs and chats related to favela tourism reveals that motivations vary considerably. Many tourists cite the film *City of God* as the impetus for their tour. Others mention the desire to connect with the 'real' people of the city and still others express the belief that their tourism would aid the community. In the numerous tours we took in researching this topic, the response of fellow tour-takers was equally varied. Perhaps predictably, internet debates reveal opinions that are largely polarized between those that consider favela tours a form of exploit-ation, a sort of 'poverty safari' and those that justify the tour as a means of local entrepreneurship and cultural contact. Likewise, favela residents interviewed in the Rocinha also had mixed feelings. Some complained that the community was not profiting from the enterprise, others thought that it was important to show the favela and demystify stereotypes. The relevant point here is that neither the official packaging of the Rocinha tours by the Riotur, nor the actions of independent entrepreneurs, nor even the efforts of the favelas' neighbourhood associations control the fabrication of social reality. Favela tours have themselves have become terrains of contested meaning which are being scripted and questioned not only by favela dwellers and tourists but by a host of anonymous voices engaged in digital debate.

Reality tours reconsidered

In contrast to the familiar terrain of middle class consumption found in the nearby towers of São Conrado and the São Conrado Fashion Mall, the foreign tourists touring the Rocinha are mostly encouraged to be interested in the cultural specificities of the socialization of the poor, and also in a certain 'shock of the real' (Jaguaribe, 2007b), made tangible by the poverty of some of the constructions and by the raw sewage and garbage that clogs many of the narrow pedestrian lanes. Given its panoramic views, relative calm, seemingly jovial and welcoming inhabitants, and proximity to the wealthy areas of the city, the tour of Rocinha can be experienced and is often packaged as an interlude, a slice of the 'real' that does not actually entail a very close range mingling with the unglamorous poverty of a housing project such as Cidade de Deus or the other favelas located in the western and northern sections of the city (Jaguaribe and Hetherington, 2004). The 'real' of the reality tours is the real of the spectacle, not because the tour necessarily camouflages or glorifies poverty, but because the relationship between the scenario of the favela and the tourist is inevitably one of protected voyeurism. The glamour of the favela rests on its being apprehended as the 'real without embellishments' nurtured by a realist encoding of the place. Distinct from the standardized hotels, from the shopping malls with their international brands, and from the contemporary high-rise buildings, the realism of the favela introduces a specific sense of place that ought to be experienced by the tourist. The selling of the 'real favela' seeks by and large to promote the dignity of poor people who encounter themselves in a mined terrain. The favela packaged as a theme park of violence and poverty, despite the cocaine stuffed nostrils exemplified by Phillips' guide do not really manage to contain the favela experience. But until recently, the sound of fireworks announcing the arrival of drug supplies or the police, the explicit prohibition that tourists are not to photograph or approach the 'bocas de fumo', the selling points of the trade, and the immediate withdrawal of the guides in the light of any forthcoming violence signal that the favela tour occurs in a ambiguous zone. Despite the presence of the local vendors that sell products of the Rocinha – products that are often standardized T-shirts or objects made from recycled materials – and aside from a display of community feeling and solidarity, the tour comes to an abrupt end if there are any signs of a real confrontation in the making.

As spectacle, reality tours of the Rocinha do not presuppose a simulation of a hidden reality but rather a montage of layers of realities that are furnished by the spatio-temporal framing of the tour itself. These are experiences negotiated by the tourists through the gaze, the camera, confinement in the jeep, by media imaginaries, and also by the unsettling sounds and smells that pervade favela spaces (Jaguaribe and Hetherington, 2004). Indeed, a central ingredient of the reality tour is rendered by the absorption of tactile and olfactory experiences. If the tourist gaze has been codified as a photographic appraisal in consonance with the modern trend of transforming the 'world into a picture' (Urry, 2002), the favela experience also destabilizes these parameters because noise, stench and climatic discomforts produced by the usual tropical heat cannot be distanced as they can in photography.

The tour, albeit briefly, also offers the possibility of mingling in instances. This experience of otherness can be cancelled by the feeling of déjà vu or by the blasé attitude of boredom but it can also provide genuine instances of discomfort, excitement or fear. In those favelas that are now instigating reality tours in the aftermath of the actions of the UPPs, the tour serves as a showcase, the favela itself as a model, of the success of the 'pacification'. Prior to the action of the UPPs' pacifying forces, reality tours were fuelled by the adrenaline of an encounter with the 'danger zone'. But while nobody truly wanted to become the victim of a stray bullet, the intrepid tourist still desired a sense of the 'real' beneath the postcard landscapes and beyond the packaged samba shows performed nightly in the tourist traps, and favela reality tours provided a viable means of taking a walk on the wild side, of entering the 'zones of risk', without really taking a risk. The motivations for embarking on a reality tour certainly vary, but the continued flow of tourists taking them attest to the fact that they deliver what they promise. The spectacle of poverty is granted and yet the sense of 'packaging' is diminished because there are no local people performing for 'foreign eyes' and foreign dollars. Rather, it is the cultural inventiveness itself that becomes a spectacle. In some of the newly 'pacified' favelas it is also the spectacle of social justice, urban improvement and architectural re-invention and the realization of some measure of social harmony that is animating the tours. It remains to be seen whether these elements will sustain continued interest.

What is sold in the favela tours is but a fleeting vision of the favela reality. Although the tours insist on the promotion of the 'real thing', this is evidently not necessarily granted. Obviously the 'real thing' in itself does not exist but is rather a compilation of views, representations and discourses. But also most crucially, the heterogeneity of the favela destabilizes the gaze and does not allow a homogeneous viewing. Poverty and violence, community values and enterprise, imaginaries of the media and politics of agency, and the new state intervention and remodelling projects signal that this is a fluid space that is perused instead of lived, that it is corporally projected but that is also asetheticized and distanced. The tourist in the favela is subject to the vigilance of the guides. The realist reality is also defamiliarized because it is not really a visit to modern popular housing projects built according to public architectural tenets, but rather it is the encounter with the 'unexpected', with improvised informality, with a scarcity that generates a culture of survival; it is an encounter with spurious instances of the meeting of the Brazilian popular that is, nevertheless, globally mediated.

Conclusion: Rio, favelas and the global imaginary

In the early decades of the twenty-first century, cities are increasingly competing with one another to capture a place in an ever-shifting urban hierarchy by positioning themselves as cultural sites, economic centres, or political capitals. The struggles over metropolitan visibility correspond to the imperative to capitalize investments, stimulate commerce, activate tourism and fortify political niches – an entire assortment of activities and services that can determine the success of cities independently of their connection to the nation–state. Cities are now bombarded by interpretative inscriptions as they seek to project themselves as cultural, political and economic referents. In renovated urban centres that sell the patina of the past, in new buildings and districts that cater to entertainment, in favelas that inaugurate museums and reality tours, in development strategies that promote cities as Olympic sites, and in the outpouring of books and images that poetize the meanings of the urban experience, cities are increasingly being packaged as consumer products.

More than any other Brazilian city, Rio de Janeiro has a guaranteed presence in the global imaginary as a city of tourist attractions. The spectacular topography framing a city of contrasts, the iconic presence of the Sugar Loaf and the Corcovado, the public arena of the city's numerous beaches, the vibrant and hybrid popular culture, sexy hedonism and the grand apotheosis of Carnival are some of the showy ingredients that highlight this global prominence. As any city that sells itself to tourist eyes, Rio has attempted to crystallize in legible forms its cultural repertoire and mythical configurations. Yet this repertoire usually bypasses its relevance as a former political capital, it dismisses its importance as a former site of public intellectuals and modernist artistic invention. Instead, it activates the usual tropical clichés of exoticism and hedonism amid a lavishly abundant topography.

In the 1930s, when Rio de Janeiro was the political and cultural capital of Brazil, the popular Carnival tune 'Cidade Maravilhosa' was coined as the official anthem of the city. The ingredients that shaped the repertoire of the 'Marvellous City' were scenic wonders, a vibrant popular culture and the belief that Rio was a synecdoche of Brazil itself. With the transfer of the capital of Brazil to Brasília in the 1960s, Rio de Janeiro lost its clout as a political center. In recent decades Rio has been overshadowed by São Paulo's spectacular economic and industrial growth. In contrast to the global city of São Paulo – a city attuned to international finance capital, a city of immigrants and massive economic power – Rio de Janeiro was seen as a decadent city burdened by years of governmental and municipal mismanagement, crippled by its leftover array of public civil servants, daunted by expanding favelas and the violence of an unchecked drug trade. By the early 1980s the festive version of the 'Marvellous City' was increasingly challenged by the unmitigated disparity between the rich and the poor and the growing violence associated with the expansion of the drug trade. Indeed, the spread of a culture of fear associated with this violence has also become a part of Rio's global representation and added another, quite different dimension to its hedonistic image. As a major site of both violence and poverty, the favelas of Rio have particularly come to the fore in a set of overlapping and competing tourist representations of the city. As a showcase of 'real' Rio, favela tours lie at the centre of this complicated nexus of contesting visions.

In recent years, there have been concerted attempts to rectify the purported decadence of Rio, endeavours connected to the discovery of major oil reserves in the state of Rio de Janeiro that has prompted economic expansion and an influx of capital into Rio and its real estate market. Further, during the second mandate of President Luiz Ignacio Lula da Silva, Rio was seen as the city that could most effectively capture a prominent place in the global imaginary by promoting a metropolis that was not only an attractive tourist site but also, once again, was a showcase of Brazilian nationality. The championing of Rio as the city that would host the World Cup in 2014 and the Olympics in 2016 is an integral part of this promotional strategy that seeks to place the city and the nation of Brazil in the fierce competition for a worthy spot in the global arena. Evidently Rio de Janeiro, despite its deep contradictions and ambiguities, it still the city that encapsulates the consolidated representations of Brazilian creativity. It is symptomatic of the change in cultural agendas that Rio's modernity is extolled on the basis of its popular culture where diversity and hybridity combine. It is precisely at the crossroads of this cultural invention that the favelas and reality tours are thrust into the limelight.

In this chapter we have explored aspects of Rio's evolving, complex and persistently ambiguous relationship with the favela. Alternately represented as the locus of an authentic and exotic 'imagined popular community' or as the terrain of violence, vice and poverty, more recently favelas have become the symbols of the city's inability to overcome its disparities and social dilemmas. The illicit and flourishing drug trade was in large part responsible for the renewed prominence of the favela as the battle between drug dealers, police and the state fomented a culture of violence that surpassed in scale and devastation all previous parameters. As the host city for the mega events of the 2014 World Cup and the 2016 Olympics, Rio de Janeiro has very recently gained extraordinary media visibility. In anticipation of the global scrutiny associated with these events, state agencies have undertaken an unprecedented series of interventions, initiated by the UPPs' massive armed police incursions into targeted favelas in an attempt to 'recapture' them for civic purpose. At the same time considerable investment is being allocated to the remodelling the favelas as part of the new Morar Carioca urbanization. In the new urban cartography created by these initiatives, those favelas that rise to prominence may feature as the 'favela chic' while others will likely have to pave their way through the politics of agency. Such politics of agency or modes of citizenship are not simply processed by the usual party politics but are formed by grassroots associations composed of mobile and fragmentary agendas that, nevertheless, are tied to the wider institutional frameworks of the organized political arena. Rio de Janeiro is embarking on a major process of urban (re) branding. It is symptomatic of a considerable shift in cultural agendas that Rio's favelas are set to become a trademark of the city itself and a crucial component in the re-launch of the 'Marvellous City' brand. The favela, at once a synecdoche of the contradictory nature of a national dilemma, emerges as the critical locus of a series of experiments which may well decide the fate of Rio's global aspirations.

At the same time, favelas themselves have surfaced in the global imaginary in their own right, in large part through their effective packaging as both reality spectacle and cultural commodity. As an antithesis of the global 'non-place', the complex space of favela offers the experience of a contrasting urbanization that is contradictory, porous and intensely negotiated. For the visitor, reality tours offer the promise of a glimpse of community life that exists in direct contrast to the isolation of the standardized high-rise, the allure of captivating views in contrast to the flatness of suburbia and the exciting new forms of reinventing favela 'cool' through music, fashion and political agency vis-à-vis the predictability of theme. Poverty in the shape of raw sewage, rampant disease, rat-infested derelict con-structions, and the vision of stricken children with rickety legs and bloated bellies is not in itself an attraction. Such images of woeful misery that shaped the imaginaries of poverty throughout and about the 'third world' are not what animate contemporary reality tours. Neither the favela poor of southern zone slums of Rio de Janeiro, nor the slum dwellers of Dhavari, nor the residents of Soweto, just to name a few 'reality sites' now routinely visited by tourists, have staked their claim to fame on just their sheer material poverty. Rather, reality tours entice because they offer a spectacle of the 'real thing' shaped as cultural practices of community belonging, collective identity, and hybridity for global tourists in search of a sporadic re-enchantment with a defamiliarized reality. It is likely that they, or some new variant, will continue to entice international tourists for some time to come.

Notes

1 This chapter builds in part on insights first articulated in Jaguaribe and Hethering-ton (2004). The authors thank Kevin Hetherington for graciously allowing us to draw upon that essay in framing some of the arguments here. The usual disclaimers apply.

2 Our use of the terms 'imaginaries' and 'imagination' are similar to what Charles Taylor (2004) defined as 'social imaginaries.' By social imaginaries, Taylor is referring to the images, stories and legends shared by a large group of people that make these the source of common practices and legitimacy. Our use of imagination and imaginaries refers to the representations spawned by the media, by artistic products, and by daily experiences that are routinely shared. Indeed, realist representations of the favela and urban life have become naturalized as a decoding of reality itself.

3 Personal interview with one of the authors, conducted on 14 August 14 2007 – but see also (Freire-Medeiros, 2009a: 587).

4 Dona Marta began laying claim to global notoriety in 1996 when Spike Lee filmed a music video of Michael Jackson dancing atop a laje, the concrete slab that features in most favela homes. At the time of the shoot, Spike Lee paid for the 'protection' of the drug lord Marcinho VP in order to make the video. Following in Jackson's footsteps a stream of musicians, including Madonna and members of The Police, have made pil-grimages to the favela. In 2010 Alicia Keys and Beyoncé were filmed giving 'impromptu' performances in Dona Marta. Newspaper coverage of the event stressed the impact of the new policies in Dona Marta and Keys was photographed greeting the female police officer in charge of the pacification of the favela.

5 For a discussion of realist aesthetics in this context see Jaguaribe (2007a, 2007b).

6 For a discussion of key favelas in Rio, the politics of removal, and the social mobility of favela residents see Janice Perlman's (2010), *Favela: Four Decades of Living in the Edge*

in Rio de Janeiro, and Licia Valladares (2005), *A Invenção da Favela: do Mito de Origem a Favela.com*.

7 Even the subsequent neo-realist socially critical films made by the Brazilian directors of the Cinema Novo in the 1960s never undermined the popularity of Camus' enchanted favela.

8 The PAC, Programa de Aceleração de Crescimento, is, among other things, a major re-urbanization and housing project initially undertaken by the federal government during the Presidency of Luiz Ignacio Lula da Silva.

9 Favela Tour: http://www.favelatour.com.br; Exotic Tours: http://www.exotictours.com.br; Be A Local: http://www.bealocal.com/flash.html. All sites last accessed 3 March 2011.

10 Personal interview with one of the authors conducted on 18 August 2010.

16

MODERN WARFARE AND THE THEORIZATION OF THE MIDDLE EASTERN CITY

Sofia T. Shwayri

> Most scholars of the urban Middle East are trained in Latin American models, and indeed we can argue that to be trained as an urbanist is to be raised on the traditions of the Latin American school.
>
> *(AlSayyad and Roy, 2004: 2)*

The opening quote to this chapter points to the ways in which development studies has been focused on literature from the non-Western city, but scholars of the Middle East have relied on a Latin America model which draws heavily on Western urban theory (Salamandra, 2004: Hanssen, 2005; Watempaugh, 2006). However, theorists such as Anthony King (1995) has expanded the geography of world cities and Jane Jacobs (1996) has unsettled the traditional Third/First world division, demonstrating the existence of the Third world city within the First, and revealed a 'Fourth world' that suggests new geographic divisions'.

Anthony King's (1995) work on the world city, its meaning and representation in an era of globalization and post-colonialism, has not only expanded the geography of world cities but the possibilities of bringing cities on the periphery closer to cities in the core. Subsequently, Jane Jacobs (1996) has unsettled the traditional Third world/First world divisions, especially as she has demonstrated the existence of the third world city within the first, and revealed a 'Fourth world' that suggests new geographic divisions. Other scholars have used these frameworks in studying cities in relation to the global economy, placing them in a hierarchical order, with some at the centre of decision-making and others at the periphery. These approaches, including some writing focused on Middle Eastern cities (Hubreychts, 2002; ElSheshtawy, 2004), have juxtaposed once obscure non-Western cities with Western cities, and attempted to render existing binaries irrelevant. However, more generally, world city and global city perspectives have kept a lot more cities 'off the map' (Robinson, 2002), particularly a majority of cities in the Middle East.

The exclusion of a majority of cities in the Middle East from urban theory can also be explained in part by the history of wars and conflict within the region since World War II. In 1948 the Arab–Israeli War broke out following the partition of Palestine. That conflict saw the establishment of the State of Israel, called *Independence* by Israelis and *Nakba* – catastrophe – by Palestinians. Hundreds of towns and villages were destroyed and hundreds of thousands of refugees settled in camps in several countries across the region. This war initiated a series of wars in a conflict that persists as the State of Israel continues to redraw its borders. Lebanon, one of those countries in the region, has suffered from wars, some with its neighbors, with repeated Israeli attacks from 1969 up to 2006. Others have been internal, such as the 15 years of fighting that took place between 1975 and 1990. Most of these wars were urban-based and caused major physical and social destruction, leaving hundreds of thousands dead and displaced. Iraq also suffered; a series of coups bought the Baathist Party to power in 1968 and a decade later their leader, Saddam Hussein, began an era of wars with neighbouring states. The Iraq–Iran war lasted the entire 1980s. Its end in 1989 saw Saddam turn on Kuwait in 1990, with the Iraqi army invading and occupying it. The failure of the regime in Baghdad to respond to UN Security calls to pull out was followed by the imposition of economic sanctions on Iraq and the subsequent first Gulf War. The sanctions lasted over a decade and only ended with the second Gulf War launched in March 2003. The official end of that war in May 2003 heralded the beginning of a series of internal wars, some launched against the Allied occupation, some by the Allied coalition forces against the locals, and others occurring among local religious factions, including Shiites, Sunnis and others.

This climate of war that has overshadowed the region since the late 1940s is symptomatic of the twentieth century globally – 'the most murderous in recorded history' – according to Eric Hobsbawm (2002). That century was marked by two world wars, scores of wars of independence, and hundreds of intra-state ethnic and religious wars, causing the displacement of millions within their own countries, leaving millions dead and millions more as refugees. In 2008, the UN Refugee Agency estimated the number of internally displaced people to be 26 million worldwide and those that crossed national borders about 11 millions; Iraqis accounted for 2.5 million of these (Koser and Dhaliwal, 2008). This number excludes 4.3 million Palestinian refugees. The huge social costs have been paralleled by massive physical destruction of cities across both the developed and the developing world.

Despite the long history of war and its role in the planning of cities, war as a subject in urban theory remained largely absent until the early 1990s (Ashworth, 1991). War was 'not on the agenda of most urban theory nor seen as relevant to urban policy' (Woollacott, 1996: 89). This absence in the literature is partially explained by the fact that no conventional war has been fought in a Western city since the end of World War II – as an area of research. Utopian ideas that marked the discipline's early beginnings still very much persist in ideas of progress that guide the development of cities, and war, as the antithesis of utopianism, progress and development, has made it unpopular as an area of research. Its omission from the literature can partially explain the absence, until 1992, of Middle Eastern cities from urban theory.

The attack on New York City on 11 September 2001 has nonetheless shifted the discourse significantly, as evidenced in the number of conferences, symposiums, seminars and workshops on the theme of war and the city held since, and also by the number of publications – books, edited volumes and journal articles – that have subsequently appeared. However, this apparent shift in response to the events of 2001 was already underway in the previous decade. The siege of Sarajevo in the mid-1990s was the longest siege of a city since the siege of Leningrad in World War II and caused the most extensive urban bomb devastation since Dresden in 1945. This was a stark reminder of the inseparability of war and cities (Woollacott, 1996). And the nature of this particular violence reminds us that war is rarely confined to one city but is often visited on other cities in acts of revenge. The importance of these urban wars produced the recognition that the city was no longer a battlefield among many but the main theatre and an awareness that the identity of the fighter had become more complicated as the nature and rules of warfare were changed.

The classical Clausewitzian war fought between two formal armies to advance political goals has been replaced in the post-Cold War era by 'new wars' among private militias, with the war in Bosnia-Herzegovina as the archetypal example, according to Mary Kaldor (1999). She further claims that despite this war not being the first post-Cold War international crisis, it is likely to be a defining event in understanding this new type of urban warfare and its meaning for foreign involvement. This was the case for the American military in the 1990s as they prepared for this inevitability, identifying types of operations and new challenges to be encountered (Press, 1998). The Bosnian war was also significant for its contribution to urban theory, re-introducing urbanists to a modified examination and interpretation of processes of destruction in times of war: *urbicide*, a term used by a group of Bosnian architects in a publication on Mostar in 1992, is the deliberate destruction of the urban environment.

The term urbicide, connoting a form of violence against the built environment, was first used by Ada Louise Huxtable in *Will they Ever Finish Bruckner Boulevard? A Primer on Urbicide* (1972), in which she examined development projects in New York City. This use of the term remained limited to New York and was subsequently echoed in the work of Marshall Berman who describes the violence caused to his neighbourhood in the Bronx by the building of the Cross Bronx Expressway by Robert Moses. Yet rather than highlighting the uniqueness of the case, Berman (1996: 172–92) argues that urbicide constitutes an essential feature of human development. This need to understand urban life in the context of urbicide during peacetime continues to be an important dimension of Berman's work in his continuing examination of the transformation of New York.

Since 2001, 'urbicide' has gained in prominence as a term used to describe the widespread violence against the built environment, especially that experienced by cities in the Middle East. The widespread destruction of Palestinian space, towns and cities by the Israeli Defense Forces (IDF) since the launching of the Second Intifada in 2000 has provided a wealth of material for scholars interested in cities and conflict (Graham, 2002, 2003, 2004; Abujidi, 2005). In fact, some scholars have used that same lens to look back at historic cases of warfare and urban destruction, as with, for example, the first two years (1975–6) of the Lebanese Civil War in the capital, Beirut (Fregonese, 2009). This shift in urban theory has encouraged diverse and rapidly evolving new areas of research, including studies on the 'war on terror', security, urban warfare and military geographies in which the cities of the Middle East play a central role.

Some of these works highlight a change in approach to the study of cities in the Middle East, notably moving away from direct parallels with the Western city. Other scholars continue to employ traditional frameworks that place cities of the Middle East in 'opposition' to the Western city. In this chapter, I focus on two key themes that have acted as intersections between some of these evolving areas of research. First, I discuss the renewed interest in the city of the Middle East as a city re-awakened by war. Second, I focus on the work of scholars who have recreated those cities, re-imagining them as 'other' to the American city. I explore the experiences of a number of cities in Iraq, Palestine and Lebanon that are currently in a state of war or have been in the recent past. Their location allows me to loosely term them cities of the 'Middle East', a region usually referred to as a block of countries stretching from Iraq to Libya and forming a political entity known as the Arab League.

The city re-awakened by war

It is the contemporary city, not the state, that has become the key strategic site for military and geopolitical conflicts. So noted Stephen Graham before September 11, 2001, as he planned a conference titled 'Cities as Strategic Sites: Militarism, Anti-globalism and Warfare', held in November 2002. It raised questions, mostly unexplored in urban theory, which have subsequently become ever more urgent after the attacks on New York City and Washington. The papers were later published in an edited volume, charting the path for a new research question: how do urban areas and organized, military conflict shape each other in these post-Cold War, post-9/11 times? (Graham, 2004: 24). It is this question that this book and other studies pose, and scholars have turned to the Middle East where the long running Israeli–Palestinian conflict has witnessed a new confrontation between the IDF and Palestinian militants in the occupied Palestinian cities of the West Bank. Moreover, as preparations for war against Iraq got underway in 2002, Iraqi cities, as its main theatres, have become the focus of the media and academics.

The ongoing Israeli–Palestinian conflict has been marked by two key wars, 1948 and 1967, and two uprisings, the first in 1988 lasting two years, the second

from 2000 and is still underway. The wars of 1948 and 1967 led to major redraw-
ing of boundaries of British Mandate Palestine, including Israeli encroachment
on West Jerusalem and consequently on hundreds of thousands of Palestinians.
The war of 1948 continues to be narrated by historians and political scientists, as
they try to make sense of the events of this war that led to the destruction of
Palestinian villages, towns and cities (Morris, 1990; Pappe, 2007; Shlaim, 2009).
Geographers and planners have instead been more focused on understanding
urban policies enacted by the Israeli government for the Judaization of Palestinian
space since 1967 especially that of East Jerusalem (Kliot and Mansfeld, 1999;
Bollens, 2000; Safier, 2001; Calame and Charlesworth, 2009). Since 2002, a shift
in the literature can be detected as scholars turned towards writing the history of
urban spaces remade by war. Some focused on analysing Israeli military strategies
in cities of the Occupied Territories during the Second Palestinian Uprising. The
bulldozing of the centre of Jenin refugee camp in the West Bank by the IDF in
April 2002 was claimed by Israeli Prime Minister Ariel Sharon to be impelled by
the dismantling of terrorist infrastructure, held responsible for the suicide attacks
within Israeli cities. Graham (2003: 63) argues that the real goal of the bulldozing
was to 'destroy the urban foundations of a proto-Palestinian State', through urbicide.
This process is part of a broader infrastructural war. That includes the systematic
bombing of Gaza's airport, port and Palestinian television and radio transmitters,
all aimed at undermining the slow process of modernizing Palestinian cities
(Graham, 2002). Together with the destruction of factories and agriculture and
the resultant effect on the economy, this has been described by Sara Roy (1999)
as a process of socio-economic re-development.

This attack on Palestinian modernity and urbanity and its very use as a weapon,
argues Eyal Weizman, goes back to 1967 when Israel seized the West Bank. It is
a two-fold process characterized by swift military actions and slower urban development, as military and civilian planners and architects translated Sharon's vision
onto the landscape through 'the use of temporary security architecture to create
permanent facts on the ground' (Weizman, 2004: 172–3). As Minister of Agriculture
in May 1977 and Head of the Ministerial Committee on Settlement Policy, Sharon
pushed to occupy the western slopes of the West Bank, limiting both the expansion of Palestinian cities and contiguity of the Palestinian territory. In 1982, as
Minister of Defence, he devised the *Masterplan for Jewish Settlement in the West Bank
through the Year 2010*. Settlements, now points on the high grounds, were envisioned
as a whole urban infrastructure and were connected by traffic arteries (ibid.:
176–7). The building of Israeli cities within Palestinian-occupied territory was
initially planned to take a couple of decades but this changed when Sharon assumed
power as Prime Minister of Israel in 2001 and the violence against the West Bank
was expedited by the opportunity provided by the global 'War on Terror'. The
introduction of a new fortified barrier, the *West Bank Wall*, was part and parcel
of this infrastructural war that aimed for the de-modernization of Palestinian
cities through the chronic lack of investment in public infrastructure, as with East
Jerusalem (Misselwtiz and Rieniets, 2009: 67).

In contrast to the slow process of de-modernization depicted above, a faster strategy was applied by the IDF in Lebanon in its 34-day war against Hizballah in the summer of 2006. Here, the strategic bombing of key infrastructure targets was intended to bring urban life to a complete standstill. Beirut International Airport was forced to close and this was followed by a systematic process of isolating cities, towns and villages, as key bridges, highways and connecting roads were knocked out. This infrastructural war extended to the bombing of power stations and media networks, hospitals, schools and housing. Its purpose was to take Lebanon back 30 years. The deliberate targeting of infrastructure integral to the sustenance of contemporary urban life had been used previously by the IDF, in the attack on Beirut International Airport in the 1960s, and since then the scale has expanded and modified to target all forms of infrastructure.

The infrastructural wars launched on cities in the Middle East have not been limited to the Arab–Israeli conflict but have become a central component of the contemporary US military strategy since the first Gulf War in 1991. Although targeting key infrastructure has been a hallmark of terrorists and state armies for a long time, Graham (2005) argues that the systematic targeting of urban networks by traditional powers like the US poses a greater threat to urban existence. The destruction of dual-use urban infrastructure systems in the first Gulf War followed by a regime of sanctions that made postwar reconstruction almost impossible has forced the de-modernization of Iraqi cities, and society, resulting in the 'the largest engineered public-health catastrophe of the late 20th century' (Graham, 2005: 182). The ground work laid during the first war and the ensuing sanctions and bombing for over a decade have been followed by a complete de-modernization of everyday urban living in 2003 causing the 'switching [of Iraqi] cities off' (ibid.).

The official end of the Second Gulf War in May 2003 marked the beginning of new types of warfare, key among them the reconstruction warfare led by US military officials. The restructuring of Iraq's state apparatus and economy postwar has been the subject of numerous chapters and articles (Harvey, 2003; Shwartz, 2007; Klein, 2007). Naomi Klein examines economic opportunities that emerge in the context of disaster, both natural and man-made, from the early 1970s, with Pinochet's Chile to the present, with Iraq constituting the main case. Klein initially focuses on the rebuilding process led by Paul Bremer, the US administrator in Iraq, and especially the laws enacted to transform Iraq into a 'capitalist dream'. Although the reconstruction process started well before the launching of the war in March 2003, it is the 'War on Terror' and especially the military landscape that has made the imposition of neo-liberal policies in Iraq possible. Neo-liberal reform in cities of the Global South has a long history, yielding a comparable pattern of social and physical disasters for locals and opportunities for foreign multinational corporations, a process of 'accumulation by dispossession' according to David Harvey (2003). The impoverishment of the population has often been manifested in the production of slum cities. Normally, this is a process that stretches over decades, but in the case of Iraq, neo-liberal reforms compounded by the violence of war has compressed the process to only few years as demonstrated by Michael

Shwartz (2007). Slum cities have not only been on the rise in Iraq but across the Middle East, especially as urban warfare – ethnic and religious – continues to be waged, resulting in population displacement that is causing a reordering of space.

The imagined versus the real city in the 'War on Terror'

> Geographies are invented to legitimate war to Americans
> *(Shears and Tyner, 2009: 224)*

Scholars have argued that the Bush Administration's post-2001 'War on Terror' has signified a process of reworking the geographies of the Middle East. However, a closer look at changes in military operations in urban terrains reveals that this change followed the first Iraq War in 1991, with mock-up cities constructed in military bases across the US from the late 1990s. Their number increased exponentially in the US and around the World as the case for the second Iraq War was being made by President George W. Bush in an address to the United Nations General Assembly on September 12, 2002. Since that date there has been little doubt that the wars of the foreseeable future were to be fought in such cities, although the form and nature of these wars remained obscure. Likewise those cities hidden from view for over a decade from the end of the first Gulf War in 1991 came more to the fore. The obscurity of the real city was countered with an imagined one, virtually and physically.

The 'War on Terror' has simply moved the Middle East region and its people into the 'colonial present', argues Derek Gregory, as Western attempts to control the region continue, linking old and new forms of intervention across time and space in Palestine and Iraq. Gregory (2004: 11) explains the spatial manifestations of the present colonial moment by employing Edward Said's Orientalist discourse, identifying contemporary tropes that took hold in the US following the September 11, 2001 attacks, in which the 'other' – Arab and Muslim Middle Easterner – was imagined as 'barbaric', 'evil' and an inhabitant of cities that are terrorist nests. Equally instrumental in this construction was the news media, in the months leading to the Iraq War, as it previewed scenarios of urban warfare in self-designed simulated cities where the only real city was Baghdad (the Iraqi capital), and later, during the war, in its minute by minute reporting of the targeting of the built environment. The media simulation of an urban battlefield may have appeared novel in 2002; however, such simulations have been around much longer, at least since the 1980s, especially in war games (Der Derian, 2001: 3).

In the 1990s, the simulation of an urban battlefield was taken to new heights as the US military transformed some of its bases into training sites. One important urban combat training site was Zussman Village MOUT at Fort Knox in Kentucky, a city covering 30 acres and consisting of a range of features including an extensive sewer system, a US embassy building, a large school, houses, apartments, a petrol station and a railway assembly line (Desch, 2001: 12–13). One aspect in particular makes Zussman interesting: the city's identity can be transformed to represent an urban setting from any part of the world through pop-up targets, fire and special effects such as providing radio and TV stations that broadcast in Arabic, Hebrew, or Russian. Odors are simulated on demand. As this is meant to simulate a battlefield, the city is 'smothered in dirt and mud', and accommodates hundreds of insurgents wearing Keffiyeh, the traditional Arab headdress for men, armed with AK-47s and rocket-propelled grenades (Graham, 2010: 189).

Following 11 September 2001, the threat of urban warfare in which the American military would fight became real. Its location, the changing identity of a generic city as a military training site, was replaced by a specific one – the city of the Middle East. It was being invented and duplicated across the world in US military bases and training grounds in the United States, the deserts of Kuwait and Israel, the downs of England, the plains of Germany, and the islands surrounding Singapore. As Graham notes, these form

> A hidden archipelago of between eighty and a hundred mini cities. The vast majority are located in the United States . . . Some such cities are replete with lines of drying laundry, wandering donkeys, Arabic graffiti, tape loops endlessly playing the call to prayer, even ersatz minarets and mosques. Others boast 'slum' or 'favela' districts and underground sewers with built-in olfactory machines capable of producing on demand the simulated smell of rotting corpses or untreated sewage. Still others are populated occasionally by itinerant groups of Arab-Americans, bussed in to wander about in Arab dress and role-play.
>
> (Graham, 2010: 183)

These training sites vary in area and design from 'a little more than air-portable set of containers, others (are) extensive sites that mimic whole city districts, with airports or surrounding countryside' (Graham, 2007: 124). With the unfolding of the Iraq war, the US army invented not merely a generic Arab city but tried to imitate the Iraqi city, specifically Baghdad, by populating these sites with dead and living animals, as well as 'Baghdad-style orange-and-white taxis' (Graham, 2010: 187). These city replicas are not exclusively a US strategy for the 'War on Terror' but one also practised by the Israelis for conflict within Palestinian cities or related to regional conflict. Built as a mock-Arab city in the Negev desert, Baladia is a training ground for Israeli soldiers' incursions into Palestinian cities and refugee camps. It is among the very first urban training sites to be built to urban scale, covering an area of 7.4 square miles, with more than half that space designated as roads and the remaining area the built urban fabric, consisting of 472 structures arranged into 4 quarters, simulating apartment buildings, a market, a mosque and a refugee camp (Graham, 2007: 123). Because this is a war zone, the streets are 'littered with burned cars, burned tires, and mock-booby traps' (Graham, 2010: 193). The city not only replicates the built environment but also the sounds of everyday activities experienced by the Palestinians: the noise of Israeli helicopters flying over, mortar rounds, calls to prayer and the like. Like Zussman MOUT, Baladia can change identity to simulate any city that the IDF intend to attack in

Gaza, West Bank or Lebanon. Its use by the US Marines for training means it can be transformed into any Iraqi city. Unlike earlier forms of the physical simulation of places to be targeted in wars that go back to World War II, Graham notes that the replicas of the city of the Middle East in the US or in Israel are novel in their relationship to political violence as they are meant to prepare those outside forces for occupation and counterinsurgency warfare (ibid.: 186).

While the recreated city of the Middle East was being established in the heart of the Western landscape, an American city has been created in the heart of the Iraqi city following the end of major combat operations in May 2003. As the US military turned from liberator to occupier, the nature of the war changed and so did the form of the urban environment. The shift changed the Iraqi city and especially the capital city Baghdad. An open city prior to March 2003, Baghdad became an occupied, segregated city demarcated by walls, hiding behind them an American city. This transformation of the capital has been widely documented primarily by the occupying army and the embedded journalists. It is their observations that have become the primary sources for academics describing this newly emerging form at the periphery of empire and its transformation by counterinsurgency warfare, the pull-out from Iraqi cities and later, in the summer of 2010, the downsizing the presence of the occupying army.

Roy Scranton (2007), who served in Iraq from May 2003 to July 2004, demonstrates how the fear for the military's own safety in a war of insurgency has shifted their focus on creating a dual city within an existing booming metropolis. It is a dual city layered on the existing one, the infrastructure of Baghdad forming its backbone, inhabited by both the US military and the administrators running the occupation. This American city is formed from a system of dispersed and heavily fortified bases connected by a road network used by military convoys, dotted with patrols and checkpoints, enabling the occupier to avoid the real city in large part. This act of avoidance includes a renaming of the roads and expressways used by the Americans, which have become Route Irish, Route Senators and Route Predators, for instance. The central base, a seven-square-mile fortified enclave, known as the Green Zone, located along the Euphrates in central Baghdad, has formed the heart of the American city, the administrative node, housing the US-run government of Iraq, the Coalition Provisional Authority, until end of June 2004, and since then the US embassy and the Iraqi government. Baghdad's Green Zone has 'its own electrical grid, its own phone and sewage systems, its own oil supply and its own state-of the-art hospital with pristine operating theaters–all protected by five-meter-thick walls' (Klein, 2007: 413). Life on the inside is a world away from its surroundings where American law is the only law that applies; drivers have been pulled over for speeding and drunk driving. The inside of the Green Zone 'feels like little America' (Chandrasekaran, 2007: 17–18).

This layering over of an existing city in the emergence of the American city in the Middle East following the end of combat in May 2003 appears novel. However, the creation of an American landscape in the outposts of the American empire has a much longer history, as detailed by Mark Gillem (2007). A close examination of the US bases in South Korea and Japan, according to Gillem, reveals the recreation of a low-density American suburb, exported from home, replete with auto dependency, isolated uses and low net densities (Gillem, 2007: xv). He adds that since 11 September 2001, as more facilities such as fitness centres, ballfields, and clubs have been built within bases to keep the army occupied and consequently reduce contact with the local population, the US military has also been relocating bases away from capital cities to remote areas (ibid.: 266). This model, according to Gillem, reveals a policy of avoidance, a divergence from the twentieth-century model of association, where bases were located within capital cities with many of the soldiers living off-base and interacting with their surroundings as they used local businesses.

This policy of association had been practised by European colonial powers, namely the French and the Italians in North Africa, since the late nineteenth century. In Algiers, the French policy shifted from an earlier policy of assimilation, with economic aims outweighing military objectives and paralleling a rise in local resistance against colonizer (Lamprakos, 1992: 189). This policy was manifested both architecturally and spatially through a variety of means ranging from integrating traditional architectural forms in government buildings to a spatial separation between colonizer and colonized by ceasing demolitions of the old city and implementing a policy of conservation while creating new and distinct quarters based on modern European urbanistic principles (ibid.: 190–1). These new quarters are what scholars refer to as the dual-city phenomenon. Prior to this phase of separation from the natives, the French military objectives of forced assimilation meant extensive destruction of the existing cities through the demolition of private and communal buildings and the transformation of others for their use. These processes of destruction have, in some cases, laid the foundations of a dual city on the footprints of an existing one, as evidenced by French intervention in Beirut on the Eastern Mediterranean in the early twentieth century. Immediately after the mandate was established in Lebanon, the French military launched three planning projects devised to modernize the traditional city centre. This resulted in extensive destruction of the existing fabric creating a tabula rasa upon which a new modern centre of wide gallery-lined avenues were laid, becoming La Place de L'Etoile (Davie, 2003: 206–7).

Clearly, dual cities built by imperial powers next to an existing city or within existing boundaries have a long history. A third, newer and more elaborate form has emerged in the 'colonial present' in the Palestinian Occupied Territories whereby the landscape is treated as a three-dimensional volume conceived in layers of infrastructure and underground resources. This multi-layered conception of occupying and controlling space is what the architect Eyal Weizman (2003) has referred to as the 'Politics of Verticality'. Although this mechanism of control has been in place since the 1970s, an aggressive application by the Israelis followed the 1993 Oslo accords. They have built an elaborate road network for civilian and military use that connects Israeli areas while controlling Palestinian movement and space. This is achieved by 'roads [that] skirt around, bridge over or tunnel under Palestinian habitation'

(Lein, 2004, cited in Pullan *et al.*, 2007: 176). The borders of the Occupied Territory form an elastic geography of a 'multitude of temporary, transportable, deployable and removable border-synonyms' that 'shrink and expand the territory at will' (Weizman, 2007: 6). This elasticity keeps the military constantly working on devising new and elaborate systems of control in addition to its many partners in this occupation – the settlers, public officials, foreign governments, 'supportive' overseas communities, and many others – as they oversee its implementation and sustenance, all to further shrink the colonized space.

The 'War on Terror' has served as a new critical moment for the Israeli government to push more intensively its occupation of the Occupied Territories and especially its control of East Jerusalem. Following the September 11, 2001 attacks on New York City, scholars studying contemporary Jerusalem have applied the lens of urban terror to connect the experiences of both cities, as well as with those of London and Madrid, and subsequent urban responses to the bombings therein (Savitch, 2005). This connectivity among cities, Savitch and Ardashev (2001) argue prior to the September 11 attacks, is due to the nature of urban terror, which, though local in character, is global in its exportability. Although the form of urban terror may not differ between cities in the developing or developed world and neither may the consequent degree of risk that urban dwellers face, studies have revealed that the impact is greater in cities of the developing world (Beall, 2007). To ignore the urban dimension of a conflict when attempting to understand urban terror, as in the case of Jerusalem, is to ignore the long history of occupation and the local dimensions of acts and, consequently, the uniqueness of these cities in their contribution to urban theory.

Conclusion

Despite the growing and varying body of scholarship on the city of the Middle East in the last few decades, the Islamic city model remains the dominant framework for studying the morphology of the contemporary city, urban society and institutions including religious ones. Its emergence dates to the early twentieth century, French Orientalists describing the colonial city in North Africa, more commonly referred to as the 'traditional city' (Bonine, 2005: 413). This French construct of the native city depicted a way of living framed in relation to the non-traditional city, the modern districts built adjacent to it, becoming one of many constructs advanced as part of a colonizer defined Orient. The preservation of the natives' quarters, the Arabization of its architecture, Hamadeh (1992: 249) argues, are French Imperial policies that contributed to the creation of a society that was frozen in time and space. Over a century later and despite widespread urbanization and development of the Middle Eastern city, the Islamic city model is again used by imperial powers to define Middle Eastern societies and as the justification for war and destruction. This time it is the history of wars and conflicts of the latter part of the twentieth century, some resulting from continuous imperial ambitions to control the region, that has shaped the Middle Eastern city. Images of chaos, lawlessness and underdevelopment caused by decade long wars overshadow images of modernization and growth. This has subsequently hampered the potential for these cities to be used as examples that might contribute to a broader theorization of the urban.

However, this all changed a decade ago. It was not 11 September 2001 but the period following the end of the Cold War that saw the city as the premier site of conflict and war. This disciplinary change was compounded with a change in the military strategy and a shift in the terrain of the battlefields of the twentieth-century that placed the Middle East at its centre, especially its cities. Shaped by their recent violent history, the US military has envisioned the city of the Middle East as a Muslim-Arab city, chaotic, lawless and underdeveloped, everything that the orderly and developed Western American city is not. The artificial creation of ersatz cities in the American heartland and across the world is part of a process that prepares for the destruction of real cities through war and occupation followed by the implantation of neoliberal practices. The layering of the American city on the existing one, set apart from it by a set of walls and fortifications and an independent infrastructure system, continues to define the Middle Eastern city.

The 'Wars of the Post-cold War Era' in general and the 'War on Terror' specifically, have brought war back to cities across the world and back to urban theory. Furthermore, despite the variation in the nature of warfare and its specific local manifestation, scholars studying war as an urban phenomenon have turned to the experiences of cities in the developing world to understand urban transformation in the contemporary city in response to violence. This has set in motion a new approach for theorizing cities in general, and a revival of interest in once-forgotten cities like those of the Middle East. Those studies have partly shifted research on the Middle Eastern city from being an area study within development studies to one increasingly situated at the core of urban theory. Whether placed in opposition to the city in the West or on the same platform, it is Western conceptualizations that are defining it, thereby hiding local agency and diverse forms of struggles.

During the writing of this chapter in May 2010, US President Barak Obama dropped the term 'War on Terror', and replaced it with 'Overseas Contingency Operation' to emphasize that the war is with Al-Qaeda and not with the Islamic World. Recentering military strategy towards Afghanistan is intended to redefine the geography of the war and consequently to shift the battlefield away from the cities of Iraq which the US forces have focused on controlling in the last few years. Moving to new theaters of war will gradually be paralleled by an academic shift towards new conflicts and cities. The place of cities of the Middle East in urban theory will undoubtedly shift but not back to obsolescence, for 'the fantasy we entertain now in the West is that some cities can remain entirely safe while others burn' (Woollacott, 1996: 91). The events of the past decade highlight that no city is immune. The increasing connectivity among and between cities, coupled with conflicts and crises that transcend local spheres, mean that the experiences of cities in the Middle East will remain key to understanding violence in cities elsewhere.

17

READING THAI COMMUNITY

The processes of reformation and fragmentation

Cuttaleeya Jiraprasertkun

As the effects of globalization and urbanization cause many cities to become seemingly indistinguishable, Bangkok and its communities are also undergoing reformation and fragmentation. Visible changes in the constituents of community over the past few decades raise questions about how Thai community is formed, reproduced and continually transformed: how has the role of these changes, with their tangible and intangible qualities, impacted upon the constitution of community? In this chapter, these metamorphoses and their characteristics are read and interpreted according to their conceptual, practical and political implications in three sites that represent different articulations of urban development in Bangkok. The study not only reveals how the changes in political approaches to the city have resulted in the modification of people's everyday lives and perceptions, but it also discusses the dilemma of reading and interpreting community in a Thai context through space and time.

Setting the scene . . .

Over the past few decades, the ideas of 'community making' and 'community culture' have been widely discussed in Thai society among scholars and local people. Many debates have focused on methods and processes for (re)creating, reinforcing, or prolonging a sense of community in localities. Accordingly, the task of identifying and (re)constructing the particular qualities of community has become crucial to Thai processes of contemporary place-making.

The concept of 'community' has only recently been adopted into Thai language as the term '*chumchon*' – the most literal translation of 'community' in Thai – did not appear in the 1950 Thai dictionary (Kanchanapan, 1992, cited in Hawanonth *et al.*, 2007). Several Thai scholars in sociology and the humanities argue that the concept of '*muban*', which translates as 'village' in English, is grounded in the notion of '*chumchon*' (Nartsupha, 1991, 1994, 1997; Hawanonth *et al.*, 2007) while, in contrast, others contend that the idea of '*muban*' might be considered as a 'myth' constructed by scholars since the spatial, administrative boundary of a village restricts understanding of complex community systems (Chareonsinoran, 1997 cited in Hawanonth *et al.*, 2007). Two questions are thus raised: how can we comprehend the concept(s) of 'community' in Thai contexts and how have ideas of '*muban*' contributed to contemporary understandings and formations of Thai communities?

While places in Bangkok have dramatically changed over the past few decades, this chapter asks how the conceptions and connotations of the term 'community' have evolved and transformed, attempting to evaluate how studies of *muban*, its notions and characteristics, might lead to better comprehension of present-day *chumchons*, and how emerging concepts of *chumchon* have contributed to the ongoing constitution of a Thai sense of place. Consequently, this chapter explores processes of reformation and fragmentation of Thai community through exploring and contrasting three places with distinctive conceptions of community, namely *muban*, *chumchon* and an 'unidentified' place.

Beginning to read urban place

Bangkok's contemporary urban scene presents a completely different image from what is familiar from earlier eras. Nostalgically held notions of Bangkok as Venice of the East and the picturesque village life along *khlongs* (waterways) no longer serve to depict the characteristics of the city today. The immense growth of the city as well as major influences from globalizing modernization over the past century, especially the last 50 years, have had huge impacts on the environment and led to the emergence of newly intermixed characteristics. The 'merging phenomena' of visual and social worlds, with the superimpositions and intermixtures of past and present scenes, old and new elements, a sense of village and city, and also spiritual and practical engagements, have created 'an intermingled essence', which is conceived as expressing distinctively Thai characteristics of modern Bangkok.

This rapid urbanization, the rise of notions of lost identity (or 'authentic' qualities and a sense of distinctiveness) and the subsequent importance of locality and community are of increasing concern in urban discourse. The emergence of *thanons* (roads) followed by the expansion of land-based settlements has continuously sprawled to the outskirts of Bangkok, intruding into and swallowing many local communities, whereas simultaneously, nostalgic images of water-focused life are seemingly declining in salience. There are, however, questions about how these phenomena have (re)formed or otherwise transformed a sense of community life and how these might be read or described, providing an opportunity to reconsider Thai concepts of community life and its constitutive processes, since prevailing discursive frameworks now appear inadequate as a basis for understanding 'modern' Bangkok with its seemingly indecipherable, disorienting and complex characteristics.

This reconceptualization and reconstitution of community in present-day Thai society is read through considering three places: Ban Bangraonok in Nonthaburi,

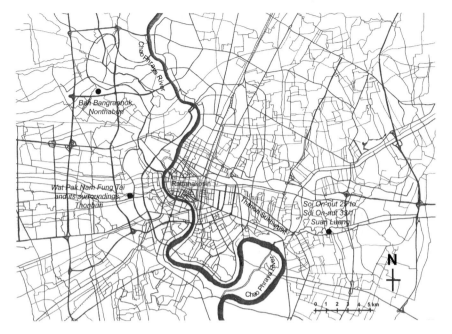

FIGURE 17.1 Map of Bangkok Region, displaying the locations
of three case study areas
Source: Author

Wat Pak Nam Fung Tai and its surroundings in Thonburi, and Soi On-nut 29 to
Soi On-nut 33/1 in Suan Luang (Figure 17.1). These three cases represent different
characteristics of urban development in Bangkok, ranging from a seemingly rural
area to a more generic urban place. Observations of these changes were carried
out between 2001 and 2004, and interviews of local people (approximately 30
for each case) were conducted in 2002. The analysis that follows discusses the
complexities and tensions inherent in exploring conceptions of community in
three different places, and under three spheres of community life: conceptual,
practical and political.

Reading place in a Thai context

Unlike in most Western[1] societies, where the terms 'the search for identity' and
'identity crisis' have long been familiar expressions, the unique qualities of 'the
Thais' had largely been taken for granted without being questioned or defined
(Barnett, 1959; Mulder, 1996). It is only recently that the new phase of turning
the traditional Siamese state into the modern nation-state of Thailand, instituted
at the end of nineteenth century, has stimulated questions about identity (Keyes,
1987).[2] The increasing impacts of globalization and related dramatic transformations have reinforced the need to redefine national identity and identify the
characteristics that make the nation unique and different from others.

Since, typically, Thais were not specifically uninterested in defining their own qualities, the subject of how Thai space and place have been characterized was more of a concern to foreign scholars. Many who simply looked for a sense of order and a spatial hierarchical system akin to that of Western cities referred only to superficial images of Thai places but were not able to describe further qualities beyond the terms 'complexity' and 'richness' (Maugham, 1995, cited in Askew, 2002; Fournereau, 1998; Crawfurd, 1967 cited in Smithies, 1993; Warren, 2002).

This legacy has meant that problems in reading and understanding Thai place have arisen, along with a need to find conceptual tools that will assist in a better understanding of the uniqueness of Thai place. The idea of a distinctive Thai 'cultural landscape' that signifies place as an expression of evolving cultural values and social behaviour is adopted here. The uniqueness of Thai culture, shaping the way Thai people see, behave in and perceive their world, necessitates the search for a specific language and vocabulary that is able to apprehend the cultural sensibilities that subsist alongside the material characteristics of Thai places. Hence existing theories, mostly originating from Western scholarship, may not be appropriate in investigating the unique social, psychological and spiritual conceptions of Bangkok.

This suggests the need to adopt suitable methods for understanding the complexity of 'modern' Bangkok, including the ways in which cultural meanings and social values are produced and considered in learned discourse and daily life. According to Winichakul (1994), local knowledge, vocabularies and conceptions are important tools in creating an in-depth understanding of Thai place. Thus, instead of using inapplicable Western concepts to explore Thai space and its constitutive places, local vocabularies and concepts are used here to develop deeper understandings of Thai place than those put forward by Western academics.

Reading community: reading place

What is it that makes each place different? The idea of reading and interpreting a particular place at a particular time, including the ways in which people perceive, shape and value their environment, was the starting point for this research. Reading the subtle and complex cultural sensibilities that subsist provides a method to identify the unique characteristics of a place, highlighting how each place is different from others.

In this study, a holistic view is acknowledged as crucial in understanding the deep meaning of place and its complex relationships. According to Alexander's semi-lattice structure, things do not exist in isolation (in his case, a tree), but are associated, overlap, or are combined in particular ways (Alexander, in Kaplan and Kaplan, 1982). Instead of concentrating on particular aspects of place, the focus here is on the interconnections of components and domains, the overall qualities and image of place in its entirety.

This chapter suggests that Wood's (1997) idea of 'community as place' provides one way to understand and describe place, focusing on processes of community constitution and the construction of community identity. I propose three fundamental themes through which to read Thai community: first, the concept of community life; second, the practices of community life; and third, the politics of community life. Each theme contributes different perspectives and forms of knowledge of community as place, and in combination they yield a more complete understanding of place.

The dimension of time is a crucial factor in the constitutive process of community formation, since meanings and affiliations are constructed through longitudinal and ongoing processes (Rivlin, 1987). Combining spatial and temporal dimensions (Tuan, 1977), this chapter proposes that one way to read changes over time is to read the different degrees of change across various contemporary city spaces, exploring how different degrees and elements of continuity and change provides insight into processes of urban diversification.

The concept of community life

The nostalgic notion of Bangkok

The imaginary formation of Thai community has been shaped by the classic, romantic scene of '*ban suan rim khlong*', the house along the waterway in the midst of an orchard, conveying a nostalgic and 'traditional' sense of a Bangkok constituted out of a series of village communities. Such an imaginary is characterized by the close connections between four fundamental components, namely *ban* (houses), *suan* (orchards), *khlong* (waterways) and *wat* (Buddhist temples).

The qualities of each element contributed to the constitution of the conventional village (*muban*) in terms of its conceptual, social and political aspects. The concentration of village life along the *khlong* and the unbounded quality[3] that applied spatially, socially, perceptually and spiritually to the village environment dominated the sense of continuity, unity, and harmony of community. The holism of a village was not only displayed through land utilization and village activities but also religious customs and spiritual beliefs. Buddhism has long been deeply integrated with the way Thai people perceive their world and carry on their lives. Accordingly, the *wat*, as central symbol of Buddhist place, expanded the conformity of the village not merely beyond a geographical pace but also beyond, across the invisible terrain of the eternal spirits.

An analysis of terminology, definitions and connotations reflects the underlying concepts of Thai place that have been formed and transformed through phenomenal changes over time). The term *muban* (village), referring to a gathering of many *bans* (houses), and the term *muang* (city or town), originally meaning an assemblage of both rural and urban (*mu*)*ban* (houses and villages), emphasizes the persistent significance of *ban* as a concept that influences the conceptualization of space constituted at a larger scale, whether community, city, or even country.[4] The idea of *ban*, meaning both the building and its compound, is attached to the sense of domesticity and enclosure of place (and the subsequent sense of security). This feeling of familiar domesticity and security, however, foregrounds distinct social perceptions of 'we' and 'they', 'local' and 'strangers', or 'inside' and 'outside'.

FIGURE 17.2 Simplified diagram of physical settings shared by the three case studies in the present day
Source: Author's observation of the three case areas

The diversification of Bangkok

Over the past 40 years, driven by the diverse forces of modernization, the urban growth and development of Bangkok has invaded indigenous settlements, over-taking the notion of '*ban suan rim khlong*' of each place to varying degrees. These changes underlie the transforming and newly emerging concepts of community in the present. The former senses of uniformity, coherence and closeness that Bangkok as a city of villages once conveyed have been shifting towards hybridization, fragmentation and glocalization.

In investigating community identity and the dynamic interrelationships between humans and their environments and the underlying qualities and meanings of place, I focused upon three seemingly similar sites. Although these three places all contain the ingredients of *ban, khlong, thanon* (roadways), *wat* and *suan*, the differing proportions and qualities of these elements, and the differing degrees of human engagement and emotional attachment with these places, shows that each place is distinctive and unique. Apart from the physical and visible setting and components, other aspects of place also contribute to the process of place-making and subtle differentiation. For the variety of community systems and characteristics, ranging from the 'traditional' or locally-concentrated to the more modernized or globally-oriented, is further articulated by concepts that connote a communal place in modern Thai language. Many of these terms and connotations are investigated through the three case studies, as briefly summarized below.

The Ban Bangraonok case exemplifies the survival of a '*muban*' (village) community, where traditional environments and lifestyles are still distinctively apparent. Such surviving villages are usually located beyond the urbanized sprawl of Bangkok, at the outer fringe of the city. Hence the notion of *muban* (village) is underlain with a rural or traditional sense of *chonnabot* (rural area) environment, which also implies impressions of underdevelopment, or to some extent, a realm of the uncivilized (Figure 17.3).

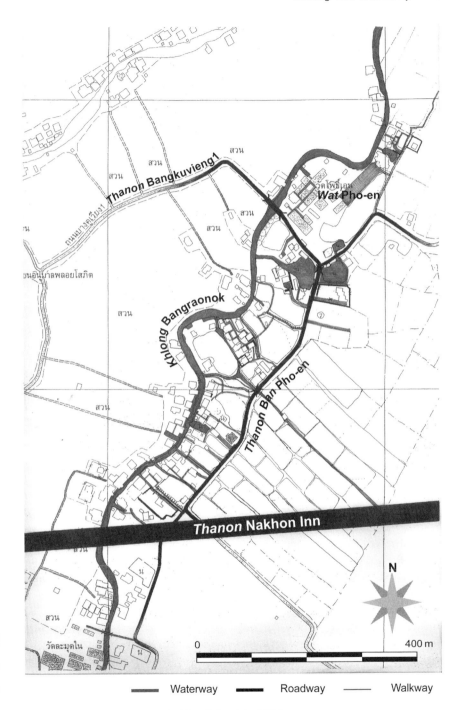

Waterway ▬▬▬ Roadway ▬▬▬ Walkway ━━━

FIGURE 17.3 Ban Bangraonok, Nonthaburi, in 2004
Source: Modified from Urban Department of Nonthaburi survey

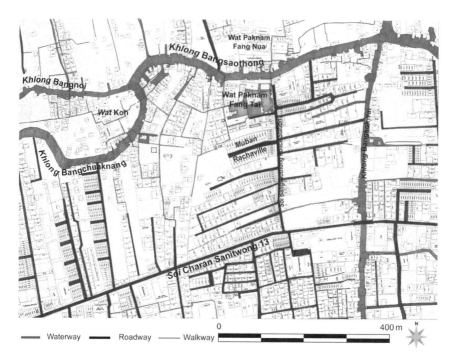

Waterway ▬▬▬ Roadway ▬▬▬ Walkway

0   400 m   N

FIGURE 17.4 Wat Paknam Fang Tai and its surroundings, Thonburi, in 2004
Source: Modified from Metropolitan Electricity Authority survey

The study in Thonburi presents the juxtaposition of a surviving traditional sense of community expressed by the concept of '*chumchon*', with the modern community concept of '*muban-chad-san*' (housing estate) and the rapid growth of urbanization. *Chumchon*, directly translated as a gathering or group of people who have something in common (*chum+chon*), embodies the nostalgic sense of *muban* (village) together with the sense of an urban community located in the *muang* (city). These two ostensibly contrasting concepts of community nevertheless represent the continuing efforts involved in establishing a sense of community in present day Thai place despite the challenges produced by urban transformation (Figure 17.4).

The third case, Suan Luang, demonstrates how the imposition of an administrative unit (*khet*, or district) affects the reconceptualization of the idea of community in the wider context of the Bangkok Metropolitan Area (BMA). This area, a so-called 'unidentified place', illustrates the erasure of a sense of local community as it becomes threatened by the recently implemented political system with its general-ized perspectives and the subsequent loss of identity with the domestic realm (Figure 17.5).

Various complicated and yet ambiguous notions of community and place in the Thai context call for further clarification of these concepts in two other domains, namely practices and the politics of community life. By observing the importance of the smallest unit of *ban* (houses), fundamental to the idea of Thai

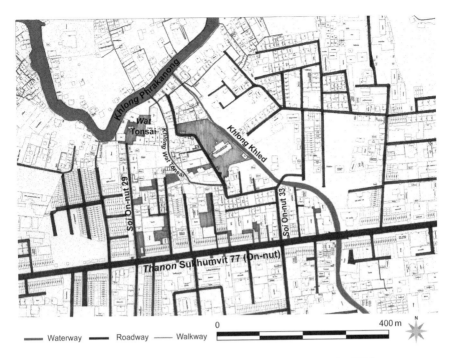

FIGURE 17.5 Soi On-nut 29 to Soi On-nut 33/1, Suan Luang, in 2004
Source: Modified from Metropolitan Electricity Authority survey

community at whatever scale it might be considered, this research therefore adopts a bottom-up view by initiating the reading of place from residents' points of view. Accordingly, the next section attempts to describe the subtlety of place as community through focusing on the local perceptions that develop through everyday life, and this is followed by a discussion about the differences produced by the influence of politics on the process of the making and conceiving of community.

The practices of community life

The different ways in which the unbounded quality (Noparatnaraporn, 2005) of the three places discussed are being transformed through multiple activities and forms of land utilization, is leading to a sense of vanishing homogeneity and is fracturing the harmony of community. This diversification of place corresponds with people's increasingly complicated lifestyles and dynamic patterns of movement, in comparison with former days when necessary commodities, social interactions and services were located within walking distance. Accordingly, it seems more difficult to find commonality among people's lives and build up a sense of shared bonds within a neighbourhood.

Community fragmentation is evident in the fracturing and hybridizing of space. The study revealed, however, that the divisions within the community were not identified simply as marked in physical space but were conceived variously in people's minds in accordance with their different experiences of everyday life. Displaying the diversity (hybridization) of place can be illustrated by people's different ways of perceiving and behaving in their neighbourhood environments.

These two ways of dealing conceptually with the boundaries of community space, namely in its physicality and in people's imaginations, help to identify how communities are contextualized at particular times. On one hand, while many forms of physical space have become increasingly bounded, a sense of unboundedness has been translated and migrated to the social, perceptual and spiritual realms of space, beyond these material boundaries. The permeability and limitlessness of space and the virtual world are instead created in people's imaginations. On the other hand, while urban expansion creates a sense of unboundedness into a 'limitless Bangkok', the human perception of place is also conceived as a smaller, lived-in world. Boundaries are thus variously confined and delimited in people's minds, and such imaginaries are aligned with the scope and form of their associations in real life.

Accordingly, two important factors in the formation of Thai community, apparent from the case studies, consisting of social interactions and mental connections are presented in the following sections. Each view contributes to a more complete picture of transforming places in the Thai contexts in identifying varying attitudes and perspectives.

Social interactions

The interviews with inhabitants from all the case study sites demonstrate how social relationships with neighbours influence how people mentally define the scope of their neighbourhood areas (*lawaek ban*). Social collaborations among residents who live in the same neighbourhood play important parts in constructing a sense of community. Many shared factors affect the possibility and degree of social interrelations among neighbours, including the degree of familiarity and closeness, length of stay, social class, ethnicity and religious beliefs. Nevertheless, local civic spaces[5] are used and perceived as a mechanism to stimulate and enhance the chances of social interaction and reinforce values of cooperation.

The continuing senses of community in Ban Bangraonok (*muban*) and in the northern part of Thonburi (*chumchon*) reflect the effective roles of neighbourhood spaces, such as *khlong*, *wat* and *ran-kha* (local store), in everyday collaboration (Plates 17.1a, 17.2a and 17.3a). In contrast, a sense of common belonging and levels of social interactions were much less in the land-based settlements, including the southern part of Thonburi and most areas of Suan Luang. Instead, public spaces and places outside the neighbourhoods, such as department stores, supermarkets, or public parks, have become more significant as substitutes for local public spaces.

It is nevertheless worthwhile analysing why the places (*khlong*, *wat* and *ran-kha*) that were once fundamental to the construction of village community, and still surviving in the Ban Bangraonok case, are no longer the major constituents of community in the other two case studies.

PLATE 17.1 Life(less?) along the khlongs in (a) Nonthaburi, (b) Thonburi
Source: Cuttaleeya Jiraprasertkun

PLATE 17.1 (*cont'd*) (c) Suan Luang

In Thonburi and Suan Luang, where many public activities have been transferred to walkways and *thanon* (roadways), the *khlong* is no longer a focus of activity in everyday community life (Plates 17.1b and 17.1c). Social interactions among people who live in the same neighbourhood were mainly conducted on the solid ground via the linked systems of walkways, roads and bridges. There is also a different physical speed – cars, or even hurrying pedestrians – that sets a different 'pace of life' that is less discursive, less contemplative, more frenetic and less able to supply the occasions for pause and interchange of concerns. Movement along *Khlong* and *thanon* produce different experiences of time.

The role of a local *wat* as a centre of a community's religious and social activities has also altered. Although Buddhist traditions are still popularly practised and attempts to provide public spaces for the neighbourhoods were observed in the two more urbanized areas, residents did not merely attach themselves to their nearest temple (Plates 17.2b and 17.2c). They may or may not follow their ancestors by going back to the family's temples, and the ones they regularly visited could be located anywhere. Various factors influence the choice of temple, such as convenience of accessibility, reputation of the temples or monks, or the attractiveness of the setting.

Similarly, 'local stores' where people used to 'hang out' and socialize with familiar neighbourhood friends are disappearing. Increasingly, *ran-kha*, or food and grocery stores, in the neighbourhood decline in their role as major communal spaces. Additionally, as people now have less leisure time, the utilization of civic space is

PLATE 17.2 Community space of the wats in (a) Nonthaburi, (b) Thonburi
Source: Cuttaleeya Jiraprasertkun

PLATE 17.2 (*cont'd*) (c) Suan Luang

increasingly limited. And as use declines and neglect takes over, the quality of such spaces deteriorates, and this in turn deters social interaction. Most grocery and/or ready-to-eat food stores now provide no sitting areas and those along the main roads, which serve both locals and outsiders, have little daily conversation (Plates 17.3a, b, c).

Overall, the changing roles and functions of spaces designed for communal purposes and the lack of neighbourhood public spaces increasingly results in people spending more time in their houses or outside the neighbourhoods altogether. The minimal opportunities for people to mingle and meet within their neighbourhood reduce social connections and reproduces tighter mental boundaries. Such material processes of fragmenting place in the perceptual world reinforce the disappearing emotional and conceptual connections with places and thereby the deteriorating sense of belonging.

Mental connections

The research suggests that residents of Bangkok have become more materialistic, concerned more with surface appearances or the physical world, its aesthetic qualities, rather than the underlying social, emotional and spiritual qualities through which place is valued. In the land-based settlements such as the southern part of Thonburi and most areas of Suan Luang, the spatial structures of settlements now follow the tree-like patterns of roadways. However, the buildings are only

PLATE 17.3 Food and grocery stores along thanons in (a) Nonthaburi, (b) Thonburi
Source: Cuttaleeya Jiraprasertkun

PLATE 17.3 *(cont'd)* (c) Suan Luang

connected at a superficial level with little relationship to the deeper levels of emotional and spiritual attachment that supports a sense of community and mental comfort amongst those who live there.

The data showed, in each place and to varying degrees, a diminishing familiarity between community members, although this remains important in establishing a sense of place and sense of home in Thai culture. Any unfamiliar changes or the advent of strangers were believed to erode or disrupt everyday practices, were perceived as destructive and harmful to the ambience of domesticity. The causes of these negative perceptions were different for each case: the emergence of *thanon* in *Ban* Bangraonok; the rapid physical changes and the increase of renters and newcomers in Thonburi; and the continuous moving in and out of residents and the rise of apartments and factories in Suan Luang.

The declining significance of the local *wat* has affected its traditional role of sustaining community and reinforcing spiritual connections between residents and the place where they live. Apart from the loss of respect for monks and Buddhist temples, it is apparent that the *wat* has changed its concentration on providing spiritual and psychological support and bringing enlightenment to people's lives. Increasingly, it conducts commercial business and earns money. The *wat* tends to interpret its new role as supporting and facilitating practical requirements, such as providing spaces for social, communal, recreational and sporting activities and parking cars! (See Plates 17.2a, b, c) There remains, however, another indication of the resilience of the human connection to the spiritual world: the belief in the existence of spirits persists not so much through the physical existence of a spirit house in every house compound, but rather in rituals and ceremonies.

The politics of community life

Community system: social versus political

Different perspectives and understandings of community, between those that were constructed in people's minds through everyday experiences and those that were invented and implemented in the political system, have challenged the sustaining of community identity during the re-formation of modern Bangkok. The political concept of community and the tensions resulting from oppositional views are now discussed.

The studies across the three areas highlighted the transformation of local municipalities of Bangkok – from the socio-political system of '*mu*' or '*ban*', abbreviated from *muban* (village), towards the more concentrated political system of *chumchon* (community) and *khet* (district). Different levels of integration between social characteristics and political forces, expressed through the different community structures and organizations, are important to the constructions of community identity and image of place in each area.

Ban Bangraonok presents a complete system of '*mu(ban)*'. Here, the political system is deeply integrated and strongly influenced by the social structures of the village. Thonburi exhibits the integration of both surviving and newly emerging concepts of community within one place, although the survival of the '*mu*' system has been supplanted by the recently applied systems of *chumchon* and *khet* and is no longer influential in community life. Suan Luang displays the fully operating political system that increasingly prevails. However, existing social networks and people's sense of community – how the social environment works in real life – should not be overlooked in the process of reconstituting community. The relationship between social networks and political organization, as exemplified in the case of Ban Bangraonok in Nonthaburi and *Chumchon* behind Wat Pak Nam Fang Tai in Thonburi, make a major contribution to the survival of the original sense of place, community and its identity in the present.

The new official administrative unit of *chumchon* is intended to modify a community structure and its bureaucratic system according to the physical and social changes of current urbanization. Until 2002, which is when the surveys were conducted, the concept of *chumchon* had been applied to both old and new settlements in Bangkok, excluding some places that are not bureaucratically defined as communities.

The government's policy of replacing the original '*mu*' system with the new concept of *chumchon* for the whole Bangkok Metropolitan Area (BMA) in 2004 does not appear to correspond with what local people perceive and prefer. Interviewees in all sites referred to the connotations of *chumchon*, or urban community, as denoting negative associations with *muang* (urbanized area), including notions of high building density, environmental pollution and social problems. Hence the term '*chumchon ae-at*', referring to a slum area, is perceived as one of the characteristics of *muang*. Interviews with residents in all three places demonstrate that most desire neither to live in *chumchon* nor to see this concept implemented in their areas. Even in the case of an official *chumchon* in the northern part of Thonburi, where the residents were forced to accept the implementation of this new political concept, people's understandings and expectations of community remain influenced by their nostalgically held notions of village (*muban*).

Community boundary: territory and authority

The imposition of political administrative systems has constructed new concepts of community boundary, territory, authorization and municipality. For example, *khlong* and *thanon*, which are perceived by locals as *uniting* elements of community life, have become *dividing* lines between districts and sub-districts in the urban community system.

The studies of Thonburi and Suan Luang illustrate that the imposed administrative systems of community (*khet*) do not correspond with local understandings of community or social organization. The apparent lack of concern for existing local communities from the political perspective impacts upon the level of social participation and sense of community within neighbourhoods. Several readjustments and resizing of political boundaries of districts (*khet*) in Bangkok over the past few decades have been carried out for administrative purposes relating to political authority and control. These changes to political boundaries have not only affected the reordering of house numbers and reorganized the bureaucratic management of localities, but also contributed to confusing locals' concepts of community.

In contrast to the progressive sense of the diversification and fragmentation of Bangkok, as illustrated in conceptual, social and perceptual realms, standardization is officially promoted as necessary for modern political structure and regulation. If the survival and recreation of community identity in each place is a priority, then existing approaches need adjustment. It seems that alongside the bureaucratic structures imposed on places there is a need for more subtle, flexible and adjustable administrative systems that suit the particular characteristics of each place.

Critical thinking . . .

This chapter examines Bangkok as it transforms into a 'modern' *muang* ('city') through investigating processes of community fragmentation and reformation. Lessons from the three areas studied reinforce the idea that the time-honoured conception of '*muban*' continues to play an important role in the understanding of Thai 'community' in the present. Although the political term (*mu*) is no longer used in Bangkok Metropolitan Area (BMA), the concept it expresses has undeniably been integrated into the more recently constructed community concept of '*chumchon*'. As I have explained, the four fundamental components of *muban* – *ban* (houses), *suan* (orchards), *khlong* (waterways) and *wat* (Buddhist temples) – still comprise key elements within the constitution of urban communities today even though green spaces have been greatly reduced. Besides, it was found that *khlong*, *wat*, and the *ran-kha* (local store) continue to perform major roles as effective neighbourhood public spaces in the construction of both *muban* and *chumchon*. More importantly, evidence of high degree of social interaction and people's deep attachment to their place confirms that the foundational element of *muban* survives in amended form, incorporated into the structure and concept of *chumchon*.

The empirical data has clarified how '*chumchon*' in Thai urban contexts today is not simply a term that can be translated to mean 'community'. Though it builds upon the conception of community from *muban*, and contains many similar physical components and social structures, *chumchon* expresses a political dimension that is qualitatively different from *muban*: it connotes a spatially bounded and registered political community that cannot simply be superimposed on every residential area. However, the notion of *chumchon* is complicated by its association with the negative concept of *chumchon(ae-at)* or slum, a place in which nobody would want to live. This reveals how concepts that connote 'community' are variously composed out of various local understandings of *muban*, *chumchon*, *muban-chad-san* (housing estate) and 'unidentified' place, highlighting the broader urban fragmentation that is expressed through the diverse, contesting conceptualizations of community that are emerging.

The second dimension of this community fragmentation is evident in the diversification of people's ways of living, and empirical evidence exposes shifts towards an independent lifestyle (*tang-khon-tang-yu*), social separation and mental detachment from place. A decreasing sense of belonging and bonding (*khwam phukphan*) to the neighbourhood results not only in a declining concern with the immediate environment but also in a waning sense of community. In contemporary contexts in which various activities and better transport facilities increasingly encourage people to travel beyond their neighbourhood space, *wat* appears to persist as the most important factor linking people's minds and spirits together through religious practices and beliefs. If the meanings and values associated with this spirituality are signified as important factors of communality, this could be crucial in reviving a sense of place and community in Thai society. This suggests that the role and meanings of the local *wat* need to be reinforced in order to stimulate a greater degree of social connection and deepen the level of emotional and spiritual attachment of people to their neighbourhood places.

The third dimension of neighbourhood fragmentation and change concerns various political concepts of community. While Bangkok seems to be shifting from communities that were originally socially orientated towards those which are constituted through politically based conceptions, the research reveals nevertheless, that social and perceptual dimensions of place remain important in the processes of constituting community. This invokes the imperative to reconsider the influence of these political agencies and search for possibilities to revive social practices that are integrated with more indigenous concepts of community and place in administrative processes. It also suggests that policies dealing with community should be designed in ways that allow the social characteristics of place to survive and constitute part of the construction of new community identities as transformation takes place.

By reading places through the transformations of communities, it is apparent that physical appearance is not a major concern for Thai people. Rather, social, psychological and spiritual values, what might be regarded as the underlying *meanings* of place, are key to achieving a deeper understanding of the cultural sensibilities that subsist in a Thai sense of place. While contributing to a more substantive understanding of community-making and place-making theory in Thai contexts, this approach also clarifies why existing theories are limited in their interpretation of Thai space and place since they tend to 'generalize' about all places, proposing the same analytical models based on similar techniques and preconceived methods of reading.

In recent times a significant body of literature has emphasized the importance of taking account of local cultural influences in the creation of the uniqueness of place. Consequently, it is imperative to develop more sensitive, specific and nuanced methods for understanding place in particular contexts and at particular times. This trend has not until now extended into research into Thai everyday space and place. However, this chapter places great importance on acknowledging the influence of Thai culture in constituting and interpreting Thai place and community. This exemplifies the imperative of synthesizing general knowledge and local wisdom in reading place, in developing theoretical and methodological approaches. The reading of Thai place through three spheres of community life – concepts, practices, and politics – reveals that social interactions and mental connections are deeply rooted in the formation of community. For Thais, mental connections, referring to deeper levels of spirituality and experiential qualities, are key to yielding a meaningful understanding of human-space relationships. Accordingly, a strengthening of these emotional attachments to places would assist processes of regenerating and reviving the qualities that all too frequently merely reside in reminiscences of the past, returning them to contemporary community life and space.

Notes

1 The use of the term the 'West' or 'Western' in this chapter reflects the idea of Orientalism being inscribed as the 'Other' of Western civilization (Winichakul, 1994: 7–8). The 'Thai' or 'Thainess' in this respect reflects the notion of 'We' or one's own self that is different or in some senses opposite to the 'Other' or the 'West'.

2 To constitute the new concept of nation-state, several ideas and policies were implemented during the King Rama VI period (1910–25); for example, the idea of nationality and the new flag of Siam were introduced, including new Western cultural importations such as a title of name, a surname and Western dressing styles. The campaign of a 'Thai' nation was seriously promoted in the King Rama VII period,

under the government of General Plaek Phibunsongkram (1938–44, 1948–57), together with the change of name from 'Siam' to 'Thailand' – the land of the free – and a new national anthem (Thanakit, 2000: 268–70; Pithipat, 2001: 51–5; Plynoy, 2000: 142–61).

3   See the explanations of unboundedness in Noparatnaraporn (2005).

4   See the connotations of '*ban*' and '*muang*' in Noparatnaraporn (2003).

5   Michael Douglass introduces the term 'civic space', which is focused neither on owner-ship (private or public) nor its organization, to the reading of community space. This term matches the concept of 'community free space', defined by King and Hustedde (1993), where people can meet for public talk and actively contribute to solving public problems (Douglass, 2002: 2).

18

URBAN POLITICAL ECOLOGY IN THE GLOBAL SOUTH

Everyday environmental struggles of home in Managua, Nicaragua

Laura Shillington

Introduction

Everyday relationships with urban natures for many inhabitants in cities of the global south[1] are very precarious. In *Planet of Slums*, Mike Davis (2006) characterizes such relationships as 'slum ecologies'. Indeed, most households deal with polluted water, raw sewage, landslides, flooding and disease on a daily basis. Yet while the majority of urban residents struggle with such 'ecologies' everyday, urban geography and urban political ecology have paid little attention to understanding the complexities of these struggles and how they shape lives and social relations. This chapter attends to such complexities by looking at the everyday relations with water, sewage and garbage in an informal settlement in Managua, Nicaragua. Inspired by work in political ecology which has examined how *rural* communities and households adapt to, rely on and create environmental changes, I ask how households in this marginalized urban area interact with, adapt to, rely on, and shape urban nature. I contend that paying attention to such everyday environmental relations assists in better understanding the role of marginalized communities in shaping socio-environmental processes of cities of the global south. I contend that paying attention to such everyday environmental relations assists in better comprehending the role of marginalized communities in shaping socio-environmental processes of cities of the global south.

The quotidian dynamics of humans and 'natures' are significant elements of the urban milieu, and accordingly need to be taken into account for more inclusive understandings of cities and urban life. This is of particular importance for cities of the global south where the majority of research on urban life has been framed through developmentalist lenses rather than being informed by the diverse actual routines of daily life. As Robinson (2006: x–xi) points out, cities of the global south have been seen as imitations of Western cities and not as sites of 'inventiveness and innovation'. Yet everyday life in cities of the global south is teeming with innovation, adaptations and mundane routines, all of which contribute to producing diverse and 'ordinary' cities. Human relations with urban natures are an integral part of shaping cities and urban life.

My examination of human–nature relations in Managua draws on scholarship about everyday (mundane) geographies, urban natures and recent attempts to re-materialize geography within urban geography. By engaging with these literatures, I aim to contribute to urban political ecology through an examination of everyday human–environment relations. While urban political ecology has made visible the ways in which cities, and thus urban natures, are produced through unequal power relations, the politics embedded in everyday activities has tended to be sidelined. For instance, the production of urban waterscapes involves inequalities at multiple scales, from its spatial distribution to its allocation in homes for daily use. And yet the majority of analyses have focused on far more grand displays of power, to the neglect of more intimate places and spaces. I aim to contribute to the urban political ecology literature by examining household scale human–nature relations, while at the same time contributing to a new critical and rematerialized urban theory of the global south.

Using a case study of households in Managua, this chapter will address two main questions. First, what does an examination of everyday human–nature relations in cities tell us about urban life in Managua? Second, what can an urban political ecological approach contribute to creating new urban theories of ordinary cities? This chapter is based on ethnographic research carried out between June 2005 and July 2006 involving in-depth interviews and participant mapping with twenty-five women as well as three focus groups in an informal settlement in Managua. Before considering the case study, I discuss recent work on the everyday, materiality, and urban nature. Next, I briefly outline main trends within urban political ecology research. Following this, I turn to the case study of Managua, examining the everyday environmental struggles in marginalized homes, using this to point to instances where an urban political ecology approach might provide more nuanced understandings of quotidian urban life and urbanity.

Everyday life, materiality, urban natures

The everyday comprises a multitude of practices, objects and places. It takes place in the home, on the streets, and at work and constantly crosses the boundaries between private and public spaces. It is the world we directly experience through material, emotional and imaginary interactions with other living beings, places and objects. Some think of the everyday has banal and mundane, yet everyday life also contains the unexpected and exciting. It is because the everyday is where life happens that it is messy, complex and contested. The everyday is informed by the past, present and future, oscillating between them through practices, thoughts and desires. In this sense, everyday life comprises repetitions of previous days as well as rehearsals for future days.

The simultaneous complexity and habitualness of everyday life has provided a rich source of knowledge for scholarship in geography and other social sciences. It is beyond the scope of this chapter to provide an overview of the work on and theorizations of everyday life.[2] Rather, I want to briefly highlight some aspects of everyday life that have been of recent interest in urban theory. In particular, I want to point to the difference between studies on everyday urban life in the global north and global south. In Anglo-American urban geography, research has focused primarily on urban life in the global north. Research that pays attention to everyday life in cities of the global south tends to occupy development studies more than urban studies. As I will show, studies in urban areas of the global north are intimately linked to material geographies while those in the global south are associated more with political economy.

There are two broad, related foci in the study of the everyday urban life: practices and relations. The first encompasses work that looks at the mundane, routine activities of urban life. In the global north, this has included examinations of the taken-for-granted acts of eating, walking, playing and working in cities. Everyday street life has been of particular interest – how the street is used in the daily activities of walking, consuming and playing. The latter has included the use of the street as a playspace which resists the normal ordering and use of urban space (Borden, 2001). Eating has also garnered attention – what people eat every day and where they consume food. Food geographies are part of a growing field of the geographies of consumption. Everyday life is made up of diverse consumptive practices (at home, in the streets, for instance), which assist individuals in constructing identities and negotiating relations and places. Intimately linked to practices of consumption are, of course, everyday relations to material goods, which I look at later in this chapter.

Part of the everyday practices of cities involve what Hubbard (2006) calls 'embodied practices', which he suggests are the daily ways that individuals negotiate the city – how they physically get by. Geographies of mobility have called attention to the ways in which differently abled bodies negotiate everyday spaces (street, home, schools, shopping malls). Feminist geographers have paid particular attention to the gendered and sexed differences in experiencing, accessing and moving about cities (Binnie and Valentine, 1999; Longhurst, 2000). In addition to examining everyday urban practices they have also interrogated the practices and relations within the home itself. The attention to the everyday routines and social relations of women both inside and outside the home has challenged dominant notions of the city and urban life. More recent research on domestic practices has looked at the material and emotional relations with objects such as computers and household appliances (Hollway and Valentine, 2001), animals (Franklin, 2006) and plants (Bhatti and Church, 2004; Hitchings, 2003). The latter two relations are part of increased attention to urban natures and materiality.

Discussions of everyday objects and natures contribute to a growing body of work in urban geography that seeks to re-materialize the city (Braun, 2005; Smith, 2004). In addition to the work on plants, animals and domestic objects noted previously, urban research has also explored urban parks (Heynen, 2003), suburban lawns (Robbins, 2007) and urban wildlife (Wolch, 2002). Such work has sought to understand how our relationships with nature shape the cultural, economic and ecological urban landscape. It also seeks to uncover the ways in which the cultural and emotional meanings of human–nature relations shape everyday spaces. This focus has also been part of a renewed material urban geography. Daily interactions with cars, computers and other objects shape our everyday practices and relations. Such material relations, Smith (2004: 89–90) suggests, might be more important for some people than social relations.

Intimate material relations have been less studied in cities of the global south. Indeed, there have been far fewer studies about everyday life; or rather, research has tended to focus on a less diverse range of everyday practices and relations, most research occurring within the field of gender and development. Early work examined household structures, gendered division of labour and livelihood strategies (Momsen and Townsend, 1987). There is a large body of research on women and work, including home-based work and informal street trade (Hays-Mitchell, 1994; Sheldon, 1996). Similar to research in the global north, the street has also been of interest in cities of the global south. In particular, research on everyday practices that oscillate between street and home have challenged Western conceptions of public and private space in looking at sites of everyday domestic activities (Drummond, 2000). Children's daily practices in street life have also garnered attention, especially their engagement in informal economic activities (Young, 2003; Beazley, 2000). Recent work has concentrated on violence in urban homes and communities (McIlwaine, 2008).

While there are similarities in the sites of research in urban studies of the global north and south (street, home, workplace), the way in which they are framed is considerably different. In the global north, studies have focused more on the cultures, emotions and materiality of everyday life, whereas in the global south, there has been more of a tendency to frame everyday life in socio-economic terms (and hence, development) (Rigg, 2007). Peake (2009) observes in her overview of gender and the city that research in the global north tends towards the development of critical urban theory, while in the global south research focuses more on explanation. Indeed, Western urban theory is largely used to understand and explain everyday urban life in the global south.

With this in mind, rather than draw on any particular theory of everyday life or human-nature relations, I want to make use of political ecology as an approach to look at intimate materials relations with nature in Managua. Political ecology has evolved as an approach to examining human–environment relations. Instead of attempting to apply any one broad theory to explain environmental change, it insists on asking particular sets of questions to better understand the complexity and interconnectedness of human relations with the environment.

Urban political ecology and cities of the global south

While early political ecology concentrated on exploring how political economic structures influenced environmental change in rural areas of the global south, current political ecology regularly engages with critical social theory such as post-structuralism, postcolonialism and feminism.[3] Political ecology has also expanded from a near exclusive focus on rural areas of the global south to include rural and urban areas in the global north. There is now a specific field of *urban* political ecology. The links between current urban political ecology and traditional political ecology are contested. According to Keil (2003, 2005), the main theoretical precursor of urban political ecology is the literature on urban sustainability, new urbanism and urban environmental movements, but not necessarily political ecology itself. Keil predominantly bases his review and conclusion on research from the global north. In spite of this claim, political ecology that draws heavily on the more 'traditional' rural-based approach tends to be situated in urban contexts of the global south.

Regardless of the division, there are common themes that currently underlie both: co-production of nature and society, unequal access to natural resources, uneven distribution of the effects of environmental change, and power relations. Within urban political ecology research in the global north, water has been of particular interest, especially the historical role of water in shaping cities and the production, distribution and consumption of water (Bakker, 2003; Swyngedouw, 2004). Other work in the global north has considered the consumption, production and distribution of 'natural resources' in cities, including those studies on material urban relations cited above.

Urban political ecology research in the global south has also given attention to production and distribution of 'natural resources' in cities; consumption has had less attention. Most prevalent in global south urban political ecologies, like that in northern research, are the growing number of studies that look at the relationship between production, distribution and sustainability of water in cities (Bakker, 2003; Loftus, 2007). Much of this work tends to focus on the politics of water at a broad urban scale. One exception is Loftus' (2006) research on everyday resistances to water meters in Durban, South Africa.

Also at a broad urban scale, Véron (2006) explores how air pollution policies differentially impact urban populations. In a different vein, Pelling (2003) looks at the ways in which political and socioeconomic processes in Georgetown, Guyana, have created uneven vulnerabilities to flooding. Njeru (2006) also explores environmental hazards, but focuses on the political ecology of plastic bag waste in Nairobi. Also looking at waste, Moore (2009) draws on environmental justice literature to show how garbage, usually seen as an environmental hazard, is deployed as a political tool by the marginalized urban community adjacent to the dumpsite. Such research is part of a growing body of work on urban political ecologies of cities of the global south, as their complex environmental and social problems are becoming recognized as critical to creating a more inclusive urban political ecological approach. Nevertheless, the everyday environmental struggles and relations in cities of both the global north and south still remain relatively unexamined. This chapter seeks to help address this gap.

I take two key elements from the above literatures. First, I understand the city as more-than-social. Building on this, and following Haraway and Latour, I see everyday spaces and relations as also fundamentally socionatural, wherein the social and natural are always intertwined and co-produced. Everyday lives in cities are tied up in diverse ways with other living and nonliving 'things'; the city is cultural-natural. In addition, I am inspired by the key political ecological questions about the uneven distribution of environmental change and access to resources. Yet, in contrast to most urban political ecological research which has focused on some form of 'chains of explanations' (Rocheleau, 2008) to examine at a broad level uneven distribution of environmental services and problems, I concentrate on the everyday. I do not seek to explain in detail how such unevenness emerges;[4] rather, I examine how individuals and households deal on a daily basis with environmental uncertainty and change. I show the different forms of human–nature relations and experiences in marginalized urban areas of Managua and the role of everyday life in shaping and internalizing environmental changes. I turn now to look specifically at everyday relations with water, sanitation and garbage.

Quotidian environmental struggles and urban life in Managua

Managua, like many cities of the global south, is scattered with informal settlements.[5] According to UN-Habitat (n.d.), 81 per cent of Managua's population live in 'slums'. As a result, the majority of Managuans have very precarious daily lives, with insecure sources of income and land-tenure. In addition, most lack basic urban services such as water, sanitation, garbage removal and adequate housing. With well over half of Managua's population living in informal settlements, their everyday lives shape, to a large degree, the urban landscape. Many of the struggles in these households are environmental – water shortages and quality, raw sewage, garbage and mosquito-born diseases. Households deal with such struggles in multiple ways. I look at three of the main environmental struggles in these marginalized Managuan homes: water, sanitation and garbage.

My case study is a large, informally settled *barrio* in the southern part of the city. The *barrio*, San Augusto, was originally settled in the early 1990s and continues to expand today.[6] The households are among the poorest in the city, with most lacking a stable source of income. Approximately 90 per cent of households are connected informally to the city's water network, thanks to a self-help project initiated by a local non-governmental agency and a Spanish organization (FUNDECI, 2005a, 2005b). However, the water they receive is not always 'safe', as I will discuss shortly. The large majority of houses in the *barrios* still have pit latrines and the area has not yet been officially connected to either the sanitary sewer or storm drainage systems. This has implications for the everyday environmental struggles. Garbage removal is another daily concern for households. Very few households in San Augusto have garbage collection; only houses located on paved or wide flat dirt roads have garbage collection. Out of a total of over 50 roads in the barrio, only six are suitable for garbage trucks to drive on. Unlike in North America and Europe where residents rarely have to think about garbage, residents in most areas of Managua have to deal with garbage removal every day.

Home spaces

Before I detail the ways that individuals and households deal with these three environmental problems, it is important to describe the everyday spaces in which struggles take place and how these are distinct from similar spaces in Western cities. Homes in San Augusto, similar to many of Managua's poor *barrios*, comprise small built structures and the surrounding patio. The patio in Nicaraguan homes is the main space in which domestic life takes place; the physical house, especially in lower income *barrios*, is small and tends to comprise only a bedroom and perhaps a common area (living room/sitting area) and, rarely, the kitchen. The patios in these homes are the primary spaces of everyday domestic activities: cooking, cleaning, bathing, watching TV, socializing, etc. (Shillington, 2008). The inside space of the house is reserved for sleeping, shelter during heavy rain and storage of clothes, food and other household items. Because of the lack of windows and the dependence on low-grade building materials (especially metal) which tend to absorb heat, the houses can be unbearably hot. As such, the patio, which is larger and more comfortable is the primary space where everyday tasks are carried out.

Patios are culturally important spaces in rural and urban houses in Nicaragua. They are sites of the traditional *huerto casero* (household garden), which contributes (at least partially) to household food security. The traditional *huerto casero* in Nicaragua is not a 'garden' in the conventional sense (especially in the Western Euro-American conception of small plots of land striped with rows of vegetables); rather, they are more commonly small-scale agroforestry arrangements, comprising a combination of large trees (usually fruit) and herbaceous plants. The large trees create a microclimate in the patios; the shade produced by the large canopies of the fruit trees protect the patios and houses from the harsh climate of Managua (the daily average temperatures regularly exceed 25 or 30 degrees Celsius year-round). The shade created by fruit trees is critical in making the home a materially comfortable and liveable space, physically and emotionally/mentally. The trees also play a role in dealing with everyday environmental struggles, which I discuss further on.

Alongside trees and plants in patios are the pit latrine, one water tap with hosepipes to a shower and washing sink, and an area for garbage disposal. All together, the composition of the patio – trees, plants, latrines, sink, garbage, etc. – creates an ecology that shapes everyday relations. I now turn to examine some of these relations in more detail. I have examined elsewhere the daily relations between plants and people (Shillington, 2008). Accordingly, I only briefly touch on those relations as they relate to the struggles of water, sanitation and garbage.

Water and sanitation

Quotidian relations between household members and water are precarious. There are two main issues: first, the supply of water is not consistent; and second, the water is not always 'safe' or clean. While the majority of Managuans might seemingly have access to safe water (95 per cent according to UN–Habitat, n.d.), most household connections are informal (self-help water connections). Pipes may not be adequately secured to the city's main (formal) networks, and as a consequence frequently have leaks. Residents and community associations seldom have funds to repair pipes and, in the case of San Augusto, the local NGO no longer has funds to assist. Furthermore, the City of Managua (*Alcaldía de la Ciudad de Managua*, ALMA) has considered such connections illegal and outside their jurisdiction. Unrepaired leaks lead to an increase in bacteria in the water as well as the loss of large quantities of water. For most Managuan residents, then, their water connection does not always provide them with safe or secure water, contrary to national and international statistics. Similarly, sanitation is also a daily struggle for most residents of Managua. The most recent national census reports that 43 per cent of households have a connection to the sewer system, while 33 per cent depend on pit latrines (INDEC, 2006).[7] Having pit latrines[8] in the yards of homes creates a host of (ecological) problems, especially given the important role of the yard – or patio – in Nicaraguan culture.[9]

Over the past six years, Managua has had serious problems distributing water. Households go hours and sometimes days without water delivered to their taps. The reasons for this are complex. In brief, however, there are two main explanations for the water shortages. Many of the water pumps in the city are old and require extensive maintenance. At times repairs to aging water pumps are slow, stopping the supply of water for hours or days. Not surprisingly the water pumps that service the wealthier residential areas and commercial districts are usually the first to be repaired. As such, wealthier households rarely experience the long water shortages that poorer households do. In other cases, the company in charge of energy in Nicaragua, Gas Natural (as well as the previous company, Unión Fenosa), has cut the electricity to ENACAL (*la Empresa Nicaragüense de Acueductos y Alcantarillados*), the entity responsible for water in Managua, because of outstanding bills. For households in barrios such as San Augusto the result is that water availability is very unpredictable. Sometimes they have water in the middle of the night and not in the day; other days it is reversed. Some days households have water for a few hours at a time. This has meant that the daily patterns of water use are constantly adapting, with *women's* water use impacted upon most significantly.

Water is used for cooking, drinking, cleaning (e.g. dishes), laundry and bathing. Most of these household tasks are the responsibility of women, who tend to structure their tasks around the availability of water or attempt to adjust water availability to meet their own schedules. Depending on the task and the quality of water desired, women will either carry out the task when there is water (even if in the middle of the night) or when water is available, collecting and storing it in containers. In addition, many women start cooking and cleaning while they have tap water available. Because it is preferable to cook both beans and rice with 'fresh' water, women have altered their cooking schedules around when water is available. Some women chose to cook an entire day's meals when water is available, and then reheat rice and beans for all three meals.

Bathing and laundry is also done preferably when the tap has water. While fresh water is preferred for bathing, it is difficult when the tap only runs for three hours in the middle of the night. Subsequently, stored water is usually used, and in most households men and school children have priority for bathing. Prioritizing water usage by gender and age creates tension between family members, reinforcing uneven intra-household power relations. Just as women's daily work patterns are affected far more than men's, so too are their personal, intimate needs. Some women will choose to bathe when the tap is on, such as when they wake to fill containers or begin soaking beans. At the same time, many women also try to clean clothes while water is available. Some women will soak dirty clothes in stored water, but prefer to rinse with clean water. This means that rinsing clothes might be done late at night, along with cooking and bathing. The daily work patterns of women constantly shift depending on when water is available, and many women collectively organize their schedules to assist each other in completing tasks or giving them time to shower. For example, neighbours will perform childcare if the water comes on during periods when the school is not in session. Although some women choose to use stored water and attempt to maintain more or less stable everyday routines, the majority of women have opted to shift their patterns to correspond to when there is 'clean', fresh water.

The decision to use 'clean' water reflects a desire not to have to store large quantities of water. Most women in San Augusto are very aware that stored water can quickly become a source for disease. If the containers used to store water do not have proper covers, they become ideal breeding environments for mosquitoes and other bacteria, creating a not-so-ideal home environment for human bodies. Outbreaks of dengue fever are common, and stagnant water around the home is the primary source. It is for this reason that women prefer to cook with water from the tap, which is presumably clean and free of bacteria. Using stored water for drinking, bathing, cleaning and cooking intensifies the possible spread of bacteria. Of course the water that comes out of taps in San Augusto is not necessarily the cleanest water: because of the self-help, informal connections that were installed without proper equipment or expertise, many pipes are in bad condition and increase the potential entry of dirt, insects and bacteria. Nonetheless, it is assumed that water from the tap is cleaner – and therefore safe – than water stored in containers which sit for long periods of time in Managua's heat. Water in containers potentially transforms from clean ('good') water to dirty ('bad').

Stored water is not the only source of so-called 'bad' water. The lack of proper drainage for grey or storm water means 'dirty' water regularly flows through the patio. Sinks drain into the patio unfiltered and flow onto streets and neighbouring patios. In addition, the dirty streams that weave throughout the barrio turn to rivers in the rainy season. Water from the roads regularly flows into the patios, carrying with it garbage, automobile oil and other run-off. Many households experience flooding inside their houses as well as patios. Alongside grey water and road run-off, households also have to deal with sewage overflow from latrines in their patios, especially in the rainy season. These latrines are generally not contained, built to eliminate seepage into the surrounding ground, and so during the heavy rains and the subsequent erosion of top soil, the water run-off from the streets, patios and causeways carries raw sewage.

There are several ways that households attempt to deal with unwanted 'dirty' water in their homes. Most common is the use of 'good' nature – namely plants and trees. As mentioned previously, the patios in San Augusto (and in other informal *barrios* of Managua) comprise a large diversity of plants and trees. Many of these plants are used as ways of dealing with the lack of services. For example, some households plant large fruit trees and hedges at the border of the patio and the road. These help reduce the amount of water that flows from the street into the patio during the rainy season. Large trees and other plants serve as soil retainers, decreasing the amount of water that runs through patios. Because many houses have dirt floors and are not raised above the ground, the increase of water during the rainy season can cause severe damage. Because of this, households have numerous plants closely surrounding their front doors as well as the base of their houses (some houses have doors several inches about the ground to prevent water from entering). Large fruit trees, in particular, impede water flow through their expansive roots systems; they essentially hold patios 'in place'. Indeed, many plants become part of patios because of their ability to 'deal with' climactic events.

Not only do plants and trees in patios help with excess seasonal water, but they are also used to minimize any raw sewage that might 'escape' during the rainy season. Small bushes and bamboo are often planted around latrine to 'seep' up any sewage as well as to provide additional privacy. Although plants and trees are used to assist households in dealing with environmental uncertainty and change, they do not, of course, radically change the social and environmental conditions of most households in Managua. Nonetheless, they are significant elements in attempts to materially produce comfortable homes, and they are important in everyday efforts to cope with the lack of basic urban services.

There is little that households can do to change the 'flow' of clean water into their houses. Few households choose to demand better quality water from the city since most have (informal) connections and few pay for the water used. Demanding pipe repairs and consistent service puts households at risk of being charged (with the installation of water meters). Indeed, many households cover their water taps and hoses by planting vines and small shrubs around them to hide them from ENACAL officials who periodically check. Rather than seek assistance from the municipality to repair pipes, the community association aligns itself with non-governmental organizations which, in the past, have been able to provide the barrio with essentially services such as water pipes to households. Because the everyday relations between water, sanitation and households are much more complex than simply lacking services, collective action to demand access to better services is equally complex and difficult to mobilize. The absence of collective movements around water and sanitation might be a reflection of the complexity of these everyday relations. It also reveals the power relations at work within the household.

In San Augusto, women are responsible for dealing with everyday environmental challenges in the majority (80 per cent) of households, reflecting the traditional role that women still perform in remaining responsible for social reproduction. Households in San Augusto are multi-familial; that is, they usually comprises more than one 'nuclear' family and are multi-generational. There are two predominant household structures: one 'nuclear' family plus one or both sets of grandparents; and two 'nuclear' families (siblings). Typically, most of the adults in the households hold jobs outside the home. Many men, for instance, drive taxis or buses or have stands in local markets; women tend to work as domestic servants, selling goods in local markets or in the streets (as *ambulantes*). In most households, it is generally a woman who remains in the home to manage the household. She is responsible for domestic tasks and in many cases also sells *tortillas*, cooked beans, or other small goods out of the home. In some households, elderly women stay home. Consequently, it is women who tend to make the decisions around how to deal with water and sanitation issues. It is their experience and understanding of relations with water and sanitation that inform household decisions. Women decide which plants and trees are used to deal with different environmental problems in the patios; which trees absorb grey- and rain water the best, which give off the strongest scent to cover up any bad smells (from grey water or the latrine). And most of their knowledge is passed down from their mothers and grandmothers or from the larger patio culture of Nicaragua. Because their daily routine depends on the availability of water, their experiences are much more immediate and intimate. Yet, because they spend most of their day coping with a lack of and poor quality water as well as seasonal changes in their domestic environmental, the time available to mobilize action around such issues is limited. Their ways of challenging environmental changes are subtle and primarily involve attempts to produce home spaces that meet their needs.

Garbage

Similar to uncertainties around water and sewage, women also have to deal with garbage in homes. Water and sewer are materially different from solid waste in that they flow. Thus, the issues and coping strategies around garbage are distinct. Garbage enters the households as waste produced by the household (includes non-organic waste – food packaging and plastic bags – and organic waste) and can also be deposited in the patio through rain and grey water run-off (garbage from other households). As discussed above, few households in San Augusto have garbage service because they are located off the main roads. The majority of the roads are not wide enough or in adequate condition for garbage trucks to pass.[10] Consequently, the houses not on paved roads – around 90 per cent of houses – discard garbage in their patios in five main ways: burning, burying, throwing in the nearby rainwater canal (if the house is located nearby), walking to the nearest road where garbage is collected, or discarding on the side of roads on the way to carry out other activities (on the way to work, market, or school).

For some people, in particular the elderly and those who live far off the main road, discarding garbage is an increasingly difficult task (given, for example, the condition of the dirt roads and distance to paved roads). Most people find it is more convenient to burn or bury garbage in their patios. The garbage that does make it to the main roads, if not collected almost immediately, begins to rapidly decompose in Managua's sweltering heat, creating an unpleasant smell for those households nearby and a breeding ground for bacteria, parasites and mosquitoes. Additionally, the daily afternoon rain for approximately five months of the year combines with the uncollected garbage and, as mentioned previously, flows through the dirt roads and patios, spreading bacteria and parasites into homes, before pouring into the open canals that empty into Lago Xototlán (Lake Managua). Many households will discard non-organic waste directly into the large open canal that passes by the eastern edge of the barrio. Most often, though, organic garbage is buried in the yard and non-organic waste in burned, usually in the patio although sometimes garbage is burned collectively on the roads.

Garbage, therefore, for most households remains in the patio creating both environmental problems and services. What does this mean for the patio and everyday domestic life? Foremost, burying and burning organic waste is convenient for households as they do not have to 'deal with' waste for long; they can bury or burn the waste soon after they produce it (the same day or several days later). Burying and burning waste in the patio also means that there is little travel involved in disposal. Of course, both burying and burning waste create short and long term problems for households.

Organic waste plays a dual role in patios. When it is buried, it contributes to the healthy production of 'good' natures (plants and trees) by providing nutrients. Since organic waste is buried in shallowly dug pits and the rates of decomposition are generally fast, nutrients from this waste are readily available. Typically, a new hole is dug every two to three days when there is enough new garbage to bury (although some households have one pile of organic waste in their patio and slowly let it decompose). In one month, there can be more than six holes of organic waste. As a result, the nutrients are to a certain extent evenly spread throughout the patio. Most households are strategic about where they bury organic waste; normally pits are dug close to fruit trees, herbs and medicinal plants, and plants that they want to grow faster, such as hedges to protect patios from rain water. Holes are rarely dug near cooking areas or the house.

Even though organic waste decomposes relatively fast, it does pose problems in patios: it can produce smells as it decomposes and attracts insects and rodents, which can bring in viruses and diseases. Part of the problem is that the organic waste can include garbage that does not necessarily compost fast. Normally, organic waste includes all food waste – raw vegetable and fruit peels, leftover cooked food (containing oil, meat, grains, sugars) and sometimes newspaper. Oil and cooked food (especially meat) tend to take much longer to decompose and attract rodents and pests (the most common problem is ant infestations).

Similar to the ways in which women attempt to deal with water and sewage, 'good' ecologies are also employed to assist with garbage. There are certain plant species which households claim assist in repelling certain pests, especially ants. These are planted around the perimeter of the yard to dissuade migrations of ants from other patios. Such strategies are not always effective. However, many women claim that since planting *hierba buena* (a mint species) they have had fewer ant problems. In addition, many households bury their yard waste, such as dried leaves, with domestic organic waste to accelerate decomposition. Yard waste is also used to help burn non-organic garbage and according to several women, helps to lessen the smoke produced (this does not, however, decrease the toxicity of burning plastics and other waste).

Non-organic garbage remains much more difficult for households to deal with because it needs to be physically removed from the patio. As I stated earlier, households deal with it by transporting the waste to other places (canal, roads, other communities). In most households, men tend to be responsible for discarding non-organic waste (if it is not burned). If they own taxis, they will take waste (sometimes both organic and non-organic) to dispose of in public trash bins, usually in parks or near public buildings. Children will also sometimes take bags of garbage on the way to school to discard. It is not uncommon to see trash bins in schoolyards, parks and streets overflowing. The household waste is taken from the so-called private space of home to the public spaces of school, park and street (and canal). In doing so, these households hold the city responsible for at least part of their waste. This is done for the most part implicitly, although some women and men knowingly state that since garbage service would not pick up their waste, they would make use of public bins. While households produce garbage and are required to think about how they dispose of it (unlike in Western cities where garbage collection is taken for granted and once removed from the home, is rarely thought of), the way in which they dispose of it shapes other urban spaces (including overflowing garbage bins in parks and schools, and rubbish in canals). In Western cities, garbage in public spaces is rarely as much of an issue. Indeed, garbage dumps are generally located outside cities, sometimes in different countries. Plastic bottles and bags regularly blow around the streets of Managua, a consequence of, in part, the everyday struggles households such as those in San Augusto have with garbage.

Everyday ways of dealing with environmental changes and uncertainties are important elements of producing urban spaces; the struggles shape home spaces in cities, which in turn shape the larger urban landscape. In San Augusto, the reliance on good natures – plants and trees – to assist in dealing with water, sanitation and garbage issues has produced domestic spaces that are green and more liveable. In turn, these patios collectively produce a *barrio* with surprisingly green canopy and cooler micro-climate. Not all *barrios* in Managua have the extensive canopy of San Augusto, but Managua's landscape is substantially greener than most cities in Central America. More importantly, the everyday relations that women and other individuals in households have with water and sanitation and the ways in which they struggle to manage such relations have to been understood as part of a collective effort to produce urban life.

Conclusion

This chapter shows that there are multiple and diverse everyday relations with nature in marginalized urban homes in Managua. At the same time that individuals, particularly women, struggle with the consequences of inadequate access to 'safe' water, proper sanitation, and other essential urban services such as garbage disposal, their relations and experiences with other natures (plants and trees) provide ways to cope with environmental uncertainties. By paying attention to the everyday human–nature relations in homes, two important aspects of urban life are rendered more visible and complex.

First, looking in detail at relations with natures shows the diverse ways of dealing with environmental change and how these are internalized through everyday practices and the production of domestic spaces. Urban life in San Augusto depends on natures (both good and bad) and in turn produces urban nature (again, both good and bad). The necessities of living life has meant that changes in, for example, water quality and supply, are dealt with by adapting daily patterns and creating alternative ecologies. Material home spaces are produced and modified according to daily (changes in water availability) and seasonal changes (the inundation of water, sewage and garbage from heavy monsoon rains is anticipated by the planting of particular species in specific places). Detailing these everyday routines (and the involuntary but necessary changes in routines) also reveals conflicts between natures and humans, and the frustrations that individuals (especially women) experience.

In this regard, the everyday exposes a more nuanced understanding of urban environmental politics. Women's experiences are not situated outside of water, sanitation or garbage politics; on the contrary, they are in integral to them since such 'natures' are central to their daily tasks. Taking into consideration everyday life shows how women's quotidian experiences and resistances give rise to larger effects at the urban scale. This is important for urban political ecology because it opens up what is considered urban environmental struggles. Struggles exist not only at the urban and community scale, but also at home, and taken together interact to rework urban space and environmental governance structures. Conversely, by examining the everyday human–nature relations in the home, it becomes possible to see how larger scale environmental processes play out in households.

The second aspect that is revealed through examining everyday human–nature relations is the multiplicity and dynamism of 'natures' in cities, especially in marginalized areas. Jones (2004: 190) writes that 'slums' are denied 'good' nature, yet are constantly threatened by 'bad' nature such as dirty water, open sewage and natural disasters. In the broader urban landscape, he argues, slums are considered 'bad' nature, and thus, "visible reminders of a city's poverty and ecological fragility". Yet while San

Augusto most certainly has its share of 'bad' nature, the homes are also alive with so-called 'good' natures. The patios are inhabited by 'natures' (numerous plants and trees) that do not always threaten bodies, homes or the urban landscape. Indeed, the production and reliance on such good natures has helped to create a barrio (and urban) landscape that is relatively green and more liveable. Most certainly, the use of certain 'natures' to deal with ecological uncertainties only partially addresses the uneven distribution of environmental problems and access to resources. Nonetheless, examining everyday relations with natures in urban areas such as San Augusto calls attention to the distinctness of such relations compared with similar studies carried out in the global north (Bhatti *et al.*, 2009; Hitchings, 2003). Relations with and productions of urban natures are shaped by politics, culture and economics. Paying attention to everyday human-natures in urban areas of the global south will tell us much about urban life, space and politics.

Notes

1 I use the term 'global south' rather than 'third world', 'developing world' or the 'south'. 'Global south' stands for 'a community of people at different geographical locations who experience a common set of problems – problems which emanate, by and large, from deep inequities of power within and between nations' (Mukherjee Reed n.d.). In this sense, 'south' does not only refer to those 'developing' or 'underdeveloped' countries located geographically in the southern hemisphere, but also communities in the 'global north' (hence Mukherjee Reed's specification 'within and between nations'). In contrast to the global south, the global north consists of spaces of power (certain nations, cities and populations). Consequently, both the global south and global north exist within a nation state. Because global north and global south do not refer to geographic locations per se, I do not capitalize either.

2 For overviews see Hubbard (2006); Jayne (2005); Whitehead (2005, 2009).

3 For an overview of political ecology, see Peet and Watts (2004) and Robbins (2005).

4 Given the space restrictions of the chapter, I only focus on everyday relations and discuss briefly how these link to larger relations of power.

5 I use the term 'informal settlement' to refer to illegally settled residential areas that are marginalized, in multiple ways, from the larger urban landscape (culturally, politically, economically and ecologically). I do not use the term 'slum' because of the negative connotations.

6 The data for this research comes from collaborative project between myself, a local non-governmental organization, and the Department of Social Work at the Universidad Centroamericana in 2004 and 2005. The data are from a general survey of the *barrio* (conducted in 2004) and interviews, participatory mapping and focus groups with 30 households (conducted in 2005).

7 Few households depend on shared facilities: only 7 per cent of households share an outdoor pit latrine and another 7 per cent share a flush toilet (FUNDECI 2005).

8 In San Augusto, the majority of houses have simple pit latrines, or outhouses in their patios. Most common are 'wet' latrines, which are simple holes dug into the earth and a seat of some sort placed over the hole. Ideally, the pit latrine have a depth of at least 3 metres so that the contents decompose and the pit does not have to be relocated. The latrines sometimes have concrete floors, but more commonly just a dirt floor. They are also enclosed by small wooden structures, although some are in prefabricated plastic structures. These wet latrines are different from a 'dry' pit where the hole is lined with concrete to prevent seepage; these pits have to be emptied regularly (WHO/ UNICEF 2006).

9 A UN-Habitat (2003) report cites that 43 per cent of households in Managua have 'improved' sanitation (from 1990 to 2000). 'Improved sanitation' refers to: 'connection to a public sewer, connection to septic system, pour-flush latrine, simple pit latrine [or] ventilated improved pit latrine. The excreta disposal system is considered adequate if it is private or shared (but not public) and if it hygienically separates human excreta from human contact' (UN-Habitat 2003: 234).

10 Garbage collection in Managua is a public service run by the City of Managua (ALMA).

19

SPECTRAL KINSHASA

Building the city through an architecture of words

Filip De Boeck

Introduction: the outpost

In 1896, some years before Joseph Conrad published *The Heart of Darkness*, he wrote *An Outpost of Progress*, a short fictional story that is part of his 'Tales of Unrest' (Conrad, 1961 [1898]) and in which he starts to use the material of his Congo years for the first time. A psychological thriller, *An Outpost* may also be read as a political statement undermining the very idea of empire. The storyline focuses on Kayerts and Carlier, two white traders who are outposted in Africa at an ivory trading station along an unnamed river, easily identifiable as the Congo River. The trading station's storehouse is called 'the fetish', 'perhaps', as Conrad remarks, 'because of the spirit of civilization it contained' (ibid.: 93).

Soon after the steamer that put them ashore disappears beyond the horizon, Kayerts and Carlier begin to feel uneasy and alone. At first, they still enjoy discussing the few novels they brought along, and from time to time they receive the visit of Gobila, the old chief of the surrounding villages, with whom they get along well. They also find some old copies of a home paper, left by the previous station master who died of a fever and lies buried in the yard ...

> That print discussed what it was pleased to call 'Our Colonial Expansion' in high-flown language. It spoke much of the rights and duties of civilization, of the sacredness of the civilizing work, and extolled the merits of those who went about bringing light, and faith and commerce to the dark places of the earth. Carlier and Kayerts read, wondered, and began to think better of themselves. Carlier said one evening, waving his hand about, 'In a hundred years, there will be perhaps a town here. Quays, and warehouses, and barracks, and – and – billiard-rooms. Civilization, my boy, and virtue – and all. And

then, chaps will read that two good fellows, Kayerts and Carlier, were the first civilized men to live in this very spot!'

(Conrad, 1961 [1898]: 94–5)

In spite of this comforting thought, it soon becomes painfully clear that Kayerts and Carlier are not really up to the job, and have no clue how to go about the heavenly mission of 'bringing light, and faith and commerce to the dark places of the earth'. All Kayerts and Carlier seem to be able to do is sit there and wait for the steamer to return. A deep silence sets in, and they sense that they are out of their element, not in control of events, and totally dependent on their 'nigger' Makola (or Henry Price, as he himself maintains his name is) who, 'taciturn and impenetrable', despises his two white bosses:

> They lived like blind men in a large room, aware only of what came in contact with them (and of that only imperfectly), but unable to see the general aspect of things. The river, the forest, all the great land throbbing with life, were like a great emptiness. Even the brilliant sunshine disclosed nothing intelligible. Things appeared and disappeared before their eyes in an unconnected and aimless kind of way. The river seemed to come from nowhere and flow nowhither. It flowed through a void.
>
> *(Conrad, 1961 [1898]: 92)*

Foreshadowing the fate of Conrad's best known fictive character, the infamous Mr Kurtz, Carlier and Kayerts are overtaken by what they call 'the unusual', and, slowly, to paraphrase Fabian (2000), they go 'out of their minds'.

The story does not end too well for Kayerts and Carlier, but they were right about one thing indeed. The river stopped flowing through a void. Their modest 'outpost of progress', adjacent to the grave of the first station master, did indeed become a town, and what a town: Kinshasa, a city that counts amongst the African continent's largest urban conglomerations today. This megalopolis, which some describe as 'the quintessential postcolonial African city' (Pieterse, 2010: 1) and 'one of the most drastic cities of the world' (Simone, 2010a: 291), is now home to a population of over nine million and keeps growing steadily. In this chapter I will first situate the three main phases of Kinshasa's expansion from the colonial era to the present day, before turning to the way in which Kinois generate alternative urban orders through the architecture of their speech.

The colonial planning of the city (1874–1960)

None of Kinshasa's unbridled growth was foreseeable at the end of the nineteenth century. Beginning in 1878, Henry Morton Stanley set up four stations along the Congo River manned by 150 European and American officials and supplied by 4 steamers. The trading post at Stanley Pool formed the meeting point between the caravans coming from the Lower Congo and the navigable upstream part of the river which reaches deep into the heart of Central Africa. In the early years, while Savorgnan de Brazza was busy claiming the land at the opposite bank of the Congo river in the name of the French, the post transformed into a thriving model station which became the administrative centre of the Stanley Pool District in 1886 and flourished into the urban conglomeration of Leopoldville/Kinshasa soon afterwards.[1] In fact, Stanley's first station was far from isolated, and the villages surrounding it far less pristine than Conrad's description of chief Gobila's village leads us to believe. Already at the beginning of the nineteenth century, these villages formed the regional core of a large market system with wide-ranging connections, a bustling place where goats, fish, salt, but also slaves and European goods changed hands and were being traded by the local Teke and Humbu populations. But all of that activity was still a far cry from the town that this settlement was soon to become.

Between 1885 and World War I, the core of present-day Kinshasa shaped up around a 10-kilometre-long axis between two sites, close to where Stanley and his men first set foot: the old military and commercial centre Ngaliema/Kintambo to the west, and what is now known as Gombe (formerly Kalina), the easternmost point of Ngaliema Bay. By the end of the nineteenth century, the Kintambo and Kalina outposts had thus developed from small trading stations into comptoir towns, Léopoldville and Kinshasa respectively. By 1900, this Léopoldville-Kinshasa agglomeration, connected by a railway and a road which became Kinshasa's main *Boulevard du 30 juin* after Congo's 1960 independence from Belgium, had already considerably expanded to engulf the former fishermen's village of Kinshasa and beyond, all the way east to Ndolo, turning that whole riverine zone along the Stanley (now Malebo) Pool into a more industrial area, a trans-shipment hub for goods and raw materials to be siphoned off to the Belgian Metropole. By 1910, the spirit of civilization that had been contained by the 'fetish', the first trading station's storehouse, had shape-shifted into its true form, the full-fledged spirit of capitalism. The riverbank was lined with at least 80 storehouses, belonging to several industrial enterprises and trading companies. Around that time, also, a railway was constructed to connect Ngaliema, Gombe and Ndolo. Meanwhile, the white population had grown to a thousand inhabitants, mostly men. And all that was left of the 'pristine' village of chief Gobila was its name. The now infrastructurally derelict but lively port of Beach Ngobila is still one of the main gateways to present-day Kinshasa.

The industrial growth of the city necessitated an increasing transfer of cheap labour from the country's rural hinterlands. The growing city also attracted people from all over the Belgian Congo and these were housed in a rapidly expanding number of labour camps and 'indigenous' living areas. These included Saint Jean, Kinshasa and Barumbu, which were spatially demarcated from the 'white' Gombe by the railway line between Gombe and Kintambo. Between the end of World War II and the end of the Belgian colonial presence in 1960, Léopoldville multiplied its population tenfold, from 40,000 in 1945 to approximately 400,000 in 1960.

In the first half of the twentieth century, Léopoldville thus rapidly grew into what essentially was a segregationist, Janus-faced city, a city with a white heart,

La Ville, the home of the city's European population, constructed along the Kintambo–Gombe axis, and a surrounding, quickly growing peripheral African city, commonly referred to as La Cité, home to an increasing number of Congolese. By 1959, these African *cités* included Dendale (currently Kasavubu), Ngiri-Ngiri, Bandalungwa, Kalamu, Lemba, Matete and Ndjili. They were the result of a large-scale housing scheme launched by the Belgian colonial administration, a plan marked by the modernist ideals that were also *en vogue* in the Metropole during the 1950s. In ten years, more than 20,000 houses were thus built by Belgian urban planners and architects in an impressive effort to respond to the demographic explosion of the city and the increasing social unrest it engendered after WWII (see De Meulder, 2000; De Boeck and Plissart, 2004). Although the white and more residential areas of Gombe were partly extended into Limete in the 1940s and 1950s, La Ville did not expand very much after 1960, caught as it was between the curbing Congo river on its western and northern side, and the growing belt of *cités* on its eastern and southern borders.

To some extent the division between La Ville on the one hand, and a growing number of townships on the other, continues to mark Kinshasa's urbanscape today. In the past, the two areas were not only separated from each other by a tangible colour bar, but they were also physically set apart by railway tracks, strategically placed army barracks (such as the notorious Camp Militaire Lt. Col. Kokolo) and other *zones tampons*, empty no man's lands which spatially drew a divisive line between these various living areas. These zones of separation were also responsible for the fact that the city became scattered over a vast distance. Even today, in the historical heart of Kinshasa many of these empty pockets of land have not yet fully densified in terms of housing and construction.

The city's postcolonial expansion (1960–2010): The random occupation of urban space

After 1960 the number of *cités* and communes *urbano-rurales* increased drastically. Existing *cités* further densified and expanded, and others were added: Selembao, Mont-Ngafula, Makala, Ngaba, Kisenso, Malweka, Masina, Kimbanseke, Kinkole and beyond. Some of these post-1960s expansions, such as Kinkole, had still been planned by the Belgian colonial administration, but many others were added on to the existing urban core in a rather unplanned and chaotic fashion. Today, the city continues to spread incessantly in western and southern direction towards the Lower Congo, and eastwards, way beyond Ndjili, Kinshasa's national airport, towards the foot of the impressive Mangengenge mountain, the eastern gateway to the city.

It is in these increasingly numerous urban areas that the city's inhabitants have started to re-territorialize and reclaim space, develop their own specific forms of urbanism, and infuse the city with their own praxis, values, moralities and temporal dynamics. In the 50 years of the post-independence period this process, which started at Kinshasa's margins, has engulfed the city as a whole, marking a move away from the physical and mental 'place' of colonialism (its spatial layout, its work ethos, its time-management and its language, French). Unhindered by any kind of formal industrialization or economic development, the city has bypassed, redefined, or smashed the colonial logics that were stamped onto its surface. It has done so spatially, in terms of its architectural and urban development, as well as in terms of its socio-cultural and economic imprint. Reaching across the formation period of high colonialism and its modernist ideals, Kinshasa rejoined, to some extent, its earlier rural roots. Aided by a never-ending political and economic crisis, the city (re-)ruralized in many ways, not only in terms of its social structures and spheres of social interaction, but also in terms of its economic survival and coping strategies, engendering a new type of agrarian urbanity.

The unused wastelands that were part of the segregationist colonial urban planning are increasingly being turned into gardens and fields, as are the empty spaces along the city's main traffic arteries. Formerly occupied spaces within the city, such as the cemeteries of Kasavubu or Kintambo, which were officially closed down by the urban authorities in the 1980s, are being occupied and converted into fields as well (De Boeck, 2008). An even more striking example is provided by the transformation which the Malebo Pool is currently undergoing. Over the past twenty years, the inhabitants of the neighbourhoods along the Congo River, from Kingabwa all the way to Nsele and Mikonga, have converted large parts of the Pool into arable land. They were inspired by the example of the Koreans, who started to develop rice paddies in the Malebo Pool near Kingabwa in the 1980s. When the Koreans abandoned these rice fields in the 1990s, the local population took over, and further expanded the farmland into the river, often with very basic tools or even with their own bare hands. By now, in certain areas such as the mouth of the river Tshangu near the Ndjili airport, the empoldered area is already reaching 10 kilometres into the Malebo Pool. In this way more than 800 out of the 6,000 hectares that make up the Malebo Pool have already been transformed from water into arable land.

Over the years, the official authorities, from the National Ministry of Agriculture down to the level of the 'commune', have made half-hearted attempts to impose a legal framework to direct, control and, above all, tax these new farming activities on previously non-existent land. In theory, the state administration has the right to allocate the land to farmers. The latter are supposed to make a payment (USD$200) to the land registry office of the province of Kinshasa, before being able to obtain a *contrat d'exploitation* from the Urban Division for Rural Development (USD$10), and a *permis d'exploitation agricole* (another USD$10) from the Inspection of Rural Development and Agriculture, a unit on the level of the commune. One then acquires the right to use the land for as long as one wants, on the condition that one can prove it is continuously cultivated. The commune is supposed to send an inspector to check on this once a year. In practice, however, none of these regulations and procedures are applied in any straightforward way. The inspector has never come, and since none of this land is on any official map, the authorities often don't even know which land should be paid for. In reality this huge new garden belt is organized outside any clearly defined form of government control on the ground. The factual 'ownership' of these gardens is, therefore, in the hands of some 80 farmers' associations. These have divided the riverine farmlands into a number of 'secteurs' (Kingabwa 1, 2 and 3, Rail 1 and 2, Tshangu, Mapela, Mafuta Kizola, Lokali 1 and 2, Tshunge-Masina, Tshunge-Nsele, Mikonga 1 and 2) which in turn are subdivided into a varying number of 'blocs', each consisting of hundreds of tiny garden plots that rarely surpass 2 to 6 acres. A '*président de secteur*' (officially representing the level of the 'commune' but in reality acting quite independently), aided by a number of '*chefs de bloc*', overlooks the farming activities of over 1,000 farmers. They also organize and oversee the contacts with the thousands of women who each day buy up the gardens' produce and ensure the vegetables' distribution over a large part of the city's numerous markets.[2]

In this way Kinshasa's inhabitants not only continue and reconnect with the city's and river's longstanding market and trading history (in which women have always played an important role (see Bouchard, 2002; Mianda, 1996)), but they also remind us of the fact that the city has not only looked into the mirror of colonialist modernity to design itself, but that it has always contained a second mirror as well.[3] This mirror is provided by the rural hinterland, Kinshasa's natural backdrop, which does not only form the city's periphery, and feed the peripheral city, but which has also deeply penetrated the city, economically but also socially (in terms of the ethnic make-up of large parts of Kinshasa), and above all, culturally and mentally. Rather than pushing the rural out, Kinshasa's urban identity has constantly been invaded and formed by blending with and depending on rural lifestyles, mentalities, moralities and modes of survival (see also De Boeck and Plissart, 2004: 40).

The small-scale modes of action that punctuate rural living – often reformulated yet continuing within the urban context – provide Kinshasa's inhabitants with urban politics of the possible. These often unsteady, provisional and constantly shifting possibilities and action schemes are perhaps not the only ones available to Kinois to give form to the making and remaking of associational life in the city (there is also the mobilizing force of Neo-Pentecostalism, for example), but as a lever for the conceptualization of collective action in the urban configuration it is impossible to underestimate their importance. It is in local zones and domains such as the one described previously, with its myriad activities and its complex web of 'informal' economies that have spun themselves around the river and Kinshasa as a whole, that the city reveals its own production and generates the possibility of economic survival and of social life in the urban context. Here the city reveals itself not as the product of careful planning or engineering, but rather as the outcome of a randomly produced and occupied living space which belongs to whoever generates, grabs and uses it.

This random occupation might, of course, in turn engender new conflicts. Again, the river-fields provide a good example of that. The creation of new arable land in the Malebo pool has led to innumerable and sometimes violent clashes concerning ownership and land rights over this previously non-existent land. On

PLATE 19.1 Slum invasion of empoldered land
Source: Filip De Boeck

one level, these conflicts are mainly played out between the farmers' associations and the Teke and Humbu *chefs de terre*, Kinshasa's original landowners. The latter have sustained their livelihood since the late nineteenth century by selling plots of ancestral land to the city and its inhabitants. At the beginning of the twenty-first century, however, they are running out of land to sell. That is the reason why they turned to these newly available plots of land, claiming ownership over them. Backed by some government officials, but without the farmers' consent, they started to sell large stretches of this new farmland to individuals and families. These in turn started to destroy the gardens to convert them into a shanty area. Hundreds of new 'landowners' constructed their shack in what essentially is a very unhealthy swamp area that does not exist on any official map of the city, lacks even the most basic infrastructure in terms of water, electricity and sanitation, and is totally unfit for habitation (Plate 19.1).

What complicates matters is that the farmers, the land chiefs, and the owners of the newly constructed houses are each backed by various administrative and judicial instances on the communal and the provincial level. This has created a highly explosive situation leading to currently ongoing violent clashes between the various parties involved. In one instance, in early 2010, the bodyguards of a local traditional Teke chief, backed by some army officials, attacked a provincial minister while the latter visited the disputed site with some policemen and ordered the destruction of what he considered an illegal occupation of farmland.

Yet, in spite of such conflicting interests, and the uncertainties and the constant renegotiations these clashes entail, it is this organic approach to the production of the city and its spaces that enables Kinois to survive at all. In many respects, Kinshasa's *cités* are conceived around architectures that remain almost invisible, and are defined by lack and absence on a material level. And many activities in the city become possible not because there is a well-developed infrastructure available to sustain them, but rather because that infrastructure is *not* there, or only exists through its paucity. People's lives in large parts of the city unfold around truncated urban forms, fragments and figments of imported urban technologies, echoes of built environments from the colonial period, and recycled levels of infrastructural accommodation. Although these infrastructures might have originated as the product of a careful engineering of the urban space, they no longer function along these lines today. Constantly banalized and reduced to its most basic function, that of a shelter, the built form is generated by a more real, living city which exists as a heterogeneous urban conglomeration through the bodies, movements, practices and discourses of urban dwellers. This embodied praxis of urban life is embedded in, as well as produces, the entanglement of a wide variety of rhizomatic trajectories, relations and mirroring realities. All of these enjoin, merge, include, fracture, fragment and re-order the urban space. They create, define, and transform new sites of transportation, new configurations of interlaced spatialities, new public spaces of work and relaxation, new itineraries and clusters of relations, new social interactions, new regimes of knowledge and power. And the more there are opportunities to short-circuit any dependence on (unstable) infrastructure and technology, and to bypass the intricate questions of maintenance, ownership and so on, the better all of these actions and transactions seem to work. In this way, the city exists beyond its architecture (see also De Boeck and Plissart, 2004).

Of course, this level of urban functioning outside of the official frameworks of formal urban planning is punctuated with precariousness and hardship, and defined by necessity. Therefore, it is often far from an ideal way to live. But yet, at least to a certain extent, it also seems to be efficient and to work for many. It generates a specific agency in a specific urban experience. It also generates the capacity or the possibility to become a wilful actor in these urban networks. And it is efficient because it allows urbanites to be local producers and controllers of infrastructure and technology, rather than local consumers of technology imported from elsewhere. It transforms city dwellers from passive victims into active participants with their own social, economic, political and religious agendas, which are often situated far beyond the level of mere survival. Concretely, it offers them a considerable freedom to capture the sudden possibilities opened up by unexpected occasions that are generated by the synergies and frictions of urban life. These energies constantly force the urban dweller to master the tricky skills of improvisation. Kinois seem to be very good at doing exactly that; at being flexible, at opening up to this 'unexpected', that often reveals itself outside the known pathways that constitute urban life as most in the global north know it. Urban residents of cityscapes such as Kinshasa are highly skilled at discovering itineraries beyond the obvious, and at exploiting more invisible paths and possibilities that lie hidden in the folds of urban domains and experiences. Often, these city dwellers have trained themselves to successfully tap into this imbroglio, and to exploit to the full the possibilities these juxtapositions offer. They are constantly busy in designing new ways to escape from the economic impositions and excesses that urban life imposes on them. They often know where

to look and what to look for in order to generate feasibility within what is seemingly unfeasible.

The new Kinshasa: The politics of erasure and spectral urbanization (2010–?)

All of this stands in sharp contrast to the official planning of the city which the urban
authorities and the Congolese government have recently committed themselves
to after decades of disinterest and *laissez faire*. For some years now, a successive series
of city governors has been engaged in 'cleaning up' the city. This cleansing basically
boils down to a hard-handed politics of erasure, destroying 'irregular', 'anarchic'
and unruly housing constructions, bulldozing bars and terraces considered to be
too close to the roadside, and banning containers, which Kinois commonly convert
into little shops, from the street. The same is happening to the small street 'restaurants'
known as *malewa* (which provide many women, and therefore whole families,
with an income), as well as many other informal structures and infrastructures
allowing urban dwellers to survive in the volatile economy of the street. The
urban authorities not only started to wage a war against these 'illegal' structures
and activities but also against the very bodies of those who perform or embody
them. Amongst those who first fell victim to the state's effort to 'sanitize' and
recolonize the city, rewrite the city's public spaces, redefine who has a right to
the street and to the city, were Kinshasa's street children and youth gangs, commonly
referred to as *bashege*, *pomba* and *kuluna*.[4] In an attempt to stamp a new material
and moral scale onto the city's surface, the urban authorities started to organize
operations such as *Kanga Vagabonds* ('Grab the Vagabonds', an operation reported
by Geenen, 2009), in order to expulse street children from the city's public eye.[5]
But this urban policy went much further than purifying the streets of unruly kids
or prostitutes. What it envisaged was a much more harmful attempt at wilfully
disrupting what is commonly referred to as the 'informal economy', the proverbial
système D or *Article 15*, which essentially refers to the entrepreneurial capacity of
urban dwellers to generate the networked agencies, coping mechanisms and survival
strategies which were discussed earlier.

In Kinshasa, every singular life is embedded in a multiplicity of relationships.
Many of these relationships are defined by family and kinship ties, but many others
have to do with the specific ways in which one inserts oneself – has to insert
oneself – in multiple complex, often overlapping, networks which include friends,
neighbours, colleagues, acquaintances, members of one's church congregation, professional relations, and so on. Within the megalopolis that Kinshasa has become, this
capacity 'to belong', to socially position oneself within as many different collectivities
as possible, and thereby to obliterate anonymity – in itself an almost unthinkable
concept – is crucial to survive and to exist beyond the raw reality of mere survival
and bare life (see also Lindell, 2010a, 2001b). The capacity at insertion constitutes
the prerequisite for a life worth living in this kind of urban environment, in
economic as well as social terms. The state's brutal destruction of citizens' material

and social environments under the guise of an urban reform which once again seems to be inspired by the earlier moral models of colonialist modernity, therefore forms a violent attack on precisely that crucial creative capacity which is a *sine qua non* to belong, and to belong together, in the city. The official urban politics 'orphans' many urban residents and in the end defines them as out of place in the contours of this newer, cleaner, 'better' and more 'modern' urban environment.

The same exclusionist dynamics are fuelling an even more outspoken attempt at redefining what a 'proper' city means today. During the campaign leading up to the 2006 presidential elections, President Kabila launched his 'Cinq Chantiers' programme, his Five Public Works. The concept summarizes Kabila's efforts to modernize education, health care, road infrastructure, access to electricity and housing accommodation in DR Congo. In 2010, the year in which Congo celebrated the 50th anniversary of its independence from Belgian colonial rule, and a year before the next presidential elections, the 'chantiers' were geared into a different speed, especially with regard to the latter three issues. Downtown Kinshasa (*la Ville*) went through a quite radical facelift, under the guidance of Chinese engineers, Indian, or Pakistani architects, and real estate firms from Dubai, Zambia, or the Emirates. Along the main boulevards and major traffic arteries all trees were cut down and adjacent gardens and fields were destroyed, while the roads and boulevards themselves were widened to become eight-lane highways leading right into the heart of the city. Some landmark buildings were embellished or restored, while others made way for new construction sites on an unprecedented scale. Plans also exist, so the city's rumour mill has it, to build a new viaduct connecting an upgraded Ndjili International airport with La Ville (and more precisely with its Grand Hotel, one of the two international hotels of downtown Kinshasa). The viaduct will follow the Congo River and run over and above the heads of the hundreds of thousands of impoverished inhabitants of the commune of Masina, commonly referred to as 'Chine Populaire', the People's Republic of China, because it is so over-populated.

Today, also, almost every main street and boulevard of Kinshasa is covered with huge billboards announcing the emergence of this new city and offering the spectral, and often spectacular though highly speculative and still very volatile, vision of Congo's reinsertion into the global ecumene. The advertisements promise to bring 'modernization' and '*un nouveau niveau de vie à Kin*' (a new standard of life to Kinshasa) (see Plate 19.2). The billboards show representations of soon-to-be-constructed conference centres, five-star hotels, and skyscrapers with names such as 'Modern Paradise', Crown Tower or Riverview Towers. Many advertisements sport a portrait of President Kabila alongside the statement that Congo will soon be 'the mirror of Africa' (see Plate 19.3). Kinshasa, in other words, is again looking into the mirror of modernity to fashion itself, but this time the mirror no longer reflects the earlier versions of Belgian colonialist modernity, but instead it longs to capture the aura of Dubai and other hot spots of the new urban Global South.[6]

The most striking billboard of all is to be found near the beach, Kinshasa's main port, close to the spot where, in Conrad's novella, Kayerts and Carlier watched over 'the fetish', the storehouse containing the capitalist spirit of civilization. Today, however, the beach offers a sorry sight. It has become an industrial wasteland. The riverbank itself is hidden from view by boats that have all sunk and do no longer offer possible lines of flight; instead cadavers of boats, in every possible shade of rust-eaten brown, just lie there, stranded, immobilized, stuck in the mud and entangled by floating carpets of water hyacinths. It is this very same setting that was chosen by a company that calls itself 'Modern Construction' to erect a new conference centre. On a huge billboard, a poster again shows a photo of a smiling Kabila (see Plate 19.3). On his left and right, one beholds a computer animated picture revealing the new international conference centre, which will be built in the form of a giant cruiser, complete with a rooftop terrace and restaurant! This building, Kabila seems to tell the Kinois, is the ultimate metaphor for the new Kinshasa and the new Congo. It offers the nation a new start and promises a prosperous voyage en route to global modernity. Even if, rather cynically, the name given to the building by the project developers is 'Modern *Titanic*', the image of the ship setting sail towards a new future for Kinshasa is powerfully seductive. Although there is no doubt in anyone's mind that the odds against the *Titanic not* sinking are overwhelming, and although many urban residents in Kin know that they will never have a right to this new city, the hope which this naval image engenders, the hope for a better future, for new and more advantageous ways to cruise through life and navigate the city, simply proves to be irresistible. Even those who count themselves amongst the President's political adversaries cannot

PLATE 19.2 A new level of life: construction in downtown Kinshasa
Source: Filip De Boeck

PLATE 19.3 Construction of a 'Modern *Titanic*'
Source: Filip De Boeck

help but exclaim, 'If only this were true', or, 'And what if it would be for real this time?' Although utopias usually remain locked within the realm of pure speculation and material impossibility, Kabila's '*chantiers*' seem to awaken new hopes, seem to have rekindled a dormant capacity to 'believe' and to dream against all odds: '*C'est beau quand-même, ça fait rêver!*' people exclaim, 'It is so beautiful that it makes one dream.'

But nowhere does the *speculum* of neoliberal global modernity conjure up the oneiric more spectacularly (and nowhere does it reveal its exclusionist logics more strongly) than in another construction project, which is currently already under-way: the Cité du Fleuve. This is the name given to an exclusive development to be situated on two artificially created islands. These will be reclaimed from sandbanks and swamp in the Congo River, a small distance from Kingabwa's Port Baramoto, another one of the city's sorry harbours, adjacent to the beach (see Plate 19.4). The Main Island, the larger of the two, will offer mixed commercial, retail and residential properties, while the smaller North Island will be reserved strictly for private homes and villas. The two islands will be connected to Kinshasa by means of two bridges[7] (see Plate 19.5).

According to the developers' website,[8] La Cité du Fleuve will provide 'a standard of living unparalleled in Kinshasa and will be a model for the rest of Africa'; it continues, '*La Cité du Fleuve* will showcase the new era of African economic development.' In reality, once more, most people currently living in the city will

PLATE 19.4 Location of the two future islands of 'Cité du Fleuve'
in the Congo River
Source: Filip De Boeck

never be able to set foot on the two islands. If all goes according to plan, the latter will be probably be accorded the administrative status of a new 'commune', and will be subject to their own special bylaws. Thus operated as a huge gated community, the Cité du Fleuve will inevitably redefine what is centre and what is edge in Kinshasa. Replicating the segregationist model of Ville and Cité that proved so highly effective during the Belgian colonial period, the islands will become the new Ville while the rest of Kinshasa, with its nine million inhabitants, will be redefined in terms of its periphery. In this way the new city map will redraw the geographies of inclusion and exclusion in radical ways, and relegate its current residents to the city's edges.

The first victims of the Cité du Fleuve project (the realization of which is planned over an eight-year period) will be a number of fishermen's villages in the Congo River, as well as hundreds of farmers who now work on the empoldered land in the river discussed above. All of them will be forced to move elsewhere to make room for the new development. Others will have to follow soon. As noted before, the emergence of the new city drastically changes the content and scale of what is deemed to be proper urban existence, and is going hand in hand with a destruction of the small-scale networked agencies and coping mechanisms that currently allow the majority of Kinois to survive in the city.

Undoubtedly, the re-urbanization process regularizes Kinshasa and ends its 'exceptionalism' in the sense that Kin's dynamics of urban growth has started, at

PLATE 19.5 3D architectural animation of the Cité du Fleuve project

Source: Filip De Boeck

last, to resemble that of other world cities in the global south such as Dubai, Mumbai, Rio, or the urban conglomerations of Southern China (see Appadurai, 2000; Cinar and Bender, 2007; Pieterse, 2008; Simone, 2010b). Simultaneously, however, Kin will also join the shadow-side of that global process of urbanization, a side revealing itself in an increasing favela-ization and an ever more difficult access and right to the city for many of its current inhabitants of which the majority is under the age of 25 (see also Hansen, 2008). Here, the spectral dimension of the marvellous inevitably combines with the dimensions of terror and the dismal. The nightmarish side of these new spectral topographies forms the tain, the back of the mirror which constantly reflects the occulted 'underneath of things' (Ferme, 2001) accompanying this process of urbanization, and bringing it back to the surface and into the daily life experience of the Kinois.

Heterotopology: dreaming/speaking the new Kinshasa

And yet, it is this very same mirror that somehow also unites Kinshasa's powerful and powerless, its *beau monde* and its *demi-monde*, its Big Men and its '*petit peuple*'. Kinshasa's residents and its leaders do not only share the same longing for a better city, but, remarkably, they often also share the same dream of what that city should look like. Upon my asking the farmers who are in danger of being relocated due to the Cité du Fleuve development whether they were well aware of what awaited them, they stated, 'Yes, we'll be the victims, but still it will be beautiful.' In other words, even though the governmental management of the urban site generates new topographies of inclusion and exclusion, of propinquity and distance, and of haves and have nots, and even if this dream of a new future for the city simultaneously generates very tangible forms of ever more pronounced segregation, even then, those who will not be granted access to the new 'Mirror of Africa' revel as much in this dream of the modern city as the ruling elites.[9]

In this sense, their commonly shared longing for a better city is not a utopia, it is something else. Unlike utopian, visionary dreams it does not generate or offer hope. Instead, it offers Kinshasa a new heterotopia, a new space[10] that escapes from the real order of things, its standard forms of classification and accumulation, if only because it conjures up the marvellous through its appeal to the imagination and oneiric. It is precisely in the specular qualities of the image of the new city, the very process of mirroring, realized in all those spaces where the interplay between real and un-real, or visible and invisible is realized, that this new heterotopology for Kinshasa is generated, allowing Kinois to overcome, even if only for a moment, the fragmented-ness, the contradictions and the ruptures that have scarred the face of the city's existence for so long now. It is not as if this new heterotopia, this other, mythic Kinshasa, doesn't have a very real relation with existing social, political, or economical processes in the city: all of these aspects are present in a very real, often material, form, but at the same time without any real or sustainable connection to place or location.

This is also the reason why, in the end, it almost doesn't seem to matter whether the new city is physically built or not. The government does not really seem to believe the new polis will emerge in any lasting way, otherwise why would it have chosen to cynically refer to it as a *Titanic*? And the Kinois themselves are not easily fooled either: they know very well from past experiences not to trust or believe in the official discourses or the outcome of its policies. The new city might well prove to be as chimerical and volatile as the speculative capital, the hedge and vulture funds, which are supposed to finance the construction of the new Kinshasa.[11] Indeed, for this kind of speculative and highly volatile capital it is inconsequential whether or not a development such as Cité du Fleuve is really built or inhabited. One might even argue that this kind of venture capital is attracted to an urban environment such as Kinshasa precisely because instability and volatility, the main characteristics and qualities of the city, also form its main asset, and generate its main financial opportunities (see also Rao, 2007).

In the end, then, short-circuiting any real and tangible roadmaps for the construction of a better urban future, the only place where the city is constantly being built is in language, in the architecture of words. More than through material infrastructures or new technologies, the sheer force of the word is perhaps the most powerful heterotopia through which the city imagines, invents and speaks itself. In Kinshasa, there is always the sneaking suspicion that the paths of transfer between language and reality have become totally unpredictable. Nevertheless words are also deeply believed in. They seem to be the ultimate weapon at one's disposal to defend oneself against an unfinished, unlivable, harsh and often hostile city. Together with the body, words also offer one of the most powerful tools, one of the most important building blocks with which to conquer, alter and erect the city over and over again.

In the Central African universe which brackets this urban world, the art of rhetorics has always been the most efficient tool for self-realization and singulariza-tion, at least for men. Words, also, have always had a tremendous power to construct or change reality, conjure up alternative orders, generate social networks and recreate public space (consider, in this respect, the word of the diviner, the ritual specialist, the sacred king and the judge during a palaver, or the speech acts of more recent urban figures of success such as the politician, the musician and the preacher). In all of these contexts, the legitimate public word always constitutes a demiurgical act of social reproduction and of world making. It is no coincidence that in the auto-chtonous Central-African cultural universe that brackets Kinshasa, colloquy, the act of palavering and of speaking together, is thought of as an act of 'weaving' the social world and as (a masculine equivalent of) giving birth to a child (see De Boeck, 1994). Words, therefore, are always charged with a lot of power, the power to make, conceive of and act upon the world in which one lives. In this socio-cultural constellation, words often seem more real than physical reality.

In a city where the built form of the house is constantly banalized and reduced to its most basic function, that of a shelter, and where the ordering and accumu-lation of things rarely works beyond the simple architectures of heaps of charcoal, loaves of bread, or white cassava flour for sale in Kinshasa's streets and markets, city dwellers use speech as a similarly potent instrument to create new urban orders.

In this respect, one is also constantly reminded of the fact that, for decades, Kinshasa was also permeated by the Word of the Dictator, and the all-pervasive aesthetics of the Mobutist *animation politique* which accompanied that Word. It is no coincidence that this totalitarian political discourse was laced with references to God and marked by a constant religious transfiguration of the political field. In today's Kinshasa, that legendary Hegemon has been replaced by the Old Testament God. With the help of the thousands of *églises de réveil* (Churches of Awakening) and prayer groups active on every street corner of Kinshasa, and propagated by the inescapable voice of preachers on every television screen in the city (see also Pype, forthcoming), the omnipresent voice of Yahweh foists itself in just as authoritarian a way upon the public urban space, in order to turn it once again, by the light of his Word and Will, into a City of God, a new Jerusalem. In such a city, where the Holy Spirit manifests itself at every moment of the day in the form of glossolalia, and where the trance-like prayers of the faithful are continuously charged with the power of the Divine, it is not all that difficult to believe in the potential of words to represent and redesign the city through the construction of rhetorical architectures. Their speech and prayers contain an unremitting attempt to subdue, to comprehend, to build and to govern the city, conjuring up new possible futures for it.

Notes

1 For a good overview of the colonial urban planning history of Kinshasa see De Meulder (2000) and Lagae (2007). See also Beeckmans (2009) for a history of Kinshasa's colonial and postcolonial planning in relation to its public markets. On Kinshasa's history of urbanization see also De Boeck and Plissart (2004), de Maximy (1984), de Saint Moulin (2007), La Fontaine (1970), Kapagama and Waterhouse (2009), Kolonga Molei (1979), Lelo Nzuzi (2008), Pain (1984) and Piermay (1997).
2 Apart from rice, many other vegetables are grown here, most notably maniangu, an aquatic tuber, reportedly imported from Bunia, in eastern Congo, in the 1980s.
3 For a more detailed treatment of the notion of the mirror and of the city's spectral structures see De Boeck and Plissart (2004). On the spectral dimensions of African city life see also Simone (2002).
4 The word *kuluna* derives from the Lingala verb kolona, which means to plant, to sow, to cultivate. The verb is, of course, itself a derivative of the French coloniser, while also referring to the military term colonne. Kuluna originated with urban youngsters from all over south-west Congo who would walk all the way to the diamond fields of the Angolan province of Lunda Norte in order to try their luck there and return with diamonds or dollars, a very common practice in the early 1990s (cf. De Boeck, 1998). Walking the small trails through the forests of northern Angola they would follow each other and form a line, like a military cohort, while penetrating and 'colonizing' new and unknown territory. (On urban street gangs in Kinshasa, see also Pype, 2007.)
5 In more than one way, street children's bodies constitute the very physicality of Kinshasa's public space. That is why they sing: '*Nzoto ya Leta, molimo ya Nzambe*', 'Our bodies belong to the state, but our souls to God'. When the state inscribes its violence onto the bodies of the bashege, it is domesticating the city's public spaces, overwriting it with its own logic.
6 At the same time, the billboards powerfully reveal the tensions and disjunctures between these mirages and images of the new city and the histories and temporalities of the

lives currently lived in Kinshasa by most. In this way, in February 2010, I was struck by two adjacent billboards on one of Kinshasa's main boulevards. One showing an image of a new shopping mall, complete with fountains and gardens, soon to be constructed along the boulevard, while a poster right next to it advertised a new brand of Aladdin lamp, still the most important tool for many Kinois to light up their nights, because large parts of the city are not, or no longer, or only at unpredictable moments, connected to the city's failing electricity network.

7 Officially the Cité du Fleuve is supposed to relocate the entire Kinshasa downtown area of Gombe. According to the current plans it will span almost 400 hectares, include 200 residential houses, 10,000 apartments, 10,000 offices, 2,000 shops, 15 diplomatic missions, 3 hotels, 2 churches, 3 daycare centers, a shopping mall, and a university, but the management is already scaling plans down somewhat (interviews by F. De Boeck with Robert Choudhury, General Manager of the Cité du Fleuve project, and Jason Meikle, its Director of Finance. Kinshasa, February 2010).

8 http://www.lacitedufleuve.com

9 In a similar fashion, Congolese subjects used to proudly sing along with Njila ya Ndolo, the famous 1954 song by Antoine Mundanda about Kinshasa poto moindo, 'Kinshasa, the African Europe', even though they were – often physically – barred from La Ville, the city's white, European heart.

10 Kinshasa's standard heterotopias are many, and include the bar, the street, the church and the body. For a full treatment of these sites, as well as for a more theoretical treatment of Foucault's notion of heterotopia, see De Boeck and Plissart (2004: 254ff). See also Dehaene and De Cauter (2008).

11 The Cité du Fleuve megaproject is being financed by a private promoter, Mukwa Investments, via Hawkwood Properties, a Lusaka based company serving US and European investors.

20

AFTERWORD

A world of cities

Tim Edensor and Mark Jayne

This book has concentrated on reviewing the ways in which urbanists have accounted for cities around the world and sought to act as a point of departure in re-imagining cities in ways that take seriously urban theory beyond 'the West'. In doing so, contributors have engaged with a diverse range of interdisciplinary theoretical work that is both urban focused and more broadly engaged with political, economic, social, cultural, and spatial practices and processes. It is abundantly clear that there is no typical or archetypal non-Western city, for the numerous urban settings featured throughout this book are extraordinarily culturally, morphologically, economically, historically and politically diverse. Moreover, chapters have variously worked at the intersection of theoretical debates concerning post-colonialism, globalization, modernity, Marxism and poststructuralism *and* considered a diverse range of topics and issues that include public space, the home, privacy, planning, squatter settle-ments, informality and everyday life, post-socialism, governance, neo-liberalization, state sponsored capitalism, exclusionary practices, urban ecologies and spectrality. Read together, these accounts offer an important redress to the failure of urban theory to keep pace with broader academic agendas that have had greater success in moving beyond theory dominated by North American and European traditions.

A clear concern of this book, then, has been to offer theoretical and empirically innovative writing in order to draw on diverse political, economic, social and cultural, and spatial theories in order to better understand urban processes. Comparative and relational approaches are foregrounded, together with a particular focus on mobilities and assemblages, and rhythmic, embodied, emotional and affective urbanisms. In this way, contributors have set about 'exploring the ways that scholars in sociology, geography, urban studies, planning and politics might make better sense of "the city" by taking its spatiality seriously' (Hubbard, 2006: 7). It is through engagement with such spatial thinking about the city that this book contributes to the progress being made in questioning the core assumptions that continue to frame urban theory. Towards this goal there are three different theoretical approaches that link the chapters in this book.

Firstly, contributors have engaged in theory building generated in cities beyond 'the West', drawing on situated concepts that emerge out of distinct cultural settings. In these accounts, there is an avoidance of the serial referencing of key thinkers who have come to dominate 'Western' urban theory, the absence of any compulsion to rearticulate well-worn contextual acknowledgments and genealogies of ideas and personalities. Moreover, a diversity of writing styles, expression and tone offer innovative ways of expressing rich, detailed and sustained theoretical engagements with urban practices and process. Most of these accounts also reveal the value of engaging in a sustained immersion in the urban fields being studied, a practice through which theory emerges rather than is superimposed with its embedded preconceptions right from the start. This is urban theory 'but not as we have come to know it', representing a different way of 'doing' theory and of theorizing cities. In challenging 'Western' epistemological and ontological conventions of how to study and represent 'the city', authors in this volume have much to offer in critiquing and supplementing dominant 'Western' modes of writing about urban theory, expanding the parameters and possibilities.

Secondly, critiques throughout the pages of this book seek to rework or refine theories that originate in 'the West' and apply them to unfamiliar contexts. This writing confronts the discursive construction of universalist thinking as well as offering a counter balance to methodological and theoretical deficiencies in both 'Western' and 'non-Western' approaches to the city. In particular, contributors highlight the ways in which theory from around the world 'travels' and can be adopted, adapted and used in multiple settings.

Thirdly, there are hybrid approaches which engage with urban theory from a variety of traditions, both from within, and beyond 'the West'. In unison, these three types of chapters draw on a mix of established and newly emerging theoretical agendas in order to respond to Ward's (2008) assertion that 'theorizing back' not only allows reflection on the geographically uneven foundations of contemporary urban scholarship but also highlights the ways in which both past successes *and* failures of urban research agendas have a vital role to play in re-imagining urban theory. We emphasize here that this theorizing back also includes the reverse flow of concepts and approaches developed and applied to 'non-Western' contexts as tools with which to explore cities anywhere, including Western cities, underlining the post-universalism we advocate. In fact, the transfer of urban models, ideals and knowledge between cities 'beyond the West' is happening anyway. Mehrotra (2008) refers to the slogan 'making Bombay Shanghai' as a spur to reproducing that Chinese city's spectacular architecture in India, and in this book Filip de Boeck shows how plans to develop Kinshasa are heavily influenced by Dubai. Urban exemplars are not necessarily any longer Paris or New York but those rapidly emerging cities outside 'the West'.

Urban Theory Beyond the West: A World of Cities ultimately outlines a significant challenge: to reflect on the past strengths and weaknesses of urban theory and research and consider how cities around the world have been conceptualized and can better be understood. This project acknowledges that while urban theorists have been relatively slow to engage with both theory and cities beyond 'the west', such an admission is only the beginning of the story and there remains much work to be done in developing critical approaches to urban geography in order to advance understanding of both large and small cities throughout the world. Nonetheless, this book highlights how studying the diversity and heterogeneity of cities can be undertaken through a productive dialogue between theoretical work within and beyond 'the west'. If the role of 'the city' is to be maintained and advanced as a key focus for interdisciplinary academic endeavour, the theoretical developments and approaches discussed throughout the pages of this book highlight how urban theory now has a significant contribution to make too, rather than lagging behind broader theoretical advances.

BIBLIOGRAPHY

Abbas, A. (2002) 'Cosmopolitan de-scriptions: Shanghai and Hong Kong', in C. Breckenridge, S. Pollock, H. Bhabha and D. Chakrabarty (eds) *Cosmopolitanism*, Durham, NC: Duke.

Abreu, M. (1997) *Evolução Urbana do Rio de Janeiro*, Rio de Janeiro: IPLANRIO.

Abu-Gazzeh, T. (1995) 'Privacy as the basis of architectural planning in the Islamic culture of Saudi Arabia', *Architecture and Behaviour*, 11: 93–112.

Abu-Lughod, J. (1980) *Rabat: Urban Apartheid in Morocco*, Princeton: Princeton University Press.

Abu-Lughod, J. (1983) 'Contemporary relevance of Islamic urban principles', in A. Germen (ed.) *Islamic Architecture and Urbanism*, Dammam: Ging Faisal University.

Abujidi, N. (2005) 'Forced to forget', paper presented at Durham University Geography Department, Conference on Urbicide, England.

African Studies Quarterly (2010) Special issue: informality and the spaces of popular agency, 11: 2/3.

Agnew, J. (1994) 'The territorial trap: the geographical assumptions of international relations theory', *Review of International Political Economy*, 1(1): 53–80.

Agostoni, C. (2003) *Monuments of Progress: Modernization and Public Health in Mexico City, 1876–1910*, Boulder: University of Colorado Press.

Agrawal, A. (2002) 'Indigenous knowledge and the politics of classification', *International Social Science Journal*, 54(173): 287–97.

Ahmed, A. (2007) 'Workers taken for a ride by Dubai taxi drivers', Gulf News, 22 September, online at http://gulfnews.com/news/gulf/uae/traffic-transport/workers-taken-for-a-ride-by-dubai-taxi-drivers-1.202184 (accessed 20/3/2010).

Ahmed, A. (2009a) 'Car pooling service fails to gather speed', Gulf News, 28 January, online at http://gulfnews.com/news/gulf/uae/traffic-transport/car-pooling-service-fails-to-gather-speed-1.47366 (accessed 20/3/2010).

Ahmed, A. (2009b) 'Dubai to have 1,800 buses', *Gulf News*, 29 June 29, online at http://gulfnews.com/news/gulf/uae/traffic-transport/dubai-to-have-1-800-public-buses-1.440110 (accessed 10/6/2011).

Ahmed, A. (2010a) 'Dubai public transport masterplan on target', *Gulf News*, 15 March, online at http://gulfnews.com/news/gulf/uae/traffic-transport/dubai-public-transport-master-plan-on-target-1.597421 (accessed 2/4/2010).

Ahmed, A. (2010b) 'Tram project progresses smoothly', *Gulf News*, 2 March: 6.

Ahmed, S. (2004) 'Collective feelings: or, the impression left by others', *Theory, Culture and Society*, 21(2): 25–42.

Alcaldía de Managua (2006) *Plan de Ejecución de Obras Mayores y Menores*, Managua: Dirección de Planificación.

Alexander, C. (1965) 'A city is not a tree', in S. Kaplan and R. Kaplan (eds) *Humanscape: Environments for People* (1982), Michigan: Ulrich's Books.

Alexander, C., Buchi V. and Humphrey, C. (2007) *Urban Life in Post Socialist Asia,* London: Routledge.

Alizadeh, H. (2006) 'The concept of Kurdish city', *The International Journal of Kurdish Studies*, 20(1/2): 167–98.

Alizadeh, H. (2007) 'Changes conceptions of women's public space in the Kurdish city', *Cities*, 24(6): 410–21.

Al-Kodmany, K. (1999) 'Residential visual privacy: traditional and modern architecture and urban design', *Journal of Urban Design*, 4(3): 283–311.

Al-Kodmany, K. (2000) 'Women's visual privacy in traditional and modern neighbourhoods in Damascus', *Architecture and Planning Research*, 17(4): 283–303.

AlSayyad, N. and Roy, A. (eds) (2004) *Informality: Transnational Perspectives from the Middle East, Latin America, and South Asia*, Maryland: Lexington Books.

AlSayyad, N. (ed.) (1992) *Forms of Dominance: On the Architecture and Urbanism of the Colonial Enterprise*, England: Avebury.

AlShafieei, S. (1997) *The Spatial Implications of Urban Land Policies in Dubai City*, Unpublished Report, Dubai Municipality.

Al-Tayer, M. (2009) 'We promised and we delivered', *Gulf News*, 2 March: 6.

Altman, I. (1975) *The Environment and Social Behaviour*, Brooks/Cole Monterey, CA.

Altman, I. (1977) 'Privacy regulation: culturally universal or culturally specific', *Journal of Social Issues*, 33(3): 66–84.

Altman, I. and Chemers, M. (1984) *Culture and Environment*, Cambridge: Cambridge University Press.

Amin, A. (2008) 'Collective culture and urban public space', *City*, 12(1): 5–24.

Amin, A. and Graham, S. (1997) 'The ordinary city', *Transactions of the Institute of British Geographers*, 22(4): 411–29.

Amin, A. and Thrift, N. (2002) *Cities: Reimagining the Urban*, Cambridge: Polity.

Amin, S. (1976) *Unequal Development*, Sussex: Harvester Press.

Amsden, A. (1989) *Asia's Next Giant: South Korea and Late Industrialization*, Oxford: Oxford University Press.

Anderson, B. (2006) 'Becoming and being hopeful: towards a theory of affect', *Environment and Planning D Society and Space*, 24: 733–52.

Anderson, B. (2009) 'Affective atmospheres', *Emotion, Space and Society*, 2(2): 77–81.

Anderson, R. (1976) *Outcasts in Their Own Land: Mexican Industrial Workers, 1906–1911*, Dekalb: Northern Illinois University Press.

Angelin, A. (2010) 'Favela picaresque and the right to the city: metropolitan tales in a miniature Rio de Janeiro', seminar presented at the Center for Latin American and Caribbean Studies, New York University, 2/12/2010.

Anjaria, J. (2009) 'Guardians of the bourgeois city: citizenship, public space and middle class activism in Mumbai', *City and Community*, 8(4): 391–406.

Appadurai, A. (1996) *Modernity at Large: Cultural Dimensions of Globalisation*, Minneapolis: University of Minnesota Press.

Appadurai, A. (2000) 'Spectral housing and urban change, notes on millennial Mumbai', *Public Culture*, 13(3): 627–51.

Appadurai, A. (2004) 'Capacity to aspire', in V. Rao and M. Walton (eds) *Culture and Public Action*, Palo Aalto: Stanford University Press.

Arias, E. (2006) *Drugs and Democracy in Rio de Janeiro: Trafficking, Social Networks and Public Security*, Chapel Hill: The University of North Carolina Press.

Arrighi, G. (1994) *The Long Twentieth Century: Money, Power and the Origins of Our Times*, London: Verso.

Ashworth, G.J. (1991) *War and the City*, London: Routledge.

Askew, M. (2002) *Bangkok: Place, Practice and Representation*, London; New York: Routledge.

Atkinson, R. and Bridge, G. (eds) (2005) *Gentrification In a Global Context: The New Urban Colonialism*, London: Routledge.

Augé, M. (1995) *Non-places: Introduction to an Anthropology of Supermodernity*, London: Verso.

Aurigi, A. and Graham, S. (1997) 'Virtual cities, social polarization and the crisis in urban public space', *Journal of Urban Technology*, 4(1): 19–52.

Auyero, J. (2000) *Poor People's Politics: Peronist Survival Networks and the Legacy of Evita*, Durham: Duke University Press.

Babb, F. (1999) '"Managua is Nicaragua": the making of a neoliberal city', *City and Society*, 1–2: 27–48.

Babb, F. (2004) 'Recycled Sandalistas: from revolution to resorts in the New Nicaragua', *American Anthropologist*, 106(3): 541–55.

Bae, Y. and Sellers, J. (2007) 'Globalization, the developmental state, and the politics of urban growth in Korea: a multilevel analysis', *International Journal of Urban and Regional Research*, 31: 543–60.

Báez Cortés, J.F. (2006) 'The problem isn't lack of resources; it's fiscal inequity and legalized pillage', *Envío in English*, 296, online at http://www.envio.org.ni/articulo/3224 (accessed 3/6/2011).

Bakker, K. (2003) 'From archipelago to network: urbanization and water privatization in the South', *The Geographical Journal*, 169: 328–41.

Ball, M. (1986) 'The built environment and the urban question', *Environment and Planning D*, 4: 447–64.

Banco Central de Nicaragua (BCN) (2007) Nicaragua en Cifras, Managua: Banco Central de Nicaragua, http://www.bcn.gob.ni/publicaciones/prensa/folletin/Folletin2006.pdf (accessed 3/6/2011).

Banco de América Central (BAC) (2006) 'Standard & Poor's financial institutions', online at https://www.bac.net/regional/pdf/BACInternationalbankIncEng0106.pdf (accessed 10/6/2011).

Banco Interamericano de Desarrollo (2005) *Apoyo a la Transformación del Transporte Público de Managua* (NI–T1019), Washington, DC: Banco Interamericano de Desarrollo.

Banerjee, P. (2006) *Politics of Time*, Delhi: Oxford University Press.

Barnett, D. (1959) *The Mask of Siam*, London: Hale.

Barsky, O. and Gelman, J. (2001) *Historia del Agro Argentino. De La Conquista hasta Fines del Siglo XX*. Buenos Aires: Editorial Mondadori.

Barth, F. (1953) *Principles of Social Organisation in Southern Kurdistan*, Oslo: Brodrene Jorgensen boktr.

Barth, F. (1969) 'Ethnic groups and boundaries', in F. Barth (ed.) *Ethnic Groups and Boundaries: The Social Organization of Culture Difference*, London: Allen and Unwin, 9–37.

Barvíková, J. (2010) 'Jak se ũije na Jiũním městě z pohledu "Husákových dětí" ', *Sociální Studia*, 3: 59–78.

Baum, V. (1986) *Shanghai '37*, New York: Oxford University Press.

Be a Local (n.d.) 'Tours', online at http://www.bealocal.com (accessed 3/3/2011).

Beall, J. (2007) 'Cities, terrorism and urban warfare of the 21st century', *Working Papers Series*, 9, London School of Economics.

Beall, J. and Fox, S. (2009) *Cities and Development*, London: Routledge.

Beazley, H. (2000) 'Home sweet home? Street children's sites of belonging', in S. Holloway and G. Valentine (eds) *Children's Geographies: Playing, Living, Learning*, Oxford: Routledge.

Beeckmans, L. (2009) 'Agency in an African city. The various trajectories through time and space of the public market of Kinshasa', paper presented at the international conference on African Perspectives: The African Inner City [Re]sourced, Pretoria, 24–28/9/2009.

Beezley, W. (1987) *Judas at the Jockey Club*, Lincoln: University of Nebraska Press.

Bell, D. and Jayne, M. (2009) 'Small cities? Towards a research agenda', *International Journal of Urban and Regional Research*, 33(3): 683–99.

Bell, D. and Jayne, M. (eds) (2006) *Small Cities: Urban Experience Beyond the Metropolis*, London: Routledge.

Bell, P.A., Greene, T.C., Fisher, J.D. and Baum, A. (2000) *Environmental Psychology*, 5th edn, Belmont: Thomson Wadsworth.

Bender, T. (2007) *The Unfinished City: New York and the Metropolitan Idea*, New York: New York University Press.

Benevolo, L. (1980) *The History of the City*, Cambridge: MIT Press.

Berman, M. (1988 [1982]) *All That Is Solid Melts Into Air: The Experience of Modernity*, New York: Penguin.

Berman, M. (1996) 'Falling towers: city life after urbicide', in D. Crow (ed.) *Geography and Identity*, Washington: Maisonneuve Press.

Beyer, J. and Wielgohs, J. (2001) 'On the limits of path dependency approaches for explaining postsocialist institution building: in critical response to David Stark', *East European Politics and Societies* 15(2): 356–88.

Bezmez, D. (2008) 'The politics of waterfront regeneration: the case of Halic (the Golden Horn), Istanbul', *International Journal of Urban and Regional Research*, 32(4): 815–40.

Bhabha, H. (1994) *The Location of Culture*, London: Routledge.

Bhatti, M. and Church, A. (2004) 'Home, the culture of nature and meanings of gardens in late modernity', *Housing Studies*, 19: 37–51.

Bhatti, M., Church, A., Claremont, A. and Stenner, P. (2009) '"I love being in the garden": enchanting encounters in everyday life', *Social and Cultural Geography*, 10: 61–76.

Bian, Y. and Logan, J. (1996) 'Market transition and the persistence of power: the changing stratification system in urban China', *American Sociological Review*, 61: 739–58.

Bianca, S. (2000) *Urban Form in the Arab World: Past and Present*, Zürich: Vdf.

Bilac, O. (1997) *Obra Reunida*, Rio de Janeiro: Nova Aguilar.

Binnie, J. and Valentine, G. (1999) 'Geographies of sexuality: a review of progress', *Progress in Human Geography*, 23: 175–87.

Binnie, J., Holloway, J., Millington, S. and Young, C. (2006) 'Introduction: grounding cosmopolitan urbanism: approaches, practices and policies', in J. Binnie, J. Holloway, S. Millington, and C. Young (eds) *Cosmopolitan Urbanism*, London: Routledge.

Bitušíková, A. (2002) 'Transformations of a city centre in the light of ideologies: the case of Banská Bystrica, Slovakia', *International Journal of Urban and Regional Research* 22(4): 614–22.

Black, G. (1981) *Triumph of the People: The Sandinista Revolution in Nicaragua*, London: Zed.

Blake, W. (1988) *The Complete Poetry and Prose of William Blake*, New York, NY: Anchor Books.

Blaney, D.L. and Inayatullah, N. (2010) *Savage Economics*, London: Routledge.

Bliss, K. (2001) *Compromised Positions: Prostitution, Public Health, and Gender Politics in Revolutionary Mexico City*, University Park: Pennsylvania State University Press.

Blunt, A. (2007) 'Cultural geographies of migration: mobility, transnationality and diaspora', *Progress in Human Geography*, 31(5): 684–94.

Bo, G. (2005) 'The role of Chinese Government in building a harmonious society', in *Network of Asia-Pacific Schools and Institutes of Public Administration and Governance (NAPSIPAG) Annual Conference 2005*, Beijing.

Bodán, O. (1999) 'La "supercarretera" de Alemán', *Confidencial*, 133, online at http://www.confidencial.com.ni (accessed 21/6/2011).

Bodnár, J. (2001) *Fin de Millénaire Budapest: Metamorphoses of Urban Life*, Minneapolis: Minnesotta.

Bois, T. (1966) *The Kurds*, Beirut: Khayats.

Bollens, S. (2000) *On Narrow Ground: Urban Policy and Ethnic Conflicts in Jerusalem and Belfast*, Albany: State University of New York Press.

Bolsa de Noticias (2011) 'Bloque de empresarios sandinistas a la orden de Daniel Ortega', 28 March, online at http://www.grupoese.com.ni/2001/bn/03/28/pt2MN0328.htm (accessed 10/6/2011).

Bonine, M. (2005) 'Islamic urbanism, urbanites, and the Middle Eastern city', in Y. Choueiri (ed.) A Companion to the History of the Middle East, Malden: Blackwell Publishing.

Borden, I. (2001) *Skateboarding, Space and the City: Architecture and the Body*, Oxford: Berg.

Bouchard, H. (2002) Commercantes de Kinshasa, Paris: L'Harmattan.

Bourdieu, P. (1994) *Raisons Pratiques*; trans. V. Dvořáková (1998) *Teorie Jednání*, Praha: Karolinum.

Boyer, C. (1996) *Cybercities: Visual Perceptions in the Age of Electronic Communication*, New York: Princeton Architectural Press.

Brabec, T. and Sýkora, L. (2009) 'Gated communities in Prague', in C. Smiegel (ed.) *Gated and Guarded Housing in Eastern Europe*, Leipzig: Leibniz–Institut für Landerkunde.

Brandtstädter, S. (2007) 'Transitional spaces: postsocialism as a cultural process: introduction', *Critique of Anthropology*, 27(2): 131–45.

Braudel, F. (1992 [1979]) *Perspective of the World: Civilization and Capitalism, 15th–18th century*, trans S. Reynolds, Berkeley: University of California Press.

Braun, B. (2005) 'Environmental issues: writing a more than urban geography', *Progress in Human Geography*, 29: 635–50.

Brenner, N. and Elden, S. (2009) 'Henri Lefebvre on state, space and territory', *International Political Sociology*, 3: 353–77.

Brosius, C. (2008) 'The enclaved gaze: exploring the visual culture of "world class-living" in urban India', in J. Jain (ed.) *India's Popular Culture: Iconic Spaces and Fluid Images*, Mumbai: MARG.

Bruinessen, M. (1978) *Agha, Shaikh and Stat: On the Social and Political Organization of Kurdistan*, Utrecht: Rijksuniversiteit.

Bruinessen, M. (1999) *The Kurds and Islam*, Islamic Area Studies Project Working Paper (13).

Bruinessen, M. (2001) 'From Adela Khanum to Leyla Zana: women as political leaders in Kurdistan history', in S. Mojab (ed.) *Women of a Non-state Nation: the Kurd*. Toronto: Mazda Publishers, Inc.: 95–112.

Bryceson, D. and Potts, D. (2006) *African Urban Economics: Viability, Vitality or Vitiation?* London: Palgrave.

Buie, S. (1996) 'Market as mandala: the erotic space of commerce', *Organisation*, 3: 225–32.

Bunnell, T. and Thompson, E. (2008) 'Unbounding area studies: Malaysian studies beyond Malaysia and other geographies of knowing', *Geoforum*, 39(4): 1517–19.

Burgess, R. (2000) 'The compact city debate: a global perspective', in M. Jenks and R. Burgess (eds) *Compact Cities: Sustainable Urban Forms for Developing Countries*, London: Taylor and Francis.

Burjanek, A. (2009) 'Sociálně vyloučené lokality města: názvosloví a charakteristiky', in S. Ferenčuhová, L. Galčanová, M. Hledíková and B. Vacková (eds) *Město: Proměnlivá ne/samozřejmost*, Červený Kostelec/Brno: Pavel Mervart/MU.

Burns, E. (1980) *The Poverty of Progress*, Berkeley: University of California Press.

Butcher, M. (2009) 'Re-writing Delhi: cultural resistance and cosmopolitan texts', in M. Butcher and S. Velayutham (eds) *Dissent and Cultural Resistance in Asia's Cities*, London: Routledge.

Brudell, P., Hammond, C. and Henry, J. (2004) 'Urban planning and regeneration: a community perspective', *Journal of Irish Urban Studies*, 3(1): 65–87.

Bruinessen, M.V. (2001) 'From Adela Khanum to Leyla Zana: women as political leaders in Kurdistan history', in S. Mojab (ed) *Women of a Non-state Nation: the Kurd*, Toronto: Mazda Publishers, Inc.

Cai, W. and Liang, Y. (2009) 'Beaten, abused and understaffed', *Shanghai Daily Online Edition*, 28 July, online at http://www.shanghaidaily.com (accessed 28/7/2009).

Calame, J. and Charlesworth, E. (2009) *Divided Cities: Belfast, Beirut, Jerusalem, Mostar and Nicosia*, Philadelphia: University of Pennsylvania Press.

Caldeira, T. (2000) *City of Walls: Crime, Segregation, and Citizenship in São Paulo*, Berkeley, CA: University of California Press.

Caldeira, T. (2008) 'From modernism to neo-liberalism in Sao Paulo: reconfiguring the city and its citizens', in A. Huyssen (ed.) *Other Cities, Other Worlds: Urban Imaginaries in a Globalizing Age*, Durham, NC: Duke.

Caldwell, M. (2004) 'Domesticating the French fry: McDonalds's and consumerism in Moscow', *Journal of Consumer Culture*, 4: 5–26.

Cameron, S. and J. Coaffee (2005) 'Art, gentrification and regeneration: from artist as pioneer to public arts', *European Journal of Housing Studies*, 5(1): 39–58.

Camín, A. (1975) *La Revolución Sonorense, 1910–1914*, Mexico: INAH.

Camus, A. (1991) *Diário de Viagem*, Rio de Janeiro: Record.

Canclini, N. (1995) *Hybrid Cultures: Strategies for Entering and Leaving Modernity*, Minneapolis: University of Minnesota Press.

Canclini, N. (2008) 'Mexico City 2010: improvising globalisation', in A. Huyssen (ed.) *Other Cities, Other Worlds: Urban Imaginaries in a Globalizing Age*, Durham, NC: Duke.

Carr, B. (1976) *El Movimiento Obrero y la Política en México*, 2 vols., Mexico.

Cartier, C. (2001) *Globalizing South China*, Oxford; London: Blackwell.

Castells, M. (2000) *The Rise of the Network Society*, Oxford: Blackwell.

Caves, R. (ed.) (2005) *Encyclopaedia of the City*, London and New York: Routledge.

Çelik, Z. (1992) *Displaying the Orient: Architecture of Islam in Nineteenth-century World's Fairs*, Berkeley: University of California Press.

Çelik, Z. (1997) *Urban Forms and Colonial Confrontations: Algiers under French Rule*, Berkeley: University of California Press.

Centre on Housing Rights and Evictions (COHRE) (2004) *Housing Rights in Nicaragua: Historical Complexities and Current Challenges*, Geneva: COHRE, online at http://www.cohre.org/store/attachments/COHRE%20Housing%20Rights%20Nicaragua%20 2004.pdf (accessed 10/6/2011).

Chakrabarty, D. (1991) 'Open space/public space: garbage, modernity and India', in *South Asia*, 16: 15–31.

Chakrabarty, D. (2000) *Provincialising Europe: Postcolonial Thought and Historical Difference*, Princeton: Princeton University Press.

Chan, K.W. (1996) 'Post-Mao China: a two-class urban society in the making', *International Journal of Urban and Regional Research*, 20(1): 134–50.

Chan, K.W. and Zhang, L. (1999) 'The *Hukou* system and rural–urban migration in China: process and changes', *The China Quarterly*, Dec.(160): 818–55.

Chan, K.W. (1994) *Cities with Invisible Walls*, Hong Kong: Oxford University Press.

Chandhoke, N. (1993) 'On the social organisation of urban space – subversions and appropriations', *Social Scientist*, 21: 63–73.

Chandrasekaran, R. (2007) *Imperial Life in the Emerald City: Inside Iraq's Green Zone*, USA: Vintage Books.

Chareonsinoran, C. (1997) *Rattasart kan Bariharn Rattakit Thrisdee: Nung Satawat Rattasart Naw Wipak* [Political Science Theory of State Management: A Century of Critical Political Science], Bangkok, Thailand: Research and Publication Center, Krerg University.

Chatterjee, P. (1993) *Nation and its Fragments*, Princeton: Princeton University Press.

Chatterjee, P. (2004) *The Politics of the Governed: Reflections of Popular Politics in Most of the World*, New York: Columbia University Press.

Chattopadhyay, S. (2005) *Representing Calcutta: Modernity, Nationalism and the Colonial Uncanny*, London: Routledge.

Chattopadhyay, S. (2009) 'The art of auto-mobility: vehicular art and the space of resistance in Calcutta', *Journal of Material Culture*, 14(1): 107–39.

Chattopadhyay, S. (forthcoming) *Unlearning the City*, Minneapolis: University of Minnesota Press.

Chen, X. (2009) 'Introduction: why Chinese and Indian megacities?', *City and Community*, 8(4): 363–8.

Chopra, P. (2011) *A Joint Enterprise*, Minneapolis: University of Minnesota Press.

Chung, J. Y. and Kirby, R. (2002) *The Political Economy of Development and Environment in Korea*, Routledge: London.

Cinar, A. and Bender, T. (eds) (2007) *Urban Imaginaries. Locating the Modern City*, Minneapolis: University of Minnesota Press.

Clark, E. (1995) 'The rent gap re-examined', *Urban Studies*, 32(9): 1,489–503.

Clemenceau, G. (1911) 'Clemenceau Pays Tribute to the Women of Argentina', *New York Times*, March 12.

Coatsworth, J. (1981) *Growth Against Development: the Economic Impact of Railroads in Porfirian Mexico*, DeKalb: Northern Illinois University Press.

Cochrane, A. and Jonas, A. (1999) 'Reimagining Berlin: world city, national capital or ordinary place?', *European Urban and Regional Studies*, 6(2): 145–64.

Colebrook, C. (2002) 'The politics and potentials of everyday life', *New Literary History* 33: 687–706.

Comisión Nacional de Zonas Francas (2010) 'Circular No. 4 de la CNZF', online at http://www. cnzf.gob.ni/index.php?option=com_content&task=view&id=105&Itemid=137&lang=es (accessed 10/6/2011).

Confidencial (2006) 'Concluyen obras de nuevo centro comercial Las Américas', *Confidencial*, 490, online at http://www.confidencial.com.ni/2006–490/negocios_490.htm (accessed 3/6/2011).

Connell, R. (2007) *Southern Theory*, Cambridge: Polity.

Conrad, J. (1961 [1898]) 'An Outpost of Progress', in *Almayer's Folly. A Story of an Eastern River* and *Tales of Unrest*, London: J.M. Dent and Sons.

Conradson, D. and Latham, A. (2005) 'Transnational urbanism: attending to everyday practices and mobility', *Journal of Ethnic and Migration Studies*, 31(2): 227–33.

Conradson, D. and Latham, A. (2007) 'The affective possibilities of London: Antipodean transnationals and the overseas experience', *Mobilities*, 2(2): 231–54.

Conradson, D. and McKay, D. (2007) 'Translocal subjectivities: mobility, connection, emotion', *Mobilities*, 2(2): 167–74.

Corbridge, S., Srivastava, M. and Veran, R. (2005) *Seeing the State: Governance and Governmentality in India*, Cambridge: Cambridge University Press.

Cornelius, W.A. (1975) *Politics and the Migrant Poor in Mexico City*, Stanford University Press.

Coutard, O. (2008) 'Placing splintering urbanism: an introduction', *GeoForum*, 39: 1,815–20.

Cowan, R. (2005) *The Dictionary of Urbanism*, London: Streetwise Press.

Cowan, M. and Shenton, R. (1995) 'The invention of development', in J. Crush, *Power of Development*, London: Routledge.

Cox, K. and Mair, A. (1989) 'Urban growth machines and the politics of local economic development', *International Journal of Urban and Regional Research*, 13: 137–46.

Crawfurd, J. (1967, c1828) *Journal of an Embassy from the Governor-General of India to the Courts of Siam and Cochin-China*, 2nd ed., Kuala Lumpur: Oxford University Press.

Cresswell, T. (2006) *On the Move: Mobility in the Modern Western World*, London: Routledge.

Cronon, W. (1991) *Nature's Metropolis: Chicago and the Great West*, New York: W.W. Norton and Co.

Crush, J. (1995) *Power of Development*, London: Routledge.

Cruz Sánchez, E. (2006) 'Desalojan a precaristas en el barrio Isaías Gómez', *La Prensa*, 2 February, online at http://www.laprensa.com.ni/archivo/2006/febrero/02/sucesos/sucesos-20060202-03.html (accessed 10/6/2011).

Cunha, E. (1902) *Os sertões*, São Paulo: Ateliê Editoria.

Curtis, G. (1999) *The Logic of Japanese Politics: Leaders, Institutions, and the Limits of Change*, New York: Columbia University Press.

Czepczyński, M. (2008) *Cultural Landscapes of Post-socialist Cities: Representation of Powers and Needs*, Hampshire, Burlington: Ashgate.

Davidson, J., Smith, M. and Bondi, L. (eds) (2005) *Emotional Geographies*, Aldershot: Ashgate.

Davie, M. (2003) 'Beirut and the Etoile area: an exclusively French project', in J. Nasr and M.Volait (eds) *Urbanism Imported or Exported: Native Aspirations and Foreign Plans*, England: Wiley-Academy.

Davies, J. (2003) 'Partnerships versus regimes: why regime theory cannot explain urban coalitions in UK?', *Journal of Urban Affairs*', 25: 570–88.

Davis, D. (2004) *Discipline and Development: Middle Classes and Prosperity in East Asia and Latin America*, Cambridge: Cambridge University Press.

Davis, M. (1990) *City of Quartz*, London: Vintage.

Davis, M. (2006) *Planet of Slums*, London: Verso.

De Boeck, F. (1994) 'Of trees and kings: politics and metaphor among the Aluund of Southwestern Zaire', *American Ethnologist*, 21(3): 451–73.

De Boeck, F. (1998) 'Domesticating diamonds and dollars: identity, expenditure and sharing in Southwestern Zaire (1984–1997)', *Development and Change*, 29(4): 777–810; reprinted in B. Meyer and P. Geschiere (eds) (1999) *Globalization and Identity. Dialectics of Flow and Closure*, Oxford: Blackwell.

De Boeck, F. (2008) '"Dead society" in a "cemetery city": the transformation of burial rites in Kinshasa', in M. Dehaene and L. De Cauter (eds) *Heterotopia and the City: Public Space in a Postcivil Society*, London: Routledge.

De Boeck, F. and Plissart, M.-F. (2004; 2nd edn 2006) *Kinshasa: Tales of the Invisible City*, Ghent/Tervuren: Ludion/Royal Museum for Central Africa.

de Certeau, M. (1984) *The Practice of Everyday Life*, Berkeley: University of California Press.

de Mattos, C. (2002) 'Transformación de las Ciudades Latinoamericanas: Impactos de la Globalización?', *EURE*, December 28, 85: 5–10.

de Maximy, R. (1984) *Kinshasa, ville en suspens: Dynamique de la croissance et problèmes d'urbanisme: Etude socio–politique*, Paris: Editions de l'Office de la Recherche Scientifique et Technique Outre–Mer (ORSTOM).

De Meulder, B. (2000) *Kuvuande Mbote: Een eeuw koloniale architectuur en stedenbouw in Kongo*, Antwerp: Houtekiet/De Singel.

de Soto, H. (2002) *The Other Path: the Economic Answer to Terrorism*, New York: Basic Books.

De Witte, M. (2008) 'Accra's Sounds and Sacred Spaces', *International Journal of Urban and Regional Research*, 23(3): 690–709.

Dear, M. (2005) 'Comparative Urbanism', *Urban Geography*, 26: 247–251.

Dear, M. and Leclerc, G. (2003) *Postborder City: Cultural Spaces of Bajalta California*, London: Routledge.

Demos (1997) 'The wealth and poverty of networks', *Demos Collection 12*. London: Demos.

De Saint Moulin, L. (2007) 'Croissance de Kinshasa et transformations du réseau urbain de la RDCongo depuis l'indépendence', in J.-L. Vellut (ed.), *Villes d'Afrique*, Paris: L'Harmattan.

Dehaene, M. and De Cauter, L. (eds) (2008), *Heterotopia and the City: Public Space in a Postcivil Society*, London: Routledge.

Degen, M. (2008) *Sensing Cities: Regenerating Public Life in Barcelona and Manchester*, London: Routledge.

Demissie, F. (2008) *Postcolonial African Cities*, London: Routledge.

Department of International Organizations, Ministry of Foreign Affairs, People's Republic of China and United Nations Country Team in China (2005) *China's Progress Towards The Millenium Development Goals 2005*, Beijing.

Der Derian, J. (2001) *The Virtuous War: Mapping the Military Industrial–Media–Entertainment Network*, Colorado: Westview Press.

Desch, M.C. (ed.) (2001) *Soldiers in Cities: Military Operations on Urban Terrain*, Carlisle: Strategic Studies Institute.

Di Tella, G. and Dornbusch, R. (1986) 'Introduction: the political economy of Argentina 1946–83', in G. Di Tella and R. Dornbusch (eds) *The Political Economy of Argentina*, London: Macmillan.

Diaz Alejandro, C. (1970) *Essays on the Economic History of the Argentine Republic*, New Haven: Yale University Press.

Diamond, L. (2008) *The Spirit of Democracy: The Struggle to Build Free Societies throughout the World*, New York: Holt.

Dick, H. and Rimmer, P. (1998) 'Beyond the third world city: the new urban geography of SE Asia', *Urban Studies*, 35(12): 2,303–21.

DiGaetano, A. and Strom, E. (2003) 'Comparative urban governance: an integrated approach', *Urban Affairs Review*, 34: 546–77.

Diouf, M. (2000) 'The Senegalese Murid trade diaspora and the making of a vernacular cosmopolitanism', *Public Culture*, 12(3): 679–702.

Diouf, M. (2003) 'Engaging postcolonial cultures: African youth and public space', *African Studies Review*, 46(2): 1–12.

Diouf, M. (2008) '(Re)imagining an African city: performing culture, arts and citizenship in Dakar (Senegal), 1980–2000', in Prakash, G. and Kruse, K. (eds) *The Spaces of the Modern City: Imaginaries, Politics, and Everyday Life*, Princeton, NJ: Princeton University Press.

Dirlik, A. (1997) 'Critical reflections on "Chinese capitalism" as paradigm', *Identities*, 3: 303–30.

Dixon, J. and Durrheim, K. (2004) 'Dislocating identity: desegregation and the transformation of place', *Journal of Environmental Psychology*, 24(4): 455–73.

D'León Masís, E. (2002) 'Entrevista con Alina Sálomon: Reactivar el viejo centro de Managua', *La Prensa*, 13 December, online at http://www-ni.laprensa.com.ni/archivo/2002/diciembre/14/literaria/comentario/ (accessed 10/6/2011).

Domański, B. (2004) 'West and East in "New Europe": the pitfalls of paternalism and a claimant attitude', *European Urban and Regional Studies*, 11: 377–81.

Dorfman, A. (1983) *Cincuenta Años de Industrialización en la Argentina 1930–1950*, Buenos Aires: Ediciones Solar.

Douglass, M. (2000) 'Mega-urban regions and world city formation: globalization, the economic crisis and urban policy issues in Pacific Asia', *Urban Studies*, 37: 2,315–35.

Douglass, M. (2002) 'Civic space, globalization and the urban future: toward a pacific Asia research agenda', unpublished paper, University of Hawaii.

Douglass, M., Ho, K.C., and Ooi, G.L. (2006) *Globalization, the City and Civil Society in Pacific Asia*, London: Routledge.

Dowling, R. (2008) 'Geographies of identity: labouring in the "neoliberal" university', *Progress in Human Geography*, 32: 812–20.

Drudy, P. and Punch, M. (2000) 'Economic restructuring, urban change and regeneration: the case of Dublin', *Journal of the Statistical and Social Inquiry Society of Ireland*, vol. XXIX: 215–87.

Drummond, L. (2000) 'Street scenes: practices of public and private space in urban Vietnam', *Urban Studies*, 37: 2,377–91.

DTZ (2009) 'The ticket to success: the impact of Dubai metro on real estate values', online at http://www.cityscapeintelligence.com/document/2333 (accessed 2/4/2010).

Duara, P. (2009) *The Global and Regional in China's Nation Formation*, London: Routledge.

Dubai Municipality (1995) *Structure Plan for the Dubai Urban Area (1993–2012)*, report prepared by Parsons Harland Bartholomew and Associates, Inc.

Dubai Statistics Center (n.d.) 'Programs & Statistical Surveys', online at http://www.dsc.gov.ae/en/Pages/Home.aspx (accessed 5/6/2011).

Duckett, J. (1998) *The Entrepreneurial State in China Real Estate and Commerce Departments in Reform Era Tianjin*, London and New York: Routledge.

Duncan, J. (1985) 'The house as symbol of social structure', in I. Altman and C. Werner (eds) *Home Environments*, New York: Plenum.

Dupont, V. (2008) 'Slum demolitions in Delhi since the 1990s: an appraisal', *Economic and Political Weekly*, 43(28): 79–87.

Dutt, A., Costa F., Aggaward, S., and Noble, A. (1994) *The Asian City*, London: GeoJournal Library.

Ebanks, G.E. and Cheng, C.Z. (1990) 'China: a unique urbanization model', *Asia-Pacific Population Journal*, 5: 29–50.

Eck, D. (1991) 'Kashi: city of all India', in T. Madan (ed.) *Religion in India*, Delhi: Oxford University Press.

Eckstein, S. (ed.) (1989) *Power and Popular Protest: Latin American Social Movements*, Berkeley: University of California Press.

Escobar, A. (2010) 'Latin America at a crossroads', *Cultural Studies*, 24(1): 1–65.

Escobar, A. and Alvarez, S. (eds) (1992) *The Making of Social Movements in Latin America*, Boulder: Westview Press.

Eckardt, F. (2005) 'In search for meaning: Berlin as national capital and global city', *Journal of Contemporary European Studies*, 13(2): 187–98.

Edensor, T. (1998) *Tourists at the Taj*, London: Routledge.

Edensor, T. (2000) 'Moving through the city', in D. Bell and A. Haddour (ed.) *City Visions*, London: Prentice Hall.

Edensor, T. (ed.) (2010) *Geographies of Rhythm*, Aldershot: Ashgate.

Eller, J.D. (2009) *Cultural Anthropology: Global Forces, Local Lives*, New York and London: Routledge.

Elsheshtawy, Y. (ed.) (2004) *Planning Middle Eastern Cities: An Urban Kaleidoscope*, London: Routledge.

Elsheshtawy, Y. (2008) 'Transitory sites: mapping Dubai's "forgotten" urban public spaces', *The International Journal of Urban and Regional Research*, 32(4): 968–88.

Elsheshtawy, Y. (2009) *Dubai: and Emerging Urbanity*, London: Routledge.

Elsheshtawy, Y. (2010) *Dubai: Behind an Urban Spectacle*, London: Routledge.

Emerson, R. (2006) *The Conduct of Life*, edited and introduced by H. Callaway, University Press of America.

Entrikin, J. (1991) *The Betweenness Of Place Towards A Geography Of Modernity*, Baltimore: The John Hopkins Press.

Enyedi, G. (1996) 'Urbanization under socialism', in G. Andrusz, M. Harloe, and I. Szelenyi (eds) *Cities After Socialism: Urban and Regional Change and Conflict in Post-socialist Societies*, Oxford: Blackwell.

Escobar, A. and Alvarez, S. (1992) *The Making of Social Movements in Latin America: Identity, Strategy, and Democracy*, Boulder: Westview Press.

Esherick, J.W. (2000) 'Modernity and nation in the Chinese City' in J.W. Esherick (ed.) *Remaking the Chinese City: Modernity and National Identity, 1900–1950*, Honolulu: University of Hawaii Press.

Everingham, M. (2001) 'Agricultural property rights and political change in Nicaragua', *Latin American Politics and Society*, 43(3): 61–93.

Exotic Tours (n.d.) 'Exotic Tours', online at http://www.exotictours.com.br/ (accessed 3/3/2011).

Fabian, J. (2000) *Out of our Minds: Reason and Madness in the Exploration of Central Africa*, Berkeley: University of California Press.

Fainstein, S. (2005) 'Cities and diversity: should we want it? can we plan for it?', *Urban Affairs Review*, 41(1): 3–19.

Falt'an, L'. (2009) 'Bratislava – problémy súčasného rozvoja', *Sociológia*, 41(4): 329–53.

Fan, C. (1995a) 'Of belts and ladders: state policy and uneven regional development in post-Mao China', *Annals of the Association of American Geographers*, 85: 421–49.

Fan, C. (1995b) 'Developments from above, below and outside: spatial impacts of China's economic reforms in Jiangsu and Guangdong provinces', *Chinese Environment and Development*, 6: 85–116.

Fan, C. (1997) 'Uneven development and beyond: regional development theory in post-Mao China', *International Journal of Urban and Regional Research*, 21: 620–39.

Fan, C. (2002) 'The elite, the natives, and the outsiders: migration and labor market segmentation in urban China', *Annals of the Association of American Geographers*, 92: 103–24.

Fan, M. (2006) 'China's Party leadership declares new priority: "Harmonious Society"', *Washington Post Foreign Service*, 12 October, online at http://www.washingtonpost.com/wp–dyn/content/article/2006/10/11/AR2006101101610.html (accessed 4/6/2009).

Fanon, F. (2004 [1963]) *The Wretched of the Earth*, trans. R. Philcox, New York: Grove.

Favela Tour (n.d.) 'What is it all about', online at http://www.favelatour.com.br/ing/whatis.htm (accessed 3/3/2011).

Fawaz, M. (2008) 'An unusual clique of city makers: social networks in the production of a neighbourhood in Beirut (1950–75)', *International Journal of Urban and Regional Research*, 32(3): 565–85.

Ferencová, M. (2008) 'Spoluūitie zaliate v bronze: Pomníky významných osobností ako prostriedok organizovaného šírenia klasifikačných schém', *Slovenský Národopis*, 56(1): 5–16.

Ferenčuhová, S. (2009) 'Coping with the past: urban planning in a post-socialist European city', in F. Eckardt and I. Elander (eds) *Urban Governance in Europe*, Berlin: BWV.

Ferguson, J. (2006) *Global Shadows: Africa in the Neoliberal World Order*, Durham, NC: Duke University Press.

Ferme, M. (2001) *The Underneath of Things: Violence, History, and the Everyday in Sierra Leone*, Berkeley: The University of California Press.

Fernandes, A.C. and Negreiros, R. (2001) 'Economic developmentalism and change within Brazilian urban systems', *Geoforum*, 32(4): 415–35.

Fournereau, L. (1998) *Bangkok in 1892*, trans. W.E.J. Tips, originally published under title *Bangkok in Le Tour du Monde* (1894), 68: 1–64, Bangkok: White Lotus Press.

Foweraker, J. and Craig, A. (eds) (1990) *Popular Movements and Political Change in Mexico*, Boulder: Lynne Rienner Publishers.

Fowler-Salamini, H. and Vaughan, M. (eds) (1994) *Women of the Mexican Countryside, 1850–1990*, Tucson: University of Arizona Press.

Franklin, A. (2006) 'Be[a]ware of dog: a post-humanist approach to housing', *Housing Theory and Society*, 23(3): 137–56.

Fregonese, S. (2009) 'The urbicide of Beirut? Geopolitics and the built environment in the Lebanese Civil War (1975–1976)', *Political Geography*, 28: 309–18.

Freire-Medeiros, B. (2007) 'Selling the favela: thoughts and polemics about a tourist destination', in *Revista Brasileira de Ciências Sociais*, 22(65): 61–72, October, São Paulo.

Freire-Medeiros, B. (2009a) 'The favela and its touristic transits', *Geoforum*, 40: 580–8.

Freire-Medeiros, B. (2009b) *Gringo na Laje*, Rio de Janeiro: Fundação Getúlio Vargas.

Friedmann, J. (2005) *China's Urban Transition*, Minneapolis: University of Minnesota Press.

Fundación Nicaragüense Pro-Desarrollo Comunitario Integral (FUNDECI) (2005a) *Informe del Barrio Memorial Sandino*, unpublished report, Managua, Nicaragua: FUNDECI.

Fundación Nicaragüense Pro-Desarrollo Comunitario Integral (FUNDECI) (2005b) *Reseña Historica del Barrio Memorial Sandino*, unpublished manuscript, Managua, Nicaragua: FUNDECI.

Fyfe, N., Bannister, J., and Kearns, A. (2006) '(In)civility and the city', *Urban Studies*, 43(5): 853–61.

Gabriel, E. (1987) *The Dubai Handbook*, Ahrensburg: Institute for Applied Economic Geography.

Gajdoš, P. (2009) 'Globalizačné súvislosti urbánneho rozvoja a jeho sociálno–priestorové špecifiká', *Sociológia* 41(4): 304–28.

Gallo, E. (1970) 'Agrarian expansion and industrial development in Argentina (1880–1930)', *Latin American Affairs*, St. Anthony Papers No 22, Oxford University Press, 45–61.

Galletti, M. (2001) 'Western images of women's role in Kurdish society', in. S. Mojab (ed.) *Women of a Non-state Nation: the Kurd*, Toronto: Mazda Publishers.

Gambetti, Z. (2009) 'Decolonizing Diyarbakir: culture, identity and the struggle to appropriate urban space', in K. Ali and M. Rieker (eds) *Re-exploring the Urban: Comparative Citiscapes in the Middle East and South Asia*, Karachi: Oxford University Press.

Gandy, M. (2002a) *Concrete and Clay: Reworking Nature in New York City*, Cambridge, MA: MIT Press.

Gandy, M. (2002b) 'Landscapes of disaster: water, modernity and urban fragmentation in Mumbai', *Environment and Planning A*, 40: 108–30.

Gans, H. (1962) *Urban Villagers: Group and Class in the Life of Italian Americans*, New York: The Free Press.

Garcia, L. (2001) 'Elitización: Propuesta en Español para el Término Gentrificación', *Revista Bibliográfica de Geografía y Ciencias Sociales*, VI, December 332.

Gayol, S. and Madero, M. (2007) *Formas de Historia Cultural*. Editorial Universidad Nacional General Sarmiento, Buenos Aires: Prometeo Libros.

Geenen, K. (2009) ' "Sleep Occupies No Space": The use of public space by street gangs in Kinshasa', *The Journal of the International African Institute*, 79(3): 347–68.

Georgiou, M. (2006) 'Architectural Privacy: A Topological Approach to Relational Design Problems', unpublished M.Sc. thesis, Bartlett School of Graduate Studies, University College.

Germani, G. (1980) *Marginality*, New Jersey: Transaction Books.

Gillem, M. (2007) *America Town: Building the Outposts of Empire*, Minnesota: University of Minnesota Press.

Gilly, A. (1971) *La Revolución Interrumpida, México 1910–1940*, Mexico.

Glass, R. (1964) *London: Aspects of Change*, London: Centre for Urban Studies and MacGillion and Kee.

Glover, W. (2007) *Making Lahore Modern: Constructing and Imaging a Colonial City*, Minneapolis: University of Minnesota Press.

Gobierno de Nicaragua (1999) Ley No. 306: Ley de Incentivos para la Industria Turística de la República de Nicaragua, online at http://www.intur.gob.ni/ley306.html (accessed 2/6/2011).

Gobierno de Nicaragua (2005) *Sistema Nacional de Inversiones Públicas: Guía de Preinversión para Proyectos de Viviendas Social*, Managua: Secretaria Técnica de la Presidencia.

Goodrich Sorensen, D. (1996) *Facundo and the Construction of Argentine Culture*, Austin: University of Texas Press.

Goodwin, P.B. (2007) *Global Studies: Latin America*, 12th edn, Dubuque: McGraw-Hill/ Contemporary Learning Series.

Gordon, D. (1984) 'Capitalist development and the history of American Cities', in W. Tabb and L. Sawyers (eds) *Marxism and the Metropolis: New Perspectives in Urban Political Economy*, New York and Oxford: Oxford University Press.

Gordon, S. (2004) *Privacy: A Study of Attitudes and Behaviors in US, UK and EU Information Security Professionals*, Symantec White Paper, Citeseer.

Graham, S. (2002) 'Bulldozers and bombs: the latest Palestinian–Israeli conflict as asymmetric urbicide', *Antipode*, 34(4): 642–649.

Graham, S. (2003) 'Lessons in urbicide', *New Left Review*, 19: 63–77.

Graham, S. (2004) *Cities, War and Terrorism: Towards and Urban Geopolitics*, Malden: Blackwell Publishing.

Graham, S. (2005) 'Switching cities off: urban infrastructure and US air power', *City*, 9(2): 169–94.

Graham, S. (2007) 'War and the city', *New Left Review*, 44: 121–32.

Graham, S. (2010) *Cities under Siege: The New Military Urbanism*, New York: Verso.

Graham, S. and Marvin, S. (2001) *Splintering Urbanism: Networked Infrastructures, Technological Mobilities and the Urban Condition*, London: Routledge.

Gramsci, A. (1971) *Selections from the Prison Notebooks*, trans Q. Hoare and G. Nowell Smith, New York: International Publishers.

Graumann, C. (1976) *The Concept of Appropriation (Aneignung) and Modes of Appropriation of Space*, in P. Korosec-Serfaty (ed.), 'Appropriation of Space (Proceedings of the 3rd International Architectural Psychology Conference' at Louis Pasteur University Strasbourg) Strasbourg (France) 21–5 June.

Graumann, C. (1983) 'On multiple identities', *International Social Science Journal*, 35(96): 309–21.

Gregory, D. (2004) *The Colonial Present: Afghanistan, Palestine, Iraq*, Malden: Wiley-Blackwell.

Guevara Jerez, F.A. (2007) 'Who's who in the new cabinet', *Envío in English*, 307, online at http://www.envio.org.ni/articulo/3460 (accessed 3/6/2011).

Gugler, J. (2004) *World Cities Beyond the West: Globalization, Development and Inequality*, Cambridge: Cambridge University Press.

Guha, R. (1983) *Elementary Aspects of Peasant Insurgency*, Delhi: Oxford University Press.

Guha, R. (1996 [1963]) *A Rule of Property for Bengal: An Essay on the Idea of Permanent Settlement*, Durham: Duke University Press.

Guha, R. (1997) *Dominance without Hegemony: History and Power in Colonial India*, Cambridge, MA: Harvard University Press.

Gunew, S. (2007) 'Can the subaltern feel? Decolonising affect/theory', keynote presentation, 'Asia and the Other', Department of English, National Taiwan Normal University, 23/6/2007.

Hackworth, J. and Smith, N. (2001) 'The changing state of gentrification', *Tijdschrift voor Eonomische en Sociale Geografi*, 22: 464–77.

Hahner, J.E. (1986) *Poverty and Politics: The Urban Poor in Brazil, 1870–1920*, Albuquerque: University of New Mexico Press.

Hakim, B.S. (1986) *Arabic-Islamic Cities: Building and Planning Principles*, London and New York: KPI.

Hakim, B. (1994) 'The "Urf" and its role in diversifying the architecture of traditional Islamic cities', *Journal of Architectural and Planning Research*, 11(2): 108–27.

Halás, M. (2007) 'Possibilities for the application of geography to land use planning', *Acta Universitas Carolinae, Geographica*, 1: 67–77.

Hall, E. (1966) *Distances in Man: The Hidden Dimension*, New York: Double Day.

Hall, P. (1998) *Cities in Civilization*, New York: Pantheon.

Hall, T. and Hubbard, P. (1996) 'The entrepreneurial city: new urban politics, new urban geographies?', *Progress in Human Geography*, 20: 153–74.

Hall, T. and Hubbard, P. (eds) (1998) The Entrepreneurial City: Geographies Of Politics, Regimes and Representation, Chichester: John Wiley and Sons.

Halperin Donghi, T. (1994) *La Larga Agonía de la Argentina Peronista*, Buenos Aires: Espasa-Ariel.

Hamadeh, S. (1992) 'Creating the traditional city: a French project', in N. AlSayyad (ed.) *Forms of Dominance: On the Architecture and Urbanism of the Colonial Enterprise*, England: Avebury.

Hamilton, K. and Hoyle, S. (1999) 'Moving cities: transport connections', in J. Allen, D. Massey, and M. Pryke (eds) *Unsettling Cities*, London: Routledge.

Hampl, M., Dostál, P., and Drbohlav, D. (2007) 'Social and cultural geography in the Czech Republic: under pressures of globalization and post-totalitarian transformation', *Social and Cultural Geography*, 8(3): 475–93.

Hann, Ch. (ed.) (2002) *Postsocialism: Ideologies and Practices in Eurasia*, London and New York: Routledge.

Hann, Ch., Humphrey, C., and Verdery, K. (2002) 'Introduction: postsocialism as a topic of anthropological investigation', in Hann, Ch. (ed.) (2002) *Postsocialism: Ideologies and Practices in Eurasia*, London and New York: Routledge.

Hannigan, J. (1999) *Fantasy City: Pleasure and Profit in the Postmodern Metropolis*, London and New York: Routledge.

Hansen, H. (1960) *Daughters of Allah*, London: Purnell and Sons.

Hansen, K.T. (2008) *Youth and the City in the Global South*, Bloomington/Indianapolis: Indiana University Press.

Hansen, T. (2001) *Wages of Violence: Naming and Idenity in Postcolonial Bombay*, Princeton, NJ: Princeton University Press.

Hanssen, J. (2005) *Fin de Siecle Beirut: the Making of an Ottoman Provincial Capital*, Oxford: Oxford University Press.

Hardoy, J. (1975) *Urbanization in Latin America: Approaches and Issues*, Garden City: Anchor Press/DoubleDay.

Hardoy, J. (1982) 'The building of Latin American cities', in A. Gilbert, J. Hardoy, and R. Ramirez (eds) *Urbanization in Contemporary Latin America*, London and New York: John Wiley and Sons.

Hardoy, J. and Gutman M. (1992) *Impacto de la Urbanización en los Centros Históricos de Iberoamérica. Tendencias y Perspectivas*, Madrid: MAPFRE.

Hardoy, J. and Satterthwaite, D. (1989) *Squatter and Citizen: Life in the Urban Third World*, London: Earthscan Publications, Ltd.

Harris, W. (1895) 'A journey in Persian Kurdistan', *The Geographical Journal*, 6(5): 453–7.

Harrison, G. (2008) 'From the global to the local? Governance and development at the local level: reflections from Tanzania', *Journal of Modern African Studies*, 46: 69–89.

Hart, J. (1987) *Revolutionary Mexico: the Coming and Process of the Mexican Revolution*, Berkeley, CA: University of California Press.

Harvey, D. (1978) 'The urban process under capitalism', *International Journal of Urban and Regional Research*, 2: 101–31.

Harvey, D. (1989) *The Urban Experience*, Baltimore: Johns Hopkins University Press.

Harvey, D. (2003) *The New Imperialism*, New York: Oxford University Press.

Hashim, A. and Rahim Z. (2006) 'Visual privacy and family intimacy: a case study of Malay inhabitants living in two-storey low-cost terrace housing', *Environment and Planning B*, 33(2): 301–18.

Hashim, A. and Rahim, Z. (2008) 'The influence of privacy regulation on urban Malay families living in terrace housing', Archnet IJAR 2.

Hassanpour, A. (1996) 'The creation of Kurdish media culture', in P. Kreyenbroek and C. Allison (eds) *Kurdish Culture and Identity*, London: Zed Books.

Hawanonth, N., Jiradechakun, P., and Padthaisong, S. (2007) *Thrisdee-tharn-rak nai reung kwam-khem-khang khong chumchon (Grounded Theory on Community Strength)*, Bangkok, Thailand: Thailand Research Fund.

Hay, W. (1921) *Two Years in Kurdistan: Experiences of a Political Officer 1918–1920*, London: Sidgwick and Jackson Ltd.

Hays-Mitchell, M. (1994) 'Streetvending in Peruvian cities: the spatio-temporal behavior of ambulantes', *The Professional Geographer*, 46: 425–38.

Hazbun, W. (2010) 'Modernity on the beach: a postcolonial reading from southern shores', *Tourist Studies*, 9(3): 225–44.

He, S. (2007) 'State-sponsored gentrification under market transition: the case of Shanghai', *Urban Affairs Review*, 43(2): 171–98.

He, S. and Wu, F. (2007) 'Socio-spatial impacts of property-led redevelopment on China's urban neighbourhoods', *Cities*, 24(3): 194–208.

Healey, P. (2007) *Urban Complexity and Spatial Strategies: Towards Relational Planning for Our Times*, London: Routledge.

Hecker, T. (2010) 'The slum pastoral: helicopter visuality and Koolhaas's Lagos', in *Space and Culture*, 13(3): 256–69.

Henn, A. (2008) 'Crossroads of Religions: Shrines, Mobility and Urban Space in Goa', *International Journal of Urban and Regional Research*, 32(3): 658–70.

Herbert, S. (2008), 'Contemporary geographies of exclusion: traversing Skid Road', *Progress in Human Geography*, 32(5): 659–66.

Heynen, N. (2003) 'The scalar production of injustice within the urban forest', *Antipode*, 35: 980–99.

Hidalgo, R. and Cáceres, G. (2003) 'Beneficencia Católica y Barrios Obreros en Santiago de Chile en la Transición del Siglo XIX y XX. Conjuntos Habitacionales y Actores Involucrados', *Scripta Nova*, VII, August 146 (100).

Higgins, B. (1990) 'The place of housing programs and class relations in Latin American cities: the development of Managua before 1980', *Economic Geography*, 66(4): 378–88.

Hill, R. and Kim, J. (2000) 'Global cities and developmental states: New York, Tokyo, and Seoul', *Urban Studies*, 37: 2,167–95.

Hillier, B. and Vaughan, L. (2007) 'The city as one thing', *Progress in Planning*, 67(3): 205–30.

Hitchings, R. (2003) 'People, plants and performance: on actor network theory and the material pleasures of the private garden', *Social and Cultural Geography*, 4: 100–13.

Hodge J.E. (1996) 'Carlos Pellegrini Argentine Nationalist 1846–1906', *Journal of Inter-American Studies*, 8(4): 541–50.

Hoffman, D. (2011) *The War Machines: Young Men and Violence in Sierra Leone and Liberia*, Durham, NC; London: Duke University Press.

Holloway, S. and Valentine, G. (2001) 'Children at home in the wired world: reshaping and rethinking home in urban geography', *Urban Geography*, 22: 562–83.

Holston, J. (2008) *Insurgent Citizenship: Disjunctions of Democracy and Modernity in Brazil*, Princeton: Princeton University Press.

Honig, E. (1992) *Creating Chinese Ethnicity: Subei People in Shanghai, 1850–1980*, New Haven, CT: Yale University Press.

Hörschelmann, K. and Stenning, A. (2008) 'Ethnographies of postsocialist change', *Progress in Human Geography*, 32(3): 339–61.

Hosagrahar, J. (2005) *Indigenous Modernities: Negotiating Architecture and Urbanism*, London: Routledge.

Howard, E. (1996) 'The town–country magnet', in R. Le Gales and F. Stout (eds) *The City Reader*, London: Routledge.

Hubbard, P. (2006) *The City*, London: Routledge.

Huyssen, A. (2008) 'Introduction: world cultures, world cities', in A. Huyssen (ed.) *Other Cities, Other Worlds: Urban Imaginaries in a Globalizing Age*, Durham, NC: Duke.

Hobsbawm, E. (2002) 'The future of war and peace', *Counterpunch*, online at http://www.counterpunch.org/hobsbawm1.html (accessed 16/6/2010).

Holston, J. (1999) 'Spaces of insurgent citizenship' in J. Holston (ed.) *Cities and Citizenship*, Durham: Duke University Press.

Hubbard, P. (2006) *City*, London: Routledge

Huxtable, A. (1972) *Will they Ever Finish Bruckner Boulevard? A Primer on Urbicide*, New York: The Maxmillan Company.

Huybrechts, E. (2002) 'Beirut: building regional circuits', in S. Sassen (ed.) *Global Networks, Linked Cities*, London: Routledge.

Illades, C. and Rodríguez Kuri, A. (eds) (1996) *Ciudad de México: instituciones, actores sociales, y conflicto político, 1774–1932*, Zamora: El Colegio de Michoacán.

Ingenieros, J. (1963) *Las Direcciones Filosóficas de la Cultura Argentina*, Buenos Aires: Editorial EUDEBA.

Instituto Nacional de Estadística y Censos (INDEC) (2001) *Población Total Nativa y No Nativa 1895–2001*, Archivo INDEC.

Instituto Nacional de Estadísticas y Censos (INDEC) (2006) *VIII Censo de población y IV de vivienda 2005, Nicaragua*, online at http://www.inec.gob.ni (accessed 23/1/2007).

Inter-American Development Bank (2002) *Nicaragua: Multi-phase Low-income Housing Program – First Phase Loan Proposal*, (NI-0064), Washington, DC: Inter-American Development Bank.

Iossifova, D. (2009a) 'Blurring the joint line? Urban life on the edge between old and new in Shanghai', *Urban Design International*, 14(2): 65–83.

Iossifova, D. (2009b) 'Negotiating livelihoods in a city of difference: narratives of gentrification in Shanghai', *Critical Planning*, 16: 98–116.

Izady, M. (1992) *A Concise Handbook: The Kurds*, London: Crane Russak (Taylor and Francis).

Jackson, I. (2002) 'Politicised territory: Nek Chand's Rock Garden in Chandigarh', *Global Built Environment Review*, 2(2): 51–68.

Jacobs, J. (1961) *The Death and Life of Great American Cities*, New York: Vintage.

Jacobs, J. (1996) *Edge of Empire: Postcolonialism and the City*, London: Routledge.

Jackson, P., Crang, P., and Dwyer, C. (eds) (2004) *Transnational Spaces*, London: Routledge.

Jaguaribe, B. (2007a) 'Cities without maps: favelas and the aesthetics of realism', in A. Çinar and T. Bender (eds) *Urban Imaginaries: Locating the Modern City*, Minneapolis: University of Minnesota Press.

Jaguaribe, B. (2007b) *O Choque do Real: Estética, Mídia e Cultura*, Rio de Janeiro: Rocco.

Jaguaribe, B. and Hetherington, K. (2004) '*Favela tours*: indistinct and maples representations of the real in Rio de Janeiro' in M. Sheller and J. Urry (eds) *Tourism Mobilities: Places to Play, Places in Play*, London: Routledge.

Jayne, M. (2005) *Cities and Consumption*, London: Routledge.

Jayne, M., Gibson, C., Waitt, G., and Bell, D. (2011) 'The cultural economy of small cities', *Geography Compass*, 4(9): 1,409–17.

Jiménez, C.M. (2001) *Making the City Their Own: Popular Groups and Political Culture in Morelia, Mexico, 1880 to 1930*, Ph.D. dissertation thesis, University of California, San Diego.

Jiménez, C.M. (2004) 'Popular organizing for public services: residents modernize Morelia, Mexico, 1880–1930', *Journal of Urban History*, 30(4): 495–518.

Jiménez, C.M. (2006) 'Performing their right to the city: political uses of public space in a Mexican City 1880–1920', *Urban History*, 33(3): 435–546.

Jiménez, C.M. (2007) 'From the lettered city to the sellers' city: vendors politics and public space in urban Mexico, 1880–1920', in G. Prakash and K. Kruse (eds) *Cities: Space, Society, History*, Princeton: Princeton University Press.

Jiménez, E. (1998) 'New forms of community participation in Mexico City: success or ailure?' *Bulletin of Latin American Research*, 7(1): 7–31.

Jiron, P. (2008) 'Mobile borders in urban daily mobility practices in Santiago de Chile', paper presented at the 38th World Congress of the International Institute of Sociology, Central European University, Budapest, Hungary, 26–30 June, 2008.

Johns, M. (1997) *The City of Mexico in the Age of Díaz*, Austin: University of Texas Press.

Johnson, C. (1982) *MITI and the Japanese Economic Miracle*, Stanford, CA: Stanford University Press.

Jonas, E. and Wilson, D. (1999) 'The city as a growth machine: critical reflections two decades later', in E. Jonas and D. Wilson (eds) *The Urban Growth Machine: Critical Perspectives Two Decades Later*, Albany: SUNY Press.

Jones, G. (2004) 'Slums', in S. Harrison, S. Pile and N. Thrift (eds) *Patterned Grounds: Entanglements of Nature and Culture*, London: Reaktion Books.

Jones, G. and Varley, A. (1999) 'The reconquest of the historic centre: urban conservation and gentrification in Puebla, Mexico', *Environment and Planning A*, 31: 1,547–66.

Jones, G. and Ward, P. (2004) 'The end of public space in the Latin American city?', Memoria of the Bi-national Conference. The Mexican Center of LILAS. University of Texas at Austin, March 4–5.

Jones, M. (2009) 'Phase space: geography, relational thinking, and beyond', *Progress in Human Geography*, 33(4): 487–506.

Joseph, G. (1982) *Revolution from Without*, Cambridge: Cambridge University Press.

Joseph, G. and Nugent, D. (1994) *Everyday Forms of State Formation: Revolution and the Negotiation of Rule in Modern Mexico*, Durham: Duke University Press.

Judin, H. (2008) 'Unsettling Johannesburg: the country in the city', in A. Huyssen (ed.) *Other Cities, Other Worlds: Urban Imaginaries in a Globalizing Age*, Durham, NC: Duke.

Kabbani, R. (1986) *Europe's Myths of Orient*, London: Pandora.

Kahler, M. and Lake, D. (2004) 'Governance in a global economy: political authority in transition', *PS: Political Science and Politics*, 37: 409–14.

Kaldor, M. (1999) *New and Old Wars: Organized Violence in a Global Era*, Stanford, CA: Stanford University Press.

Kanchanapan, A. (1992) 'Khwam Pen Chumchon' [Community] in *Research Direction and Cutural Development in Northern Region*, 28–29 August, Chiangmai, Thailand.

Kapagama, P. and Waterhouse, R. (2009) *Portrait of Kinshasa: a City on (the) Edge*, London: London School of Economics and Political Science (Crisis States Research Centre working papers series 2, 53).

Kaplan, S. and Kaplan, R. (eds) (1982) *Humanscape: Environments for People*, reprinted edn, Ann Arbor, Michigan: Ulrich's Book, Inc.

Kasraian, N. and Arshi, Z. (1990) *Kurdistan*, Östersund, Sweden: Oriental Art.

Katz, C. (2004) *Growing Up Global: Economic Restructuring and Children's Everyday Lives*, Minneapolis: University of Minnesota Press.

Kaufman, V., Bergman, M., and Joye, D. (2004) 'Motility: mobility as capital', *International Journal of Urban and Regional Research*, 28(4): 745–56.

Keil, R. (2003) 'Progress report 1: urban political ecology', *Urban Geography*, 24: 723–38.

Keil, R. (2005) 'Progress report 2: urban political ecology', *Urban Geography*, 26: 640–51.

Keller, M. (2005) 'Needs, desires and the experience of scarcity', *Journal of Consumer Culture*, 5(1): 65–85.

Kent, S. (1984) *Analyzing Activity Areas: An Ethnoarchaeological Study of the Use of Space*, University of New Mexico Press.

Keyes, C.F. (1987) *Thailand, Buddhist Kingdom as Modern Nation-state*, Boulder: Westview Press.

Khatib-Chahidi, J. (1981) 'Sexual prohibitions, shared space and fictive marriage in Shi'ite Iran', in S. Ardener (ed.) *Women and Space*, London, Croom Helm.

Khazanov, A. and Crookenden, J. (1994) *Nomads and the Outside World*, Madison, WI: University of Wisconsin Press.

Kheirabadi, M. (1991) *Iranian Cities: Formation and Development*, Texas: University of Texas Press.

Khilnani, S. (1997) *The Idea of India*, London: Hamish Hamilton.

Kidambi, P. (2007) *The Making of an Indian Metropolis: Colonial Governance and Public Culture in Bombay 1890–1920*, Aldershot: Ashgate.

King, A. (1976) *Colonial Urban Development*, London: Routledge.

King, A. (1985) 'Colonial cities: global pivots of change', in R. Ross and G.J. Telkamp (eds) *Colonial Cities: Essays on Urbanism in a Colonial Context*, Dordrecht: Martinus Nijhof.

King, A. (1990) *Urbanism, Colonialism, and the World Economy*, New York: Routledge.

King, A. (1991) *Culture, Globalisation and the World System*, London: Palgrave.

King, A. (1995) 'Re-presenting world cities: cultural theory/social practice', in P. Knox and P. Taylor (eds) *World Cities in a World-system*, Cambridge: Cambridge University Press.

King, A. (1996) 'Worlds in the city: Manhattan transfer and the ascendance of spectacular space', *Planning Perspectives*, 11: 97–114.

King, A. (2004) *Spaces of Global Cultures: Architecture, Urbanism, Identity*, Routledge, London.

King, B. and Hustedde, R. (1993) 'Community free spaces', *Journal of Extension* [Online], 31: 4.

Kirby, J. (1985) *Urbanization in China*, New York: Columbia University Press.

Klein, N. (2008) *The Shock Doctrine: The Rise of Disaster Capitalism*, New York: Picador.

Klein, S. (1983) 'The integration of Italian immigrants into the United States and Argentina: a comparative analysis', *The American Historical Review*, 88(2): 306–29.

Kliot, N. and Mansfeld, Y. (1999) 'Case studies of conflict and territorial organization in divided cities', *Progress in Planning*, 52: 167–225.

Knight, A. (1990) *The Mexican Revolution*, vols 1 and 2, Lincoln: Nebraska University Press.

Knight, A. (1994) 'Popular culture and the revolutionary state in Mexico, 1910–1940', *Hispanic American Historical Review*, 74(3): 393–444.

Kolonga Molei (1979) *Kinshasa, ce village d'hier*, Kinshasa: SODIMCA.

Koser, K. and Dhaliwal, D. (2008) 'Surge in the number of Iraqi refugees: causes and implications', *Brookings*, online at http://www.brookings.edu/multimedia/video/2008/0703_iraq_refugees_koser.aspx (accessed 8/6/2010).

Kostof, S. (1992) *The City Assembled: Elements of Urban Form through History*, Boston: Little, Brown.

Kothari, U. (2008) 'Global peddlers and local networks: migrant cosmopolitanisms', *Environment and Planning D: Society and Space*, 26(3): 500–16.

Kreyenbroek, P. (1996) 'Religion and religious in Kurdistan', in P. Kreyenbroek and C. Allison (eds) *Kurdish Culture and Identity*, London: Zed Books.

Kudva, N. (2009) 'The everyday and the episodic: the spatial and political impacts of urban informality', *Environment and Planning A*, 41: 1,614–28.

Kurian, A. (2010) 'Dubai Metro: a glimpse of life', *Gulf News*, 9 March: 8.

Kusno, A. (2000) *Behind the Postcolonial: Architecture, Urban Space and Political Cultures*, London: Routledge.

Kuus, M. (2004) 'Europe's eastern expansion and the reinscription of otherness in East–Central Europe', *Progress in Human Geography*, 28: 472–89.

La Fontaine, J. (1970) *City Politics. A Study of Léopoldville 1962–1963*, Cambridge: Cambridge University Press.

Lagae, J. (2007) 'Léopoldville – Bruxelles, villes miroirs? L'architecture et l'urbanisme d'une capitale coloniale', in J.L. Vellut (ed.) *Villes d'Afrique. Explorations en histoire urbaine*, Paris: L'Harmattan (Cahiers Africains 73).

Lagerkvist, A. (2010) 'The future is here: media, memory and futurity in Shanghai', *Space and Culture*, 13(3): 220–38.

Laguiazf.org (n.d.) online at http://www.laguiazf.org/zfafiliadas/DB_ZFAfiliadas_14.xls (accessed 21/2/2008).

Lambright, A. and Guerrero, A. (eds) (2007) *Unfolding the City: Women Write the City in Latin America*, Minneapolis: University of Minnesota Press.

Lamprakos, M. (1992) 'Le Corbusier and Algiers: the plan obus as colonial urbanism', in N. AlSayyad (ed.) *Forms of Dominance: on the Architecture and Urbanism of the Colonial Enterprise*, England: Avebury.

Langervang, T. (2008) '"We are managing!" Uncertain paths to respectable adulthoods in Accra', Ghana' *Geoforum*, 39(6): 2,039–47.

Laquain, A. (2005) *Beyond Metropolis: the Planning and Governance of Asia's Mega-Urban Regions.* Washington, DC: Woodrow Wilson Center Press; Baltimore, MD: Johns Hopkins University Press.

Larkin, B. (2008) *Signal and Noise*, Durham: Duke University Press.

Latham, A. and McCormack, D. (2004) 'Moving cities: rethinking the materialities of urban geographies', *Progress in Human Geography*, 28(6): 701–24.

Lawlor, S. (2005) 'Disgusted subjects: the making of middle class identities', *Sociological Review*, 53(3): 429–46.

Lawrence, R. (1990) 'Public collective and private space: a study of urban housing in Switzerland', in S. Kent (ed.) *Domestic Architecture and the Use of Space*, Cambridge: Cambridge University Press.

Lawson, B. (2001) *The Language of Space*, Oxford: Architectural Press.

Le Corbusier (1925; republished 1987; translated by Frederick Etchells) *The City of To-morrow and its Planning*, New York: Dover Publications.

Le Corbusier (1931; Dover Edition, 1986) *Towards a New Architecture*, New York: Dover Publications.

Leach, E. (1940) 'Social and economic organisation of the Rowanduz Kurds: monogaphs' in *Social Anthropology*, 3, London: London School of Economics.

Lear, J. (1993) *Workers, Vecinos, and Citizens: the Revolution in Mexico City, 1910–1917*, Ph.D. dissertation thesis, University of California, Berkeley.

Lear, J. (1996) 'Mexico City: space and class in the Porfirian capital, 1884–1910', *Journal of Urban History*, 22(4): 454–92.

Lee, S.J. and Arrington, C. (2008) 'The Politics of NGOs and democratic governance in South Korea and Japan', *Pacific Focus*, 23(1): 75–96.

Lees, A. (1985) *Cities Perceived: Urban Society European and American Thought, 1820–1940*, New York: Columbia University Press.

Lees, L., Slater, T., and Wyly, E. (2008) *Gentrification*, London: Routledge.

Lefebvre, H. (1991) *Critique of Everyday Life*, vol. 1, London: Verso.

Lefebvre, H. (1995) *Writings on Cities*, Oxford: Blackwell.

Lefebvre, H. (2003) *The Urban Revolution*, Minneapolis: University of Minnesota Press.

Legg, D. and McFarlane, C. (2008) 'Ordinary urban spaces: between postcolonialism and development', *Environment and Planning A*, 40(1): 6–14.

Legg, S. (2007) *Spaces of Colonialism: Delhi's Urban Governmentalities*, Malden, MA: Wiley-Blackwell.

Lein, Y. (2004) 'Forbidden roads: Israel's discriminatory road regime in the West Bank', *B'Tselem*, online at http://www.btselem.org (accessed 10/6/2011).

Leitner, H. and Sheppard, E. (1999) 'Transcending interurban competition: conceptual issues and policy alternatives in the European Union', in E. Jonas and D. Wilson (eds) *The Urban Growth Machine: Critical Perspectives Two Decades Later*, Albany: SUNY Press.

Lelo Nzuzi, F. (2008) *Kinshasa, Ville et Environnement*, Paris: L'Harmattan.

Ley, D. (1996) *The New Middle Class and the Remaking of the Central City*, Oxford: Oxford University Press.

Ley, D. (2004) 'Transnational spaces and everyday lives', *Transactions of the Institute of British Geographers*, 29(2): 151–64.

Light, D. (2004) 'Street names in Bucharest, 1990–1997: exploring the modern historical geographies of post-socialist change', *Journal of Historical Geography*, 30: 154–72.

Lin, G. (1994) 'Changing theoretical perspectives on urbanization in Asian developing countries', *Third World Planning Review*, 16: 1–23.

Lin, G. (1997) *Red Capitalism in South China: Growth and Development of the Pearl River Delta*, Vancouver: University of British Columbia Press.

Lin, G. (1999) 'State policy and spatial restructuring in post-reform China, 1978–95', *International Journal of Urban and Regional Research*, 23: 670–96.

Lin, G. (2002) 'Changing discourses in China geography: a narrative evaluation', *Environment and Planning A*, 34: 1,809–31.

Lin, G. (2009) *Developing China: Land, Politics and Social Conditions*, London and New York: Routledge.

Lindell, I. (2010a) 'Between exit and voice: informality and the spaces of popular agency', *African Studies Quarterly*, 11(2–3): 1–11.

Lindell, I. (2010b) *Africa's Informal Workers: Collective Agency, Alliances and Transnationional Organizing in Urban Africa*, London/Uppsala: Zed Books/Nordic Africa Institute.

Lloyd, D. (2005) 'The subaltern in motion: subalternity, the popular, and Irish working class history', in S. Chattopadhyay and B. Sarkar (eds) *PostColonial Studies*, 8(4): 421–38.

Lo, C.P. (1994) 'Economic reforms and socialist city structure: a case study of Guangzhou, China', *Urban Geography*, 15: 128–49.

Loáisiga Mayorga, J. (2003) 'Fininsa era del "grupo financiero" del FSLN', *La Prensa*, 11 February, online at http://www-ni.laprensa.com.ni/archivo/2003/febrero/11/nacionales/nacionales-20030211-12.html (accessed 10/6/2011).

Loáisiga Mayorga, J. (2005) 'Sombra de Bayardo Arce tras fastuoso edificio', *La Prensa*, 16 May, online at http://www-ni.laprensa.com.ni/archivo/2005/mayo/16/nacionales/nacionales-20050516-11.html (accessed 10/6/2011).

Locatelli, F. and Nugent, P. (2009) *African Cities: Competing Perspectives on Urban Spaces*, Leiden: Brill.

Loftus, A. (2006) 'Reification and the dictatorship of the water meter', *Antipode*: 1,023–45.

Loftus, A. (2007) 'Working the socio-natural relations of the urban waterscape in South Africa', *International Journal of Urban and Regional Research*, 31: 41–59.

Logan, J. and Molotch, H. (1987) *Urban Fortunes: The Political Economy of Place*, Berkeley: University of California Press.

Logan, J. (ed.) (2002) *The New Chinese City Globalization and Market Reform*, Malden, MA: Blackwell.

Logan, J. (ed.) (2007) *Urban China in Transition*, Malden, MA: Blackwell.

Longhurst, R. (2000) *Bodies: Exploring Fluid Boundaries*, London: Routledge.

Low, S. (2000) *On the Plaza: the Politics of Public Space and Culture*, Austin: University of Texas Press.

Lu, F. (2009) 'Local customs baffle some overseas visitors', *Shanghai Daily Online Edition*, 6 July, online at http://www.shanghaidaily.com (accessed 6/7/2009).

Ludden, D. (1998) 'Area studies in age of globalization', online at http://www.frontiersjournal.com (accessed on 1/6/2010).

Luxemburg, R. (2003 [1913]) *The Accumulation of Capital*, trans. A. Schwarzwild and C. Lo (1994) 'Economic reforms and socialist city structure: a case study of Guangzhou, China', *Urban Geography*, 15: 128–49.

Lungo, M. and Baires, S. (2001) 'Socio-spatial segregation and urban land regulation in Latin American cities', conference paper, Lincoln Institute of Land Policy.

Lyotard, J. (1984) *The Postmodern Condition: a Report on Knowledge*, trans. from French by G. Bennington and B. Massumi, Minneapolis: University of Minnesota Press.

Ma, L. (1971) *Commercial Development and Urban Change in Sung China (960–1279)*, Ann Arbor: University of Michigan Press.

Ma, L. (2002) 'Urban transformation in China, 1949–2000: a review and research agenda', *Environment and Planning A*, 34: 1,545–69.

Ma, L. and Fan, M. (1994) 'Urbanisation from below: the growth of towns in Jiangsu, China', *Urban Studies*, 31: 1,625–45.

McCann, E. and Ward, K. (2010) 'Relationality/territoriality: towards a conceptualization of cities in the world', *Geoforum*, 41: 175–184.

McCann, E. and Ward, K. (2011) 'Introduction: urban assemblages, territories, relations, practices and power' in E. McCann and K. Ward (eds) *Mobile Urbanism: City Policy-*

making in the Global Age, Minneapolis: University of Minnesota Press.

MacCannell, D. (1976) *The Tourist: a New Theory of the Leisure Class,* New York: Schocken Books.

McFarlane, C. (2008) 'Postcolonial Bombay: decline of a cosmopolitan city?', *Environment and Planning D: Society and Space,* 26: 480–99.

McFarlane, C. (2011; forthcoming) 'The city as assemblage: dwelling and urban space', *Environment and Planning D: Society and Space.*

McGee, T., Lin, G.C.S., Wang, M.Y.L., Marton, A., and Wu, J. (2007) *China's Urban Space,* Routledge: London.

McGinley, S. (2010) 'No major hike in rental rates near Dubai Metro stations', *Arabian Business,* 14 March, online at http://www.arabianbusiness.com/583668-no-major-hike-in-rental-rates-near-dubai-metro-stations (accessed 20/3/2010).

McIlwaine, C. (2008) 'Gender- and age-based violence', in V. Desai and R. Potter (eds) *The Companion to Development Studies,* London: Arnold.

Madanipour, A. (1998a) *Design of Urban Space: an Inquiry into a Socio-spatial Process,* Chichester and New York: Wiley.

Madanipour, A. (1998b) 'Social exclusion and space', in A. Madanipour, G. Cars, and J. Allen (eds) *Social Exclusion in European Cities: Processes, Experiences and Responses,* London: Jessica Kingsley Publishers.

Madanipour, A. (1998c) *Tehran: The Making of a Metropolis,* Chichester: John Wiley and Sons.

Maffitt, K. (2003) 'From the ashes of the poet kings: exodus, identity formation, and the new politics of place in Mexico City's industrial suburbs, 1948–1970', *International Labour and Working-Class History,* 64: 202–28.

Mairena Martínez, M. (1999) 'Cuadras Shultz reclaman ingenio que endeudaron', *El Nuevo Diario,* 21 August, online at http://archivo.elnuevodiario.com.ni/1999/agosto/21-agosto-1999/nacional/nacional5.html (accessed 10/6/2011).

Mallon, F. (1995) *Peasant and Nation: The Making of Postcolonial Mexico and Peru,* Berkeley: University of California Press.

Mann, M. (1984) 'The autonomous power of the state: its origins, mechanisms and results', *European Journal of Sociology,* 25(2): 185–213.

Marenco Tercero, E. (2005) 'Daniel creó el liderazgo político de los empresarios sandinistas', *La Prensa,* 13 February, online at http://www-usa.laprensa.com.ni/archivo/2005/febrero/13/enfoque/enfoque-20050213-02.html (accessed 10/6/2011).

Margulis, S. (2003) 'Privacy as a social issue and behavioural concept', *Journal of Social Issues,* 59(2): 243–61.

Martínez-Vergne, T. (1999) *Shaping the Discourse on Space,* Austin: University of Texas Press.

Martins, L. (2005) *João do Rio: Uma Antologia,* Rio de Janeiro: José Olympio.

Marx, K. (2004) *The Eighteenth Brumaire of Louis Napoleon,* Whitefish, MT: Kissinger.

Massey, D. (1984) 'Introduction: geography matters', in D. Massey and J. Allen (eds) *Geography Matters,* Cambridge, UK: Cambridge University Press.

Massey, D. (1986) 'Nicaragua: some reflections on socio-spatial issues in a society in transition', *Antipode,* 18(3): 322–31.

Massey, D. (2007) *World City,* Cambridge: Polity.

Matlovič, R. (2004) 'Tranzitívna podoba mesta a jeho intraurbánnych štruktúr v ére post-komunistickej transformácie a globalizácie', *Sociológia,* 36(2): 137–58.

Maugham, S.W. (1995) *The Gentleman in the Parlour* (first published 1930), Bangkok: White Orchid Press.

Mayer, T. and Mourad, S.A. (2008) *Jerusalem,* London: Routledge.

Mayorga, F. (2007) *Megacapitales de Nicaragua,* Managua: Ediciones Magnus.

Mazumdar, S. and Mazumdar, S. (2001) 'Rethinking public and private space: religion and women in Muslim society', *Journal of Architectural and Planning Research*, 18(4): 302–24.

Mbembe, A. (2001) *On the Postcolony*, Berkeley: University of California Press.

Mbembe, A. (2004) 'Aesthetics of superfluity', *Public Culture*, 16(3): 373–405.

Mbembe, A. and Nuttall, S. (2004) 'Writing the world from an African metropolis', in *Public Culture*, 16(3): 347–72.

Meade, T. (1997) *'Civilizing' Rio: Reform and Resistance in a Brazilian City, 1889–1930*, University Park: Pennsylvania State University Press.

Mehrotra, R. (2008) 'Negotiating the static and kinetic cities: the emergent urbanism of Mumbai', in A. Huyssen (ed.) *Other Cities, Other Worlds: Urban Imaginaries in a Globalizing Age*, Durham, NC: Duke.

Mehta, U. (1999) *Liberalism and Empire: a Study in Nineteenth-century British Liberal Thought*, Chicago: University of Chicago Press.

Memarian, G. and Brown, F. (2003) 'Climate culture and religion: aspects of the traditional courtyard house in Iran', *Architecture and Planning Research*, 20(3): 180–98.

Menin, S. (2003) 'Introduction', in S. Menin (ed.) *Constructing Place: Mind and Matter*. London and New York: Routledge: 1–37.

Metcalf, T. (1989) *An Imperial Vision: Indian Architecture and Britain's Raj*, Berkeley: University of California Press.

Mexican Government (1857) *Seventh Congreso Constitucional Constitutión Federal de Los Estados de México por el Congreso Constituyente el Día 5 de Febrero de 1857*, Mexico: Imprenta del Gobierno Federal.

Meyer, M. and Beezley, W. (eds) (2000) *The Oxford History of Mexico*, New York: Oxford University Press.

Mianda, G. (1996) *Femmes Africaines et Pouvoir: Les Maraicheres de Kinshasa*, Paris: L'Harmattan.

Mignolo, W. (2002) 'The many faces of cosmo-polis; border thinking and critical cosmopolitanism', in C. Breckenridge, S. Pollock, H. Bhabha, and D. Chakrabarty (eds) *Cosmopolitanism*, Durham: Duke University Press.

Mignolo, W. (2005) 'Subalterns and other agencies', in S. Chattopadhyay and B. Sarkar (eds) *PostColonial Studies*, 8(4): 381–408.

Milgram, S. (1970) 'The experience of living in cities', *Science*, 167: 1,461–8.

Miraftab, F. (2009) 'Insurgent planning: situating radical planning in the Global South', *Planning Theory*, 8: 32–50.

Misselwitz, P. and Rieniets, T. (2009) 'Jerusalem and the Principles of Conflict Urbanism', *Journal of Urban Technology*, 16(2): 61–78.

Mitchell, D. (1997) 'The annihilation of space by law: the roots and implications of anti-homeless laws in the United States', *Antipode*, 29(3): 303–35.

Mitchell, T. (1991) *Colonizing Egypt*, Berkeley: University of California Press.

Mohan, D. (2005) *Public Transportation Systems for Urban Areas: a Brief Review*, Transportation Research and Injury Prevention Programme, Indian Institute of Technology, Delhi.

Mollenkopf, J. and Castells, M. (1991) *Dual City: the Restructuring New York*, New York: Russell Sage Foundation.

Molotch, H. (1976) 'The city as a growth machine', *American Journal of Sociology*, 82: 309–55.

Momsen, J. and Townsend, J. (eds) (1987) *Geography of Gender in the Third World*, London: Hutchinson.

Monteguin, F-A. (1983) 'The essence of urban existence in the world of Islam', in A. Germen (ed.) *Islamic Architecture and Urbanism*, Dammam: Ging Faisal University.

Montgomery, M., Stren, B., Cohen, B., and Reed, H. (2004) *Cities Transformed: Demographic Change and its Implications in the Developing World*, London: Earthscan.

Moore, S. (2009) 'The excess of modernity: garbage politics in Oaxaca, Mexico', *The Professional Geographer*, 61: 426–37.

Morris, B. (1990) *1948 and After: Israel and the Palestinians*, Oxford: Clarendon Press.

Morse, R. and Hardoy, J. (Eds) (1992) *Rethinking the Latin American City*, Baltimore: John Hopkins Press.

Mugerauer, R. (1985) 'Language and the emergence of environment', in Seamon, D. and Mugerauer, R. (ed.) *Dwelling, place, and environment: towards a phenomenology of person and world*, Dordrecht [Netherlands]: Martinus Nijhoff Publishers: 51–70.

Mukherjee Reed, A. (n.d.) 'The Global South: politics, policy and development', course syllabus, York University, Toronto, Canada, online at www.yorku.ca/ananya/Globalsouthhome.htm (accessed 15/6/2007).

Mulder, N. (1996; 5th edn 2000) *Inside Thai Society: an Interpretation of Everyday Life*, Amsterdam, Kuala Lumpur: Pepin Press.

Mulíček, O. (2002) 'Suburbanizace v Brně a jeho okolí', in *Suburbanizace a její sociální, ekonomické a ekologické důsledky*, Praha: Ústav pro ekopolitiku, o.p.s.

Mulíček, O. (2009) 'Prostorové vzorce postindustriálního Brna', in S. Ferenčuhová, L. Galčanová, M. Hledíková and B. Vacková (eds) *Město: Proměnlivá ne/samozřejmost*. Červený Kostelec/Brno: Pavel Mervart/MU.

Mumford, L. (1961) *The City in History*, New York: Harcourt.

Mumford, L. (1984) The Urban Prospect, Peter Smith Publisher Inc.

Municipalidad de Providencia (2007) 'Plan de Desarrollo Comunal de Providencia 2006–2012' Available in http://www.providencia.cl.

Municipalidad de Recoleta (2005) 'Proyectos Relevantes Proyectados in Bellavista–Recoleta Periodo 2001–2005', Departamento de Edificación. Architect Rodrigo Barros report.

Muramatsu, M. (1997) *Local Power in the Japanese State*, Berkeley: University of California Press.

Musil, J. (2003) 'Proměny urbánní sociologie ve Spojených státech a Evropě 1950–2000', *Sociologický Časopis*, 39(2): 137–67.

Musil, J. (2005) 'Jak se formovala sociologie bydlení', *Sociologický Časopis*, 41(2): 207–25.

Myers, G.A. (2003) *Verandahs of Power: Colonialism and Space in Urban Africa*, Syracuse: Syracuse University Press.

Nartsupha, C. (1991) *Watthanatham thai kab krabuankarn plianplang thang sangkhom* [Thai culture and the transforming process of Thai society], 5th (2004) reprint ed., Bangkok, Thailand: Chulalongkorn Press.

Nartsupha, C. (1994) *Watthanatham muban thai (The Culture of Thai Village)*, 2nd edn 1998, reprinted Bangkok, Thailand: Sangsan Publishing.

Nartsupha, C. (1997) *Ban kab muang (Ban and muang)*, 2nd reprint edn, Bangkok, Thailand: Chulalongkorn Press.

Nash, C. (2004) 'Post-colonial geographies', in P. Cloke, P. Crang, and M. Goodwin (eds) *Envisioning Human Geographies*, London: Arnold.

Nee, V. (1989) 'A theory of market transition: from redistributive to markets in state socialism', *American Sociological Review*, 54: 663–81.

Needell, J. (1987) 'The revolta contra Vacina of 1904: the revolt against modernization in Belle–Époque Rio de Janeiro', *Hispanic American Historical Review*, 67(2): 233–69.

Needell, J. (1995) 'Public place and consciousness in Latin America', *Comparative Studies in Society and History*, 37(3): 519–40.

Neuwirth, R. (2005) *Shadow Cities: a Billion Squatters, a New Urban World*, London: Routledge.

New York Times (1857) 'The Central Park: proposed plan of improvement', report of Egbert L. Viele, 20 February.

Newell, P. (1998) 'A cross-cultural comparison of privacy definitions and functions: a systems approach', *Journal of Environmental Psychology*, 18(4): 357–71.

Newman, D. (2003) 'On borders and power: a theoretical framework', *Journal of Borderlands Studies*, 18(1): 13–25.

Newman, D. and Paasi, A. (1998) 'Fences and neighbours in the postmodern world: boundary narratives in political geography', *Progress in Human Geography*, 22(2): 186–207.

Newman, P. and Thornley, A. (2005) *Planning World Cities: Globalization and Urban Politics*, New York: Palgrave Macmillan.

Nijman, J. (2007a) 'Introduction: comparative urbanism', *Urban Geography*, 28: 92–107.

Nijman, J. (2007b) 'Place particularity and "deep analogies": a comparative essay on Miami's rise as a world city', *Urban Geography*, 28: 92–107.

Nikitine, V. (1987) *Kurd and Kurdistan*, trans. M. Qazi, Tehran: Nilufar.

Nitlapán-Envío team (1999) 'In the vortex of hurricane "Corruption"', *Envío in English*, 212, online at http://www.envio.org.ni/articulo/2225 (accessed 3/6/2011).

Njeru, J. (2006) 'The urban political ecology of plastic bag waste problem in Nairobi, Kenya', *Geoforum*, 37: 1,046–58.

Noparatnaraporn, C. (2003) 'Living place and landscape in Bangkok: the merging character', paper presented at Hawaii International Conference on Arts and Humanities, in Hawaii: University of Hawaii – West Oahu.

Noparatnaraporn, C. (2005) 'Transforming "unbounded" nature: the evolution of a Thai cultural landscape', *Raneang: Journal of the Faculty of Architecture*, Kasetsart University, 4, 10th anniversary edition, Bangkok, Thailand: Text and Journal Publication Co. Ltd.: 204–15.

Núñez, O. (2006) *La Oligarquía en Nicaragua*, Managua: CIPRES.

O'Meara, R. (2007) *Space and Muslim Urban Life*, London: Routledge.

OED Online (2009) Oxford University Press, University of Texas Libraries.

Oi, J. (1992) 'Fiscal reform and the economic foundations of local state corporatism', *World Politics*, 45: 99–126.

Oi, J. (1995) 'The role of the local state in China's transitional economy', *China Quarterly*, 144: 1132–49.

Oldenburg, V. (1984) *The Making of Colonial Lucknow, 1856–77*, Princeton: Princeton University Press.

Olds, K. (2001) *Globalization and Urban Change: Capital, Culture and Pacific Rim Mega–Projects*, London and New York: Oxford University Press.

Oliver, P. (ed.) (1997) *Encyclopaedia of Vernacular Architecture of the World*, Cambridge, Cambridge University Press.

Oliveira, N. (1996) 'Favelas and ghettos: race and class in Rio de Janeiro and New York City', *Latin American Perspectives*, 90(23): 71–89.

Organization for Economic Cooperation and Development (OECD) (2005) *OECD Territorial Reviews: Japan*, Paris: OECD.

Orum, A. and Chen, X. (2003) *The World of Cities: Places in Comparative and Historical Perspective*, Oxford: Blackwell.

Ouředníček, M. (2003) 'Suburbanizace Prahy', *Sociologický Časopis*, 39(2): 235–53.

Ouředníček, M. and Temelová, J. (2009) 'Twenty years after socialism: the transformation of Prague's inner structure', *Studia Universitatis Babes–Bolyai, Sociologia*, 54(1): 9–30.

Ouředníček, M., Temelová, J., Brabec, T., and Vyhnánková, M. (2009) 'Praŭské předměstí, sídliště Máj a suburbium Kodetka: případové studie proměňujících se lokalit Českobudějovické aglomerace', in J. Kubeš a kol *Urbánní geografie Českých Budějovic a Českobudějovické aglomerace II*, Banská Bystrica: UMB.

Oushakine, S.A. (2000) 'The quantity of style: imaginary consumption in the New Russia', *Theory, Culture and Society*, 17(5): 97–120.

Paasche, T. and Sidaway, J. (2010) 'Transecting security and space in Maputo', in *Environment and Planning A*, 42: 1,555–76.

Pacione, M. (1990) *Urban Problems. An Applied Urban Analysis*, London: Routledge.

Pain, M. (1984) *Kinshasa, la Ville et la Cite*, Paris: Editions de l'Office de la Recherche Scientifique et Technique Outre–Mer (ORSTOM).

Panadero M. (2001) 'El Proceso de Urbanización de América Latina Durante el Periodo Científico-Técnico', *Revista Bibliográfica de Geografía y Ciencias Sociales*, 298: 742–98.

Panelli, R. (2008) 'Social geographies: encounters with Indigenous and more-than-White/Anglo geographies', in *Progress in Human Geography*, 32(6): 801–11.

Pannell, C. (2002) 'China's continuing urban transition', *Environment and Planning A*, 34: 1,571–84.

Pappe, I. (2007) *The Cleansing of Palestine*, Oxford: Oneworld.

Parish, W.L. (1990) 'What model now', in Y. Kwok, W. Parish, A. Yeh and X. Xu (eds) *Chinese Urban Reform: What Model Now?* New York and London: M.E. Sharpe Inc.

Park, C. (2000) 'Local politics and urban power structure in South Korea', *Korean Social Science Journal*, 27: 41–67.

Park, R., Burgess, E. and McKenzie, R. (1925) *The City: Suggestions for the Study of Human Nature in the Urban Environment*, Chicago: University of Chicago Press.

Parker, S. (2004) *Urban Theory and the Urban Experience: Encountering the City*, London: Routledge.

Parry, B. (1993) 'The contents and discontents of Kipling's imperialism', in E. Carter, J. Donald and J. Squires (eds) *Space and Place: Theories of Identity and Location*, London: Lawrence and Wishart.

Patel, S. (2009) 'Comments and reflections', *City and Community*, 8(4): 467–71.

Patico, J. (2009) 'Spinning the market: the moral alchemy of everyday talk in postsocialist Russia', *Critique of Anthropology*, 29: 205–24.

Peake, L. (2009) 'Gender in the city', in *The International Encyclopedia of Human Geography*, London: Elsevier.

Peck, J., Theodore, N., and Brenner, N. (2009) 'Neoliberal urbanism: models, moments, mutations', *SAIS Review*, 29: 49–66.

Peet, R. and M. Watts (2004) *Liberation Ecologies: Environment, Development, Social Movements*, London: Routledge.

Pekkanen, R. (2006) *Japan's Dual Civil Society: Members without Advocates*, Stanford: Stanford University Press.

Pelling, M. (2003) 'Toward a political ecology of urban environmental risk: the case of Guyana', in K. Zimmerer and T. Bassett (eds) *Political Ecology: an Integrative Approach to Geography and Environmental-development Studies*, New York: Guilford Press.

Pempel, T. (1999) *Regime Shift: Comparative Dynamics of the Japanese Political Economy*, Ithaca: Cornell University Press.

Percival, T. and Waley, P. (2010) 'Exporting urbanisation: Northeast Asian planners and developers in Southeast Asian cities', conference presentation at 'Global Urban Frontiers: Asian Cities in Theory, Practice and Imagination', 8–9 September, National University of Singapore.

Perlman, J. (1976) *The Myth of Marginality: Urban Poverty and Politics in Rio de Janeiro*, Berkeley: University of California Press.

Peter-Smith, M. (2000) *Transnational Urbanism: Locating Globalization*, London: Wiley-Blackwell.

Peter-Smith, M. (2005) 'Transnational urbanism revisited', *Journal of Ethnic and Migration Studies*, 31(2): 235–44.

Petherbridge, G. (1984) 'The house and society', in G. Michell (ed.) *Architecture of the Islamic World: Its History and Social Meaning*, London: Thames and Hudson.

Phillips, T. (2003) 'How Favelas went chic', *Brazzil* – Poverty, December, online at http://www.brazzil.com/2003/html/articles/dec03/p105dec03.htm (accessed 3/6/2011).

Phillips, T. (2008) 'Gangsters, guns and drugs – now tourists can see the "real Rio"', *The Guardian*, 7 May.

Phillips, T. and Smith, P. (2006) 'Rethinking Urban Incivility Research: strangers, bodies and circulations', *Urban Studies*, 43(5–6): 879–901.

Piccato, P. (1995) 'El Paso de Venus el disco del Sol: criminality and alcoholism in the late Porfiriato', *Mexican Studies/Estudios Mexicanos*, 11(2): 203–41.

Pickles, J. and Smith, A. (2007) 'Post-socialist economic geographies and the politics of knowledge production', in A. Tickell, E. Sheppard and J. Peck (eds) *Politics and Practice in Economic Geography*, London: Sage.

Piermay, J-L. (1997) 'Kinshasa: a reprieved mega–city?', in C. Rakodi (ed.) *The Urban Challege in Africa: Growth and Management of its Large Cities*, Tokyo/New York: United Nations University Press.

Pierre, J. (2005) 'Comparative urban governance: uncovering complex causalities', *Urban Affairs Review*, 40: 446–62.

Pieterse, E. (2008) *City Futures. Confronting the Crisis of Urban Development*, London/New York: Zed Books.

Pieterse, E. (2010) 'Cityness and African urban development', *Urban Forum*, 21(3): 205–19.

Pile, S. (2010) 'Emotions and affect in recent human geography', *Transactions of the Institute of British Geographers*, 35(1): 5–20.

Pineo, R. and Baer, J. (eds) (1998) *Cities of Hope: People, Protests and Progress in Urbanizing Latin America, 1870–1930*, Boulder: Westview Press.

Pithipat, S. (2001) *Khwam pen khon thai (Being Thai People)*, Knowledge of Thai History II, Bangkok: Amarin.

Plynoy, S. (2000) *Wan kon khun kao (The former days, the old nights)*, Bangkok: Pimkam Press.

Porter, S. (2003) *Working Women in Mexico City: Public Discourses and Material Conditions, 1879–1931*, Tucson: University of Arizona Press.

Portes, A. (1989) 'Latin American urbanization in the years of the crisis', *Latin American Research Review*, XXIV(2): 7–44.

Potter, S. and Potter, J. (1990) *China's Peasants: The Anthropology of a Revolution*, Cambridge: Cambridge University Press.

Pousada Favelinha (n.d.) 'Home', online at http://www.favelinha.com (accessed 3/3/2011).

Pow, C. (2009) *Gated Communities in China: Class, Privilege and the Moral Politics of the Good Life*, London and New York: Routledge.

Prakash, G. (2008) 'Mumbai: the modern city in ruins', in A. Huyssen (ed.) *Other Cities, Other Worlds: Urban Imaginaries in a Globalizing Age*, Durham, NC: Duke.

Press, D. (1998) 'Urban warfare: opinion, problems and the future', summary of a conference on urban warfare, Bedford, Massachusetts, online at http://web.mit.edu (accessed 22/5/2010).

Pullan, W., Misselwitz, P., Nasrallah, R., and Haim, Y. (2007) 'Jerusalem's Road 1', *City*, 11(2): 176–98.

Pype, K. (2007) 'Fighting boys, strong men and gorillas: notes on the imagination of masculinities in Kinshasa', *Africa*, 77(2): 250–71.

Pype, K. (forthcoming) *The Making of the Pentecostal Melodrama: an Ethnography of Media, Religion and Mimesis in Kinshasa*, New York/Oxford: Berghahn Publishers.

Quijano, A. (1975) 'The urbanization of Latin American society', in J. Hardoy (ed.) *Urbanization in Latin America: Approaches and Issues*, Garden City: Anchor Press/DoubleDay.

Rabinow, P. (1989) *French Modern: Norms and Forms of the Social Environment*, Cambridge: MIT Press.

Ramírez Mercado, S. (1999) *Adiós Muchachos: Una Memoria de la Revolución Sandinista*, Mexico: Aguilar.

Rao, V. (2006) 'Slum as theory', *International Journal of Urban and Regional Research*, 30(1): 225–32.

Rao, V. (2007) 'Venture capital', *Public Culture*, 19(3): 593–7.

Rapoport, A. (1962) *House Form and Culture*, New Jersey: Prentice-Hall.

Rapoport, A. (1977) *Human Aspects of Urban Form: Towards a Man–Environment Approach to Urban Form and Design*, Oxford: Pergamon Press.

Rapoport, A. (1990) *History and Precedent in Environmental Design*, New York: Plenum Press.

Read, S. (2005) 'Flat city: a space syntax derived urban movement network model', in A. van Nes (ed.) *Proceedings*, 5th International Space Syntax, Symposium, TUDelft, Amsterdam: Techne Press.

Redfield, R. and Singer, M. (1954) 'The cultural role of cities', *Economic Development and Cultural Change*, 3: 53–73.

Rich, C. (1836) *Narrative of a Residence in Koordistan, and on the Site of Ancient Nineveh: with Journal of a Voyage Down the Tigris to Baghdad and an Account of a Visit to Shirauz and Persepolis*, London: J. Duncan.

Richardson, B. (1998) *The Political Culture of Japan*, Berkeley: University of California Press.

Rigg, J. (2007) *An Everyday Geography of the Global South*, London: Routledge.

Rio, J. (1911) *Vida Vertiginosa*, Rio de Janeiro: Garnier.

Rivera Reynaldos, L.G. (1996) *Desamortización y Nacionalización de Bienes Civiles y Eclesiáticos en Morelia, 1856–1876*, Morelia: Universidad Michoacana de San Nicolás de Hidalgo, Instituto de Investigaciones Históricos.

Rivlin, L. (1987) 'The neighborhood, personal identity, and group affiliations', in I. Altman and A. Wandersman (eds) *Neighborhood and Community Environments*, New York: Plenum Press.

Robins, S., Cornwall, A., and von Lieres, B. (2008) 'Rethinking "citizenship" in the post-colony', *Third World World Quarterly*, 29: 1,069–86.

Robbins, P. (2005) *Political Ecology: an Introduction*, Malden, MA: Blackwell.

Robbins, P. (2007) *Lawn People: How Grasses, Weeds and Chemicals Make Us Who We Are*, Philadelphia, PA: Temple University Press.

Robinson, J. (2002) 'Global and world cities: a view from off the map', *International Journal of Urban and Regional Research*, 26(3): 531–54.

Robinson, J. (2003) 'Postcolonialising geography: tactics and pitfalls', *Singapore Journal of Tropical Geography*, 24(3): 273–89.

Robinson, J. (2005) 'In the tracks of comparative urbanism: difference, urban modernity and the primitive', *Urban Geography*, 25(8): 709–23.

Robinson, J. (2006) *The Ordinary City: Between Modernity and Development*, London: Routledge.

Robinson, J. (forthcoming) 'Cities in a world of cities: the comparative gesture', *International Journal of Urban and Regional Research*.

Robinson, W. (1998) '(Mal)development in central America: globalization and social change', *Development and Change*, 29(3): 467–97.

Rocha, J., Martínez, T., and Rocha, X. (1999) 'Summing up hurricane Mitch: the good, the bad and the ugly', *Envío in English*, 221, oneline at http://www.envio.org.ni/articulo/2289 (accessed 3/6/2011).

Rocheleau, D. (2008) 'Political ecology in the key of policy: from chains of explanation to webs of relation', *Geoforum*, 39: 716–27.

Rochovská, A., Blaūek, M., and Sokol, M. (2007) 'Ako zlepšit' kvalitu geografie: o dôleūitosti kvalitatívneho výskumu v humánnej geografii', *Geografický Časopis*, 59(1): 323–58.

Rock, D. (1985) *Argentina 1516–1982: from the Spanish Colonization to the Falklands War*, Berkeley: University of California Press.

Rockwell, E. (1994) 'Schools or the revolution: enacting and contesting state forms in Tlaxcala, 1910–1930' in G. Joseph and D. Nugent (eds) *Everyday Forms of State Formation: Revolution and the Negotiation of Rule in Modern Mexico*, Durham: Duke University Press.

Rodgers, D. (2004) 'Disembedding the city: crime, insecurity, and spatial organisation in Managua, Nicaragua', *Environment and Urbanization*, 16(2): 113–24.

Rodríguez, A. and Winchester, L. (2001) *Santiago Report: Governance and Urban Poverty*, London: University of Birmingham.

Rojas, E. (2002) 'Urban heritage conservation in Latin America and the Caribbean: a task for all social actors', Sustainable Development Department Technical Paper Series, Inter-American Development Bank.

Rojo, A. (1999) ' "Gordoman" en el país de la piñata', *El Mundo*, 4 March, online at http://www.elmundo.es/1999/03/04/internacional/04N0053.html (accessed 10/6/2011).

Roniger, L. and Waisman, C. (eds) (2000) *Globality and Multiple Modernities: Comparative North American and Latin American Perspectives*, Portland: Sussex Academic Press.

Rose, G., Degen, M., and Basdas, B. (2010) 'More on "big things": building events and feelings', *Transactions*, 35(3): 334–49.

Rosenzweig, R. and Blackmar, E. (1992) *The Park and the People*, Ithaca: Cornell University Press.

Ross, R. and Telkamp, G. (1985) 'Introduction', in Ross, R. and G. Telkamp (eds) *Colonial Cities: Essays on Urbanism in a Colonial Context*, Dordrecht: Martinus Nijhof.

Round, J. and Williams, C. (2010) 'Coping with the social costs of "transition": everyday life in post-Soviet Russia and Ukraine', *European Urban and Regional Studies*, 17: 183–96.

Roy, A. (2005) 'Urban informality: towards an epistemology of planning', *Journal of the American Planning Association*, 71(2): 147–58.

Roy, A. (2008) 'The 21st century metropolis: new geographies of theory', *Regional Studies*, 43(6): 819–30, online at http://rsa.informaworld.com/10.1080/00343400701809665 (accessed 3/6/2011).

Roy, A. and AlSayyad, P. (2004) *Urban Informality: Transnational Perspectives from the Middle East, Latin America and South Asia*, Oxford: Lexington Books.

Roy, S. (1999) 'De-development revisited: Palestinian economy and society since Oslo', *Journal of Palestine Studies*, XXVIII(3): 64–82.

Rubino, S. (2005) 'A curios blend? City revitalisation: gentrification and commodification in Brazil', in R. Atkinson and G. Bridge (eds) *Gentrification in a Global Context: The New Urban Colonialism*, London: Routledge.

Růūička, M. (2010) 'Urbanizace chudoby a etnicity v (post)socialistickém městě', in B. Vacková, S. Ferenčuhová and L. Galčanová (eds) *Československé město včera a dnes*. Červený Kostelec/Brno: Pavel Mervart/MU.

Sa, L. (2007) *Life in the Megalopolis: Mexico City and Sao Paulo*, London: Routledge.

Sabatini, F. (1997) 'Liberalización de los mercados de suelo y segregación social en las ciudades Latinoamericanas: el caso de Santiago', draft, Cambridge: Lincoln Institute of Land Policy, Urban Land Markets in Latin America Project.

Sabato, H. (2001) *The Many and the Few: Political Participation in Republican Buenos Aires*, Stanford: Stanford University Press.

Sachs, J. and Woo, T. (2000) 'Understanding China's economic performance', *Journal of Policy Reform*, 4: 1–50.

Sadler, S. (1999) *The Situationist City*, Cambridge, Mass.: MIT Press.

Saegert, S. (2010) 'Environmental psychology', in R. Hutchison (ed.) *Encyclopedia of Urban Studies*, Thousand Oaks, CA: Sage: 250–4.

Safier, M. (2001) 'The Struggle for Jerusalem: arena of nationalist conflict or crucible of cosmopolitan co-existence', *City*, 5(2): 135–68.

Said, E. (1993) *Culture and Imperialism*, London: Verso.

Salaff, J. (2004) 'Singapore: forming the family for a World City', in J. Gugler (ed.) *World Cities beyond the West: Globalization, Development and Inequality*, New York: Cambridge University Press.

Salamandra, C. (2004) *A New Old Damascus: Authenticity and Distinction in Urban Syria*, Bloomington: Indiana University Press.

Saleh, M. (1997) 'Privacy and communal socialization: the role of space in the security of traditional and contemporary neighbourhoods in Saudi Arabia', *Habitat International* 21(2): 167–84.

Saleh, M. (1998) 'The impact of Islamic and customary laws on urban form development in south western Saudi Arabia', *Habitat International*, 22(4): 537–56.

Sánchez, R. (2008) 'Seized by the spirit: the mystical foundation of squatting among Pentecostals in Caracas (Venezuela) today', *Public Culture*, 20(2): 267–305.

Sánchez Campbell, G. (2004), 'Poca oferta para pobres', *La Prensa*, 9 February, online at http://www-ni.laprensa.com.ni/archivo/2004/febrero/09/nacionales/nacionales-20040209-07.html (accessed 10/6/2011).

Sánchez, Díaz, G. (1991) *Pueblos, Villas y Cuidades de Michoacán en el Porfiriato*, Morelia: Universidad Michoacana de San Nicolás de Hidalgo.

Sandercock, L. (1998) *Towards Cosmopolis: Planning for Multicultural Cities*, London: Wiley.

Sargatal, M. (2000) 'El sstudio de la gentrificacion', Biblio 3W, *Revista Bibliográfica de Geografía y Ciencias Sociales*, May, 228.

Sarlo, B. (2008) 'Cultural landscapes: from integration to fracture', in A. Huyssen (ed.) *Other Cities, Other Worlds: Urban Imaginaries in a Globalizing Age*, Durham, NC: Duke.

Sarmiento, D. (2010) [1845].*Civilizacion y Barbarie*. Stockcero.

Sassen, S. (1996) *Losing Control? Sovereignty in an Age of Globalization*, New York: Columbia University Press.

Savitch, H. (2005) 'An anatomy of urban terror: lessons from Jerusalem and elsewhere', *Urban Studies*, 42(3): 361–96.

Savitch, H. and Ardashev, G. (2001) 'Does terror have an urban future', *Urban Studies*, 38(13): 2,515–33.

Schaefer, F.K. (1953) 'Exceptionalism in geography: a methodological examination', *Annals of American Geographers*, 43: 226–46.

Schneider, C. (1995) *Shantytown Protest in Pinochet's Chile*, Philadelphia: Temple University Press.

Schnore, L. (1965) 'On the spatial structure of cities in the two Americas', in P. Hauser and L. Schnore (eds) *The Study of Urbanization*, New York: Wiley.

Schoeman, F. (1984) *Philosophical Dimensions of Privacy: an Anthology*, Cambridge: Cambridge University Press.

Scott, A., Agnew, J., Soja, E., and Storper, M. (2002) 'Global city-regions', in A. Scott (ed.) *Global City-regions: Trends, Theory, Policy*, New York: Oxford University Press.

Scott, I. (1982) *Urban and Spatial Development in Mexico*, Washington, DC: The International Bank for Reconstruction and Development.

Scranton, R. (2007) 'Walls and shadows: the occupation of Baghdad', *City*, 11(3): 277–92.

Seabrook, J. (1996) *In the Cities of the South: Scenes from a Developing World*, New York: Verso.

Seddon, G. (ed.) (1997) Landprints: reflection on place and landscape, Cambridge, UK: University of Cambridge.

Sellers, J. (2005) 'Replacing the nation: an agenda for comparative urban politics', *Urban Affairs Review*, 40: 419–45.

Sennett, R. (1974) *The Fall of Public Man*, New York: W.W. Norton.

Sennett, R. (1994) *Flesh and Stone*, London: Faber.

Shapin, S. (1998) 'Placing the view from nowhere: historical and sociological problems in the location of science', *Transactions of the Institute of British Geographers*, 23: 5–12.

Shears, A. and Tyner, J. (2009) '(De)constructing the geography of America's surge in Iraq', *Antipode*, 41(2): 221–5.

Sheldon, K. (ed.) (1996) *Courtyards, Markets, City Streets: Urban Women in Africa*, Boulder, CO: Westview Press.

Sheller, M. and Urry, J. (2003) 'Mobile transformations of "public" and "private" life', *Theory, Culture and Society*, 20(3): 107–25.

Sheller, M. and Urry, J. (2000) 'The city and the car', *The International Journal of Urban and Regional Research*, 24: 737–57.

Shevchenko, O. (2002) ' "In case of fire emergency": consumption, security and the meaning of durables in a transforming society', *Journal of Consumer Culture*, 2: 147–70.

Shillington, L. (2008) 'Being(s) in relation at home: socio-natures of patio "gardens" in Managua, Nicaragua', *Social & Cultural Geography*, 9(7): 755–6.

Shlaim, A. (2009) *Israel and Palestine: Reappraisals, Revisions, Refutations*, London: Verso.

Short, J. (2006) *Urban theory: a critical assessment*, London: Palgrave Macmillan.

Shwartz, M. (2007) 'Neoliberalism on crack: cities under siege in Iraq', *City*, 11(1): 21–69.

Sibley, D. (1995) *Geographies of Exclusion: Society and Difference in the West*, London: Routledge.

Silver, C. (2007) *Planning the Megacity: Jakarta in the Twentieth Century*, London: Routledge.

Simone, A. (2002) 'Spectral selves: practices in the making of African cities', unpublished paper presented at the Wits Institute for Social and Economic Research (WISER), Johannesburg: University of the Witwatersrand.

Simone, A. (2004) *For the City Yet to Come: Urban Change in Four African Cities*, Durham NC: Duke University Press.

Simone, A. (2005) 'Urban circulation and the everyday politics of African urban youth: the case of Douala, Cameroon', *International Journal of Urban and Regional Research*, 29(3): 516–32.

Simone, A. (2010a) '2009 Urban Geography plenary lecture – on intersections, anticipations, and provisional publics: remaking district life in Jakarta', *Urban Geography*, 31(3): 285–308.

Simone, A. (2010b) *City Life from Jakarta to Dakar: Movements at the Crossroads*, New York: Routledge.

Simpfendorfer, B. (2009) *The New Silk Road: How a Rising Arab World is Turning Away from the West and Rediscovering China*, London: Palgrave Macmillan.

Singh, R.P.B. (1987) 'Time and Hindu rituals in Varanasi', in R.L. Singh and R.P.B. Singh (eds) *Trends in the Geography of Pilgrimages*, Varanasi: Banares Hindu University.

Singh, R.P.B. (1990) 'The pilgrimage manadala of Varanasi (Kashi): a study in sacred geography', in L. Gopal and D. Dubey (eds) *Pilgrimage Studies: Text and Context*, Allahabad: Society of Pilgrimage Studies.

Sit, V. (2010) *Chinese City and Urbanism: Evolution and Development*, New Jersey: World Scientific Pub.

Skeggs, B. (2005) 'The making of class and gender through visualising moral subject formation', *Sociology*, 39(5): 965–82.

Skidmore, T. (1974) *Black into White: Race and Nationality in Brazilian Thought*, Oxford: Oxford University Press.

Slater, D. (ed.) (1985) *New Social Movements and the State in Latin America*, Amsterdam: CEDLA.

Smart, J. (1995) *Local Capitalism: Situated Social Support for Capitalist Production in China*, Hong Kong: Chinese University of Hong Kong.

Smith, A. (2002) 'Culture/economy and spaces of economic practice: positioning households in post-communism', *Transactions of the Institute of British Geographers*, 27: 232–50.

Smith, A. and Rochovská, A. (2007) 'Domesticating neo-liberalism: everyday lives and the geographies of post-socialist transformations', *Geoforum*, 38(6): 1,163–78.

Smith, A. and Stenning, A. (2006) 'Beyond household economies: articulations and spaces of economic practice in postsocialism', *Progress in Human Geography*, 30(2): 190–213.

Smith, A. and Timár, J. (2010) 'Uneven transformations: space, economy and society 20 years after the collapse of state socialism', *European Urban and Regional Studies*, 17: 115–25.

Smith, J. and Jehlička, P. (2007) 'Stories around food, politics and change in Poland and the Czech Republic', *Transactions of the Institute of British Geographers*, 32: 395–410.

Smith, N. (1979) 'Toward a theory of gentrification: a back to the city movement by capital not people', *Journal of the American Planning Association*, 45: 538–48.

Smith, N. (1996) *The New Urban Frontier: Gentrification and the Revanchist City*, London and New York: Routledge.

Smith, N. (2000) 'El Nuevo urbanismo: ¿De Quién? "La época de Giuliani" y el Revanchismo de los 90', Universidad de Guadalajara 19.

Smith, N. (2002) 'New globalism, new urbanism: gentrification as global urban strategy', *Antipode*, 34(3): 434–57.

Smith, N. and Herod, A. (1991) 'Gentrification: a comprehensive bibliography', discussion paper, department of Geography, new series, 1, Rutgers University.

Smith, S. (2004) 'Editorial: living room', *Urban Geography*, 25: 89–91.

Smithies, M. (1993) *Old Bangkok*, Singapore: Oxford University Press.

Sooner Real Estate (n.d.) online at http://www.soonerrealestate.com.ni/index.php?option=com_content&task=view&id=24&Itemid=28 (accessed 21/2/2008).

Sorkin, M. (2009) *20 Minutes in Manhattan*, New York: Reaktion Books.

Sorkin, M. (ed.) (1992) *Variations on a Theme Park: The New American City and the End of Public Space*, New York: Hill and Wang.

Spivak, G. (2005) 'Scattered speculations on the subaltern and the popular', in S. Chattopadhyay and B. Sarkar (eds) *PostColonial Studies*, 8(4): 381–408.

Spurr, D. (1993) *The Rhetoric of Empire*, London: Duke.

Staples, A. (1994) 'Policia y Buen Gobierno: municipal efforts to regulate Public Behaviour, 1821–1857', in W. Beezley, C. Martin and W. French (eds) *Rituals of Rule Rituals of Rule, Rituals of Resistance*, Wilmington: Scholarly Resources.

Steinführerová, A. (2003) 'Sociálně prostorové struktury mezi setrvačností a změnou. Historický a současný pohled na Brno', *Sociologický Časopis*, 39(2): 169–92.

Stenning, A. (2005a) 'Post-socialism and the changing geographies of the everyday in Poland', *Transactions of the Institute of British Geographers*, 30(1): 113–27.

Stenning, A. (2005b) 'Re-placing work: economic transformations and the shape of community in post-socialist Poland', *Work Employment Society*, 19: 235–59.

Stenning, A. and Hörschelmann, K. (2008) 'History, geography and difference in the post-socialist world: Or, do we still need post-socialism?', *Antipode*, 40(2): 312–35.

Stenning, A., Smith, A., Rochovská, A., and Świątek, D. (2010) *Domesticating Neo-liberalism. Spaces of Economic Practice and Social Reproduction in Post-socialist Cities*, Malden, Oxford, Chichester: Wiley-Blackwell.

Stern, A. (1999) 'Buildings, boundaries, and blood: medicalization and nation building on the U.S.–Mexico Border, 1910–1930', *Hispanic American Historical Review*, 79(1): 41–81.

Stewart, D.J. (2001) 'Middle East urban studies: identity and meaning', *Urban Geography*, 2: 157–83.

Stoker, G. (1998) 'Theory and urban politics', *International Political Science Review*, 19: 119–29.

Stone, C. (2008) 'Urban regimes and the capacity to govern: a political economic approach', in M. Orr and V. Johnson (eds) *Power in the City: Clarence Stone and the Politics of Inequality*, Lawrence: Kansas University Press.

Stone, C. (1989) *Regime Politics: Governing Atlanta, 1946–1988*, Lawrence: Kansas University Press.

Su, S. and Feng, L. (1979) 'On the stages of social development after the proletariat has seized power' (in Chinese) [Wuchanjieji qude zhenquan hou de shehui fazhan wenti], *Economic Research* [*Jingji Yanjiu*], 5: 14–19.

Suessmuth-Dyckerhoff, C., Hexter, J., and St-Maurice, I. (2008) 'Marketing to China's new traditionalists', *Far Eastern Economic Review*, 171(3): 28.

Sunega, M. (2005) 'Efektivnost vybraných nástrojů bytové politiky v České republice', *Sociologický Časopis*, 41(2): 271–99.

Surazska, W. (1996) 'Local revolutions in Central Europe, 1990 to 1994: memoirs of mayors and councillors from Poland, Slovakia and the Czech Republic', *Publius: The Journal of Federalism*, 26(2): 121–40.

Swyngedouw, E. (2004) *Social Power and the Urbanization of Water*, Oxford: Oxford University Press.

Sýkora, L. (2002) 'Global competition, sustainable development and civil society: three major challenges for contemporary urban governance and their reflection in local development practices in Prague', *Acta Universitas Carolinae Geographica*, 2: 65–83.

Sýkora, L. (2003) 'Suburbanizace a její společenské důsledky', *Sociologický Časopis*, 39(2): 217–33.

Sýkora, L. (2006) 'Urban development, policy and planning in the Czech Republic and Prague', in U. Altrock, S. Güntner, S. Huning, and D. Peters (eds) *Spatial Planning and Urban Development in the New EU Member States: from Adjustment to Reinvention*, Hampshire, Burlington: Ashgate.

Sýkora, L. (2007) 'Office development and post-communist city formation: the case of Prague', in K. Stanilov (ed.) *The Post-socialist City: Urban Form and Space Transformations in Central and Eastern Europe after Socialism*, Dordrecht: Springer.

Sýkora, L. (2008) 'Idiografická nebo nomotetická koncepce v geografii: kontraproduktivní spor o povahu a podstatu poznání', *Geografický Časopis*, 60(3): 299–315.

Sýkora, L. (2009a) 'Post-socialist cities', in R. Kitchin and N. Thrift (eds) *International Encyclopedia of Human Geography* 8: 387–395.

Sýkora, L. (2009b) 'Revolutionary change, evolutionary adaptation and new path dependencies: socialism, capitalism and transformations in urban spatial organizations', in W. Strubelt and G. Gorzelak (eds) *City and Region: Papers in Honour of Jiří Musil*, Budrich, Germany: Budrich UniPress.

Sýkora, L. and Ouředníček, M. (2007) 'Sprawling post-communist metropolis: commercial and residential suburbanisation in Prague and Brno, the Czech Republic', in M. Dijst, E. Razin and C.Vázquez (eds) *Employment Deconcentration in European Metropolitan Areas: Market Forces versus Planning Regulations*, New York: Springer.

Sýkora, L. and Posová, D. (2007) 'Specifika suburbanizace v postsocialistickém kontextu: nová bytová výstavba v metropolitní oblasti Prahy 1997–2005', *Geografie – Sborník České Geografické Společnosti*, 112(3): 334–56.

Szaló, C. and Hamar, E. (2006) 'Váš Trianon, náš holokaust: segregace a inkluze kultur vzpomínání', in R. Marada (ed.) *Etnická Různost a Občanská Jednota*, Brno: CDK.

Szelényi, I. (1996) 'Cities under socialism – and after', in G. Andrusz, M. Harloe, and I. Szelenyi (eds) *Cities after Socialism: Urban and Regional Change and Conflict in Post-socialist Societies*, Oxford: Blackwell.

Taylor, C. (2004) *Modern Social Imaginaries*, Durham, Duke University Press.

Taylor, D. (1999) 'Central Park as a model for social control: urban parks, social class and leisure behavior in nineteenth-century America', *Journal of Leisure Research*, 31(4): 420–77.

Taylor, P. (1996) 'Embedded statism and the social sciences: opening up to new spaces', *Environment and Planning A*, 28: 1917–28.

Taylor, P. (2004) 'God invented wars to teach Americans Geography', *Political Geography*, 23: 487–92.

Téfel Vélez, R. (1974) *El Infierno de los Pobres: Diagnóstico Sociológico de los Barrios maRginales de Managua*, Managua: El Pez y la Serpiente.

Temelová, J. (2007) 'Flagship developments and the physical upgrading of post-socialist inner city: the Golden Angel project in Prague', *Geografiska Annaler*, 89B(2): 169–81.

Temelová, J. (2009) 'Urban revitalization in central and inner parts of (post-socialist) cities: conditions and consequences', in T. Ilmavirta (ed.) *Regenerating Urban Core*, Helsinki: Helsinki University of Technology.

Tenorio-Trillo, M. (1996) *Mexico at the World's Fairs: Crafting a Modern Nation*, Berkeley: University of California Press.

Terada, R. (2001) *Feeling in Theory: Emotion after the 'Death of the Subject'*, UK: Harvard University Press.

Thanakit (ed.) (2000) *Phraratchaprawat kao ratchakan lae phra boromarachini (The stories of the nine kings and queens)*, reprinted edn, Bangkok: Pyramid Publishing.

Thrall, G. (1987) *Land Use and Urban Form: the Consumption Theory of Land Rent*, London: Methuen.

Thrift, N. (2000) ' "Not a straight line but a curve", or, cities are not mirrors of modernity', in D. Bell and A. Haddour (ed.) *City Visions*, London: Prentice Hall.

Thrift, N. (2004) 'Movement–space: the changing domain of thinking resulting from the development of new kinds of spatial awareness', *Economy and Society*, 33: 582–604.

Thrift, N. (2004) 'Intensities of feeling towards a spatial politics of affect', *Geografiska Annaler Series B Human Geography*, 86B(1): 57–78.

Thrift, N. (2008) *Non-representational Theory: Space, Politics, Affect*, London: Routledge.

Tiebout, C. (1956) 'A pure theory of local expenditures', *Journal of Political Economy*, 64: 416–24.

Tilly, C. (1988) 'Misreading, then rereading, nineteenth-century social change', in B. Wellman and S. Berkowitz (eds) *Social Structures: a Network Approach*, Cambridge: Cambridge University Press.

Tilly, C. (1994) 'Entanglements of European cities and states', in C. Tilly and W. Blockmans (eds) *Cities and the Rise of States in Europe: AD 1000 to 1800*, Boulder, CO: Westview.

Tilly, C. (1996) 'What good is urban history?', *Journal of Urban History*, 22(6): 702–19.

Timár, J. (2004) 'What convergence between what geographies in Europe? A Hungarian perspective', *European Urban and Regional Studies*, 11: 371–5.

Time magazine (1953) 'Argentina: Night of Fire', April 27.

Tolia-Kelly, D. (2009) 'The geographies of cultural geography: identities, bodies and race', *Progress in Human Geography*, 34(3): 358–67.

Tomba, L. (2004) 'Creating an urban middle class: social engineering in Beijing', *The China Journal*, 51(1): 1–26.

Tonda, J. (2005) *Le Souverain Moderne: Le Corps Du Pouvoir En Afrique Centrale (Congo, Gabon)*, Paris: Karthala.

Transparency International (2004) *Global Corruption Report 2004*, London: Pluto Press.

Tuan, Y.-F. (1977) *Space and Place: The Perspective of Experience*, Minneapolis: University of Minnesota Press.

Tyler, I. (2006) 'Chav scum: the filthy politics of social class in contemporary Britain', *Media/Culture*, 9(5), online at http://journal.media–culture.org.au/0610/09–tyler.php (accessed 3/6/2011).

Ulusoy, Z. (1998) 'Housing rehabilitation and its role in neighbourhood change: a framework for evaluation', *Journal of Architectural and Planning Research*, 15(3): 243–57.

UN-Habitat (n.d.) 'Country Profile: Nicaragua', online at http://www.unhabitat.org/categories.asp?catid=158 (accessed 12/3/2010).

UN-Habitat (2003) *The Challenge of Slums: Global Report on Human Settlements 2003*, Nairobi: UN-Habitat.

Uribe Salas, A. (1993) *Morelia, los pasos a la modernidad*, Morelia: Centro de Investigaciones Históricas, Universidad Michoacana de San Nicholas de Hidalgo.

Urry, J. (2002) *The Tourist Gaze*, London: Sage.

Urry, J. (2007) *Mobilities*, Cambridge: Polity Press.

US Census Bureau (1920) 'Abstract of the Fourteenth Census of the United States', Washington, DC: GPO, 1923.

Valenca, M. (2006) 'Culture, politics, faith and poverty in Belem (Brazil)', *Geoforum*, 37(2): 159–61.

Valentine, G. (1999) 'A corporeal geography of consumption', *Environment and Planning D: Society and Space*, 17: 329–51.

Valentine, G. (2008) 'Living with difference: reflections on geographies of encounter', *Progress in Human Geography*, 32(3): 323–37.

Valentová, B. (2005) 'Vývoj sociálního bydlení s důrazem na zacílení na určité sociální vrstvy obyvatelstva', *Sociologický Casopis*, 41(2): 301–15.

Valladares, L. (2005) *A Invenção da Favela: do Mito de Origem a Favela.com*, Rio de Janeiro: Fundação Getúlio Vargas.

van Houtum, H. and van Naerssen, T. (2002) 'Bordering, ordering and othering', *Tijdschrift voor Economische en Sociale Geografie*, 93(2): 125–36.

Vaughan, M. (1982) *The State, Education and Social Class in Mexico, 1880–1928*, DeKalb: Northern Illinois University Press.

Ventura, Z. (1994) *A Cidade Partida*, São Paulo: Companhia das Letras.

Véron, R. (2006) 'Remaking urban environments: the political ecology of air pollution in Delhi', *Environment and Planning A*, 38: 2,093–2,109.

Vianna, H. (1995) *O Mistério do Samba*, Rio de Janeiro: Zahar/UFRJ.

Vich, V. (2004) 'Popular capitalism and subalternity: street comedians in Lima', *Social Text*, 81, 22(4): 47–64.

Vilas, C. (1992) 'Family affairs: class, lineage, and politics in contemporary Nicaragua', *Journal of Latin American Studies*, 24(2): 309–41.

Violich, F. (1987) *Urban Planning for Latin America: the Challenge of Metropolitan Growth*, Boston: Oelgeschlager, Gunn and Hain in association with the Lincoln Institute of Land Policy.

Virno, P. (2009) 'Angels and the general intellect: individuation in Duns Scotus and Gilbert Simondon', *Parrhesia*, 7: 58–67.

Visser, G. (2003) 'Gay men, leisure space and South African cities: the case of Cape Town', *Geoforum*, 34(1): 123–37.

Vogel, E. (1991) *The Four Little Dragons: The Spread of Industrialization in East Asia*, Harvard: Harvard University Press.

Wall, D. (1996) 'City profile: Managua', *Cities*, 13(1): 45–52.

Wang, Y.P. and Murie, A. (2005) 'Social and spatial implications of housing reform in China', *International Journal of Urban and Regional Research*, 24: 397–417.

Wang, Y., Wang, Y. and Bramley, G. (2005) 'Chinese housing reform in state-owned enterprises and its impacts on different social groups', *Urban Studies*, 42(10): 1,859–78.

Ward, K. (2008) 'Commentary: towards a comparative (re)turn in urban studies? Some reflections', *Urban Geography*, 29: 405–10.

Ward, K. (2010) 'Towards a relational comparative approach to the study of cities', *Progress in Human Geography*, 34(6): 471–87.

Ward, K. and McCann E. (2011) 'Conclusion: cities assembled, space neoliberalization, (re)territorialisation and comparison', in E. McCann and K. Ward (eds) *Mobile Urbanism: City Policymaking in the Global Age*, Minneapolis: University of Minnesota Press.

Warner, S. (1974) 'If all the world were Philadelphia: a scaffolding for urban history, 1774–1930', in C. Tilly (ed.) *An Urban World*, Boston: Little, Brown and Company.

Warren, J. and Fethi I. (1982) *Traditional Houses in Baghdad*, Horsham, England: Coach.

Warren, W. (2002) *Bangkok*, London: Reaktion.

Watenpaugh, K. (2006) *Being Modern in the Middle East: Revolution, Nationalism, Colonialism, and the Arab Middle Class*, Princeton: Princeton University Press.

Watson, S. (2006) *City Publics: the (Dis)enchantments of Urban Encounters*, London: Routledge.

Weber, M. (1958) *The City*, New York: Free Press.

Webster's Encyclopedic Unabridged Dictionary of the English Language (2001), Berkeley, CA: Thunder Bay Press.

Weisenthal, L. (2010) 'Producing spatial pluralism: town-planning, house building and everyday practices in a South African township', paper presented at conference: 'Non-human in Anthropology', Charles University, Prague, 4–5 December.

Weiss, L. (2000) 'Developmental states in transition: adapting, dismantling, innovating, not "normalizing"', *The Pacific Review*, 13: 21–55.

Weizman, E. (2007) *Hollow Land: Israel's Architecture of Occupation*, London: Verso.

Weizman, E. (2003) 'The politics of verticality: the West Bank as an architectural construction', in A. Franke (ed.) *Territories*, Berlin: K.W. Institute of Contemporary Art.

Weizman, E. (2004) 'Strategic points, flexible lines, tense surfaces, and political volumes: Ariel Sharon and the geometry of occupation', in S. Graham (ed.) *Cities, War and Terrorism: Towards an Urban Geopolitics*, Malden: Blackwell Publishing.

Wells, A. (1985) *Yucatan Gilded Age: Haciendas, Henequen, and International Harvester, 1860–1915*, Albuquerque: University of New Mexico Press.

Wendall Cox (2004) 'The public purpose, a national journal', online at http://www.publicpurpose.com (accessed 5/6/2011).

Wheatley, P. (1971) *The Pivot of the Four Quarters: a Preliminary Enquiry into the Origins and Character of the Chinese City*, Chicago: Aldine.

Whisnant, D. (1995) *Rascally Signs in Sacred Places: the Politics of Culture in Nicaragua*, Chapel Hill, NC: University of North Carolina Press.

White, P. (1984) *The West European City: a Social Geography*, London: Longman.

Whitehand, J. and Gu, K. (2006) 'Research on Chinese urban form: retrospect and prospect', *Progress in Human Geography*, 30(3): 337–55.

Whitehead, M. (2005) 'Between the marvellous and the mundane: everyday life in the socialist city and the politics of the environment', *Environment and Planning D: Society and Space*, 23: 273–94.

Whitehead, M. (2009) 'The wood for the trees: ordinary environmental justice and the everyday right to urban nature', *International Journal of Urban and Regional Research*, 33: 662–81.

Williams, R. (1973) *The Country and the City*, New York: Oxford University Press.

Wilson, D. (2007) 'City transformation and the global trope: Indianapolis and Cleveland', *Globalisations*, 4(1): 29–44.

Wimmer, A. and Glick-Schiller, N. (2002) 'Methodological nationalism and beyond: nation-state building, migration and the social sciences', *Global Networks*, 2(4): 301–34.

Winchester, L., Cáceres, T. and Rodríguez, A. (2001) 'Bellavista: La defensa de un barrio. Activismo político local', *Proposiciones*, 28: 36–76.

Winichakul, T. (1994) *Siam Mapped: A History of the Geo-body of a Nation*, Honolulu: University of Hawaii Press.

Winter, T. (2009) 'Conclusion: recasting tourism theory towards and Asian future', in T. Winter, P. Teo and T. Chang (eds) *Asia on Tour: Exploring the Rise of Asian Tourism*, London: Routledge.

Wirth, L. (1938) 'Urbanism as a way of life', *American Journal of Sociology*, 44(1): 1–24.

Wise, A. (2005) 'Hope and belonging in a multicultural suburb', *Journal of Intercultural Studies*, 26(1–2): 171–86.

Witte, N. (2003) *Privacy: Architecture in Support of Privacy Regulation*, Architectural thesis, Cincinnati, University of Cincinnati. Master: 79.

Wittfogel, K.A. (1957) *Oriental Despotism: a Comparative Study of Total Power*, New York: Vintage Books.

Wolch, J. (2002) 'Anima urbis', *Progress in Human Geography*, 26: 721–42.

Wolfe, T. (1981) *From Bauhaus to our House*, New York: Pocket Books.

Woo, W. (2001) 'Recent claims of China's economic exceptionalism: reflections inspired by WTO accession', *China Economic Review*, 12(2–3): 107–36.

Woo-Cumings, M. (1999) *The Developmental State*, Ithaca: Cornell University Press.

Wood, A. (1996) 'Analysing the politics of local economic development: making sense of cross-national convergence', *Urban Studies*, 33: 1,281–95.

Wood, J. (1997) *The New England Village*, London: Johns Hopkins University Press.

Woollacott, M. (1996) 'From Dresden to Sarajevo: cities, war and political will', *City*, 1(1): 89–91.

World Bank (1993) *The East Asian Miracle: Economic Growth and Public Policy*, New York: Oxford University Press.

World Bank (2000) 'World Development Report 2000/2001: Attacking Poverty', Mundi Prensa Libros.

World Bank (2008) *Decentralization and Local Democracy in the World*, Washington DC: The World Bank.

World Health Organisation (WHO) and United Nations Children Fund (UNICEF) (2006) *Meeting the Millenium Development Goals' Drinking Water and Sanitation Target: the Urban and Rural Challenge of the Decade*, Geneva: WHO Press.

Wright, F.L. (1996) 'Broad-Acre city: a new community plan', in R. Le Gates and F. Stout (eds) *The City Reader*, London: Routledge.

Wright, G. (1991) *The Politics of Design in French Colonial Urbanism*, Chicago: Chicago University Press.

Wu, F.L. (1996) 'Changes in the structure of public housing provision in urban China', *Urban Studies*, 33: 1,601–27.

Wu, F.L. (1997) 'Urban restructuring in China's emerging market economy: towards a framework for analysis', *International Journal of Urban and Regional Research*, 21: 640–63.

Wu, F.L. (2002) 'China's changing urban governance in the transition towards a more market-oriented economy', *Urban Studies*, 39: 2,071–93.

Wu, F.L. (2004) 'Transplanting cityscapes: the use of imagined globalization in housing commodification in Beijing', *Area*, 36(3): 227–34.

Wu, F.L. (2006) *Globalization and the Chinese City*, London: Routledge.

Wu, F.L. (2007) *China's Emerging Cities*, London: Routledge.

Wyly, E. and Hammel, D. (2001) 'Gentrification, housing policy, the new context of urban development', in K. Fox Gotham (ed.) *Critical Perspectives on Urban Development*, 6: 211–76, Research in Urban Sociology, London: Elsevier.

Yan, Y. (2006) 'CPC promotes "core value system" to lay moral foundation for social harmony', *China View*, 18 October, online at http://news3.xinhuanet.com/english/2006–10/18/content_5220576.htm (accessed 4/6/2009).

Yan, Z. (2007) 'Divided rooms raise concerns', *Shanghai Daily Online Edition*, 20 June, online at http://www.shanghaidaily.com (accessed 9/11/2008).

Yang, L. (2008) 'Snack vendors get booted off city streets', *Shanghai Daily Online Edition*, 13 October, online at http://www.shanghaidaily.com (accessed 22/10/2008).

Yarwood, D. (1974) *The Architecture of Europe*, London: Chancellor Press.

Yeoh, B. (1996) *Contesting Space: Power Relations and the Urban Built Environment in Colonial Singapore*, Oxford: Oxford University Press.

Young, C. and Kaczmarek, S. (2008) 'The socialist past and postsocialist urban identity in Central and Eastern Europe: the case of Łódź', *European Urban and Regional Studies* 15(1): 53–70.

Young, L. (2003) 'The "place" of street children in Kampala, Uganda: marginalisation, resistance, and acceptance in the urban environment', *Environment and Planning D: Society and Space*, 21: 607–27.

Zaluar, A. and Alvito, A. (1999) *Um século de favela*, Rio de Janeiro: Fundação Getúlio Vargas.

Zang, X. (2007) *Ethnicity and Urban Life in China*, London: Routledge.

Zhang, L. (2002) 'Spatiality and urban citizenship in late socialist China', *Public Culture*, 14(2): 311–34.

Zhang, L. and Ong, A. (eds) (2008) *Privatizing China Socialism from Afar*, Ithaca, NY: Cornell University Press.

Zhang, T. (2002) 'Urban development and a socialist pro-growth coalition in Shanghai', *Urban Affairs Review*, 37: 475–99.

Zhang, Y. (2008) 'Remapping Beijing: polylocality, globalization, cinema', in A. Huyssen (ed.) *Other Cities, Other Worlds: Urban Imaginaries in a Globalizing Age*, Durham, NC: Duke.

Zhou, Y. and Ma, L. (2000) 'Economic restructuring and suburbanization in China', *Urban Geography*, 21: 205–36.

Zhu, J.M. (2002) 'Urban development under ambiguous property rights: a case of China's transition economy', *International Journal of Urban and Regional Research*, 26: 41–57.

Zukin, S. (1988) *Loft Living: Culture and Capital in Urban Change*, London: Radius.

Zukin, S. (1993) *Landscapes of Power: From Detroit to Disney World*, Berkeley, CA: University of California Press.

Zukin, S. (1995) *The Cultures of Cities*, Cambridge, MA: Blackwell.

INDEX